D1196740

FREE RINGS AND
THEIR RELATIONS

L.M.S. MONOGRAPHS

Editors: P. M. COHN *and* G. E. H. REUTER

1. Surgery on Compact Manifolds *by* C. T. C. Wall, F.R.S.
2. Free Rings and Their Relations *by* P. M. Cohn

*Published for the London Mathematical Society
by Academic Press Inc. (London) Ltd.*

FREE RINGS
AND THEIR RELATIONS

P. M. COHN

Department of Mathematics,
Bedford College,
London

1971

ACADEMIC PRESS · LONDON & NEW YORK

ACADEMIC PRESS INC. (LONDON) LTD.
24/28 Oval Road
London, N.W.1

U.S. Edition published by
ACADEMIC PRESS INC.
111 Fifth Avenue
New York, New York 10003

Library of Congress Catalog Card Number: 76–185210

ISBN: 0–12–179150–5

Printed in Great Britain by
ROYSTAN PRINTERS LIMITED
Spencer Court, 7 Chalcot Road
London N.W.1

*To the memory
of my parents*

Preface

I have had a dream,—
   past the wit of man to say what dream it was:
man is but an ass,
   if he go about to expound his dream.

Shakespeare, A Midsummer Night's Dream

In the development of ring theory there are two strands, each with its own problems and methods, and although with many points of contact, they have never merged completely. One of them is the theory of algebras; this was a non-commutative (indeed sometimes even non-associative) theory from the beginning but there were always heavy finiteness restrictions, which are only gradually being relaxed. Thus while there is a fairly substantial theory of Artinian rings, the theory of Noetherian rings is still in its early stages, and although it is being developed vigorously, it is clear that some type of maximum condition is essential for its development.

A second and quite distinct line originated with the study of arithmetic in algebraic number fields. In the hands of Kummer, Dedekind and E. Noether this led to the abstract notion of a Dedekind ring. Meanwhile, algebraic geometers found the need for affine rings in the study of algebraic varieties; here Dedekind rings appear again, but as a rather special case (essentially the 1-dimensional case). Now in the last few years our way of describing the geometrical notions has changed quite radically, with the result that the correspondence (rings)→(varieties) has been extended and made precise: There is a contravariant functor from the category of commutative rings to the category of affine schemes, which is an equivalence. The effect of this connexion has been profound in both directions; we are concerned here particularly with the help afforded by the geometrical notions in studying rings. Although some hypotheses may need to be imposed to obtain particular results, there is only one overall restriction left: the rings must be commutative. Moreover, it is not at all clear that this restriction is essential, but simply that there is so far not even a proper beginning of 'non-commutative algebraic geometry'.

To make such a beginning one would need to study solutions of algebraic equations in non-commuting indeterminates with coefficients in skew fields.

This in itself is a daunting task, raising many problems of its own, and if one is to attack them successfully, one will need some knowledge of non-commutative polynomial rings over skew fields, i.e. free associative algebras. Such knowledge is still rather scanty, in fact since free algebras come under neither of the two main branches, they are not even mentioned in most books on ring theory.

It is the object of this book to present an account of free algebras and related rings, as far as they are known today. Our basic objects of study will be firs (free ideal rings) and their variants; this is a class of rings arising rather naturally and including free algebras as well as many other types. In fact it stands in the same relation to free algebras as commutative principal ideal domains do to polynomial rings in one variable over a field. Much of the theory of firs is a natural generalization of the commutative theory, although there is a wider range of possibilities here. Let us briefly indicate the topics covered:

Chapter 1 is devoted to the definition and basic properties of firs. Chapter 2 treats analogues of the Euclidean algorithm; Chapter 3 does the same for unique factorization and Chapter 5 for torsion modules. The fact that such generalizations exist may seem less than noteworthy to the observer of modern trends. But what gives these generalizations their real interest is the fact that they apply to such diverse rings as free algebras, group algebras of free groups, free products of skew fields etc. A number of new phenomena (without significant parallel in the commutative case) which occur in the study of factorizations, are dealt with in Chapter 4. Subrings of firs satisfying commutativity or finiteness conditions are examined in Chapter 6, and the final Chapter 8 returns to the case of principal ideal domains, to see what additional information is available in this case.

By contrast Chapter 7 is concerned with quite general rings. The problem posed and solved here is to obtain necessary and sufficient conditions for a ring to be embeddable in a (skew) field. The conditions obtained lead to an easy proof that any semifir can be embedded in a field, which justifies their inclusion.

Although on the face of it, the topic of this book is rather specialized, our aim has been to bring out the connexion with both of the major branches of ring theory, and to emphasize that all is part of the same subject. Secondly it is hoped that the information on free rings provided here will help to further the study of 'non-commutative algebraic geometry'. Thirdly, the methods used here on firs may give some indication of what is needed to study more general rings.

Since most of the material has not appeared in book form before, it was necessary to start at the beginning. In fact a good deal of background material not readily available in the literature has been included in Chapter 0, and the only prerequisite is some acquaintance with the basic notions of ring theory, as exposed in Jacobson's *Structure of Rings* or Lambek's *Rings and Modules*. Some parts of the book use lattice theory and homological algebra, and the relevant facts have been summarised in an appendix, to make the book more widely accessible. Throughout the book a substantial number of exercises and open problems has been included. The former range from simple questions to test the reader's understanding to brief versions of further developments and counter-examples. The open problems range from open-ended exercises to deep problems whose solution would materially advance the subject. It goes without saying that the author would be glad to hear from any readers who discover errors in the text or exercises. He would even more like to hear from readers who succeed in solving an open problem, but it should be observed that the presentation of the solution of a problem marked 'open' does not constitute a prima facie case for publication.

Finally a word about the sources. Early results in this field, although scanty, are widely scattered; we have tried, in the bibliography at the end, to include most papers of interest. The systematic development of the subject dates from about 1960; a landmark was the appearance of G. M. Bergman's thesis in 1967, in which some outstanding problems were solved, and which simplified and generalized much of what was then known. At that time Bergman and the author planned a joint book on the subject, but as the work progressed it became clear that there was material for (at least) two books, and that the most logical arrangement was to have one author for each. The present work therefore concentrates on the structure of firs, and does not go into detail over methods of constructing firs and *n*-firs, or questions involving free products. The material presented here includes roughly the first half of Bergman's thesis; that source has been drawn on freely. It is a pleasure to acknowledge the very great debt I owe to Professor Bergman for allowing me to use his thesis and other unpublished work so extensively for this book, as well as offering much helpful criticism and advice during its preparation.

I have also benefitted greatly from the advice of friends on whom I tried my ideas, and the comments of auditors at my lectures on the topic over the last ten years in three continents. Mr Warren Dicks has read most of the proofsheets and made some valuable comments. My thanks go to all of them, as well as to the staff of the Academic Press for their helpfulness and their efficiency in getting the book published.

Bedford College,   P. M. Cohn
London
November 1971

Contents

Note to the reader

Chapter **0** consists of background material from ring theory that may not be entirely standard. The main subject matter of the book is introduced in Chapter **1** and the reader may wish to start here, referring back to Chapter **0** only when necessary. Chapters **2** and **3** are used throughout, but Chapters **4**, **5**, and **6** are largely (though not entirely) independent and may be read in any order. The subject matter of Chapter **8** is likely to be more familiar to the reader, and this chapter can be read at any stage, if the reader is willing to turn back for the occasional definition. Chapter **7** is quite independent of the rest, except for the application to semifirs at the end, and it can also be read at any stage.

All theorems, propositions and lemmas are numbered consecutively in a single series in each section, thus Prop. 2.3 follows Lemma 2.2 in Section **4.2** (and outside Chapter **4** they are referred to as Prop. 4.2.3 and Lemma 4.2.2). The end (or absence) of a proof is indicated by ∎.

References to the bibliography are by author's name and last two digits of year of publication, thus Ore [31], with primes to distinguish publications by the same author in the same year.

There are exercises at the end of each section; the harder ones are marked * and open-ended (or open) problems are marked °.

Generally even terms in current use have been defined, but not always on their first occurrence. In all such cases the definition can be traced through the index.

Added in proof. Exercises **1**.1.9, p. 47 and **2**.2.5, p. 75 should not be marked 'open'; they are in fact rather easy to do.

Some terminology, notation and conventions used throughout the book

All rings occurring are associative, but not necessarily commutative (in fact much of the book reduces to trivialities in the commutative case). Every ring has a unit-element, denoted by 1, which is preserved by homomorphisms, inherited by subrings and acts as the identity operator on modules. The same conventions apply to semigroups. A ring may consist of 0 alone; this is so precisely if $1 = 0$.

In any ring R the set of non-zero elements is denoted by R^*. A ring R, not necessarily commutative, such that R^* is a group under multiplication is called a *field*; occasionally the prefix 'skew' is used, to emphasize that our fields need not be commutative. An element u in a ring or semigroup is *invertible* or a *unit* if it has an *inverse* u^{-1} satisfying $uu^{-1} = u^{-1}u = 1$; such an inverse is unique if it exists at all. The units of a ring (or semigroup) R form a group, denoted by $U(R)$. The ring of all $n \times n$ matrices over R is written R_n and instead of $U(R_n)$ one also writes $GL_n(R)$.

An element u in a ring is called a *left zero-divisor* if $u \neq 0$ and $uv = 0$ for some $v \neq 0$; if u is neither zero nor a left zero-divisor, it is called a *left non-zerodivisor*. Corresponding definitions hold with 'left' replaced by 'right'. A left or right zero-divisor is called a zero-divisor, and by a *non-zerodivisor* we understand an element that is neither zero nor a zero-divisor. A non-zero ring without zero-divisors is called an *integral domain*.

An element of a ring is called an *atom* if it is a non-unit and cannot be written as a product of two non-units. A *complete* or *atomic* factorization is one in which all factors are atoms, and an integral domain is called *atomic* if every element other than zero or a unit has an atomic factorization.

Two elements a, b of a ring (or semigroup) R are *associated* if $a = ubv$ for some u, $v \in U(R)$. If $u = 1$ ($v = 1$) they are right (left) associated; if $u = v^{-1}$, they are *conjugate* under $U(R)$. A polynomial in one variable (over any ring) is said to be *monic*, if the highest coefficient is 1. Two elements a, b of a ring R are *left coprime* if they have no common left factor apart from units; they are *right comaximal* if $aR + bR = R$. Clearly two right comaximal elements are left coprime; but not necessarily conversely. Two elements a, b are *right commensurable* if there exist a', b' such that $ab' = ba' \neq 0$.

Let K be a commutative ring; by a *K-algebra* we understand a ring R which is a K-module such that multiplication is bilinear. Sometimes we shall want a non-commutative coefficient ring K; this means that our ring R is a K-bimodule such that $x(yz) = (xy)z$ for any x, y, z from R or K. This will be called a *K-ring*. To rephrase the definitions, a K-ring is a ring R with a homomorphism $\alpha \mapsto \alpha . 1$ of K into R, while a K-algebra is a ring R with a homomorphism of K into the centre of R. Moreover, the use of the term '*K*-algebra' implies that K is commutative.

Let R be K-ring; a family (a_i) of elements of R is *right linearly dependent over K,* or *right linearly K-dependent* if there exist $\lambda_i \in K$ almost all (i.e. all but a finite number) but not all zero, such that $\Sigma a_i \lambda_i = 0$. In the contrary case (a_i) is *right linearly K-independent.* Occasionally we speak of a *set* being linearly dependent; this is to be understood as a family indexed by itself. For example, two elements of an integral domain K are right commensurable if and only if they are right linearly K-dependent and both non-zero.

Corresponding definitions hold with right and left interchanged.

To save space, columns will sometimes be written as rows, with a T to indicate transposition.

The letters **N**, **Z**, **GF**(p), **Q**, **R**, **C** stand as usual for the set (respectively ring) of non-negative integers, all integers, all integers mod p, rational, real and complex numbers, respectively.

0. Preliminaries

This Chapter collects some facts on rings which form neither part of our subject proper, nor part of the general background (described in the appendix) By its nature the contents is rather mixed, and the reader may well begin with Chapter **1** and only turn back when necessary.

In **0**.1 we describe the special form taken by Morita theory in the case of interest to us, the projective trivial rings. These rings are defined and characterized in **0**.2, which also describes the IBN and weak finiteness, and which includes Kaplansky's theorem on projective modules. In **0**.3 we discuss an important special case of projective trivial rings, namely local rings.

0.4 deals with eigenrings and centralizers; this is mainly used in the study of factorization in Chapter **3**.

The Ore construction of rings of fractions is behind much of the later development even when it does not appear explicitly. In **0**.5 we outline this construction for semigroups and its application to rings, as well as some other methods of embedding semigroups in groups, used later. This is followed by a discussion of modules over Ore rings (**0**.6); it turns out that the (left or right) Ore condition has some unexpected consequences.

In **0**.7 we define free algebras, the main object of study in this book, and then (in **0**.8) consider some variants of the 1–generator case, which form a useful source of examples.

0.1 Matrix rings

We shall adopt the convention of writing (as far as practicable) homomorphisms of *right* modules on the *left* and vice versa. Mappings will be composed accordingly; this applies in particular to multiplication in endomorphism rings and it obviates in many cases the need for passing to the opposite ring.

Let R be any ring, M a right R-module and write M^n for the direct sum of n copies of M. If μ_i is the natural inclusion map from the i-th summand into M^n and π_i the natural projection from M^n to the i-th factor, we have the equations (where composition is from right to left, in accordance with the above conventions)

$$\pi_i \mu_j = \delta_{ij}, \qquad \Sigma \mu_i \pi_i = 1, \tag{1}$$

1

where δ_{ij} denotes the Kronecker delta: $\delta_{ij} = 0$ or 1 according as $i \neq j$ or $i = j$. It follows that the n^2 R-endomorphisms ε_{ij} of M^n defined by

$$\varepsilon_{ij} = \mu_i \pi_j \tag{2}$$

satisfy the familiar equations for a matrix basis:

$$\varepsilon_{ij}\varepsilon_{kl} = \delta_{jk}\varepsilon_{il}, \qquad \Sigma\varepsilon_{ii} = 1. \tag{3}$$

Let us write

$$S = \text{End}_R(M). \tag{4}$$

Then each element (α_{ij}) of the $n \times n$ matrix ring S_n defines an endomorphism α of M^n by the equation

$$\alpha = \Sigma\mu_i\alpha_{ij}\pi_j,$$

and the correspondence $(\alpha_{ij}) \mapsto \alpha$ is easily seen to be a ring-homomorphism $S_n \to \text{End}(M^n)$, with inverse $\alpha \mapsto (\pi_i\alpha\mu_j)$. This proves

LEMMA 1.1 *Let M be any R-module and $S = \text{End}_R(M)$, then for any $n \geqslant 1$,*

$$\text{End}_R(M^n) \cong S_n. \ \blacksquare \tag{5}$$

When we considered M^n as R-module, its endomorphism ring turned out to be S_n. But we can also consider M^n as R_n-module; in that case its endomorphism ring, i.e. the centralizer of R_n in $\text{End}(M^n)$, is the centralizer of the matrix basis $\{\varepsilon_{ij}\}$ in S_n, i.e. S itself. Thus we have

$$\text{End}_{R_n}(M^n) \cong S. \tag{6}$$

In the two cases (5) and (6) M^n may be visualized as consisting of column vectors and row vectors respectively, over M. We shall distinguish these cases by writing the set of column vectors, i.e. $n \times 1$ matrices, as ^nM, and the set of row vectors, i.e. $1 \times n$ matrices, as M^n. As an example take R considered as right R-module over itself; we shall write this as R_R and correspondingly write $_RR$ for R as left R-module. It is well known (and easily verified) that $\text{End}_R(R_R) \cong R$, $\text{End}_R(_RR) \cong R$. Therefore R^n, the set of row vectors, has a natural (R, R_n)-bimodule structure and there is no need to indicate the action by suffixes. Likewise ^nR, the set of column vectors, has a natural (R_n, R)-bimodule structure, and (5) and (6) become in this case

$$\text{End}_R(^nR) \cong R_n, \qquad \text{End}_{R_n}(^nR) \cong R. \tag{7}$$

The vectors $e_1 = (1, 0, ..., 0)$, $e_2 = (0, 1, 0, ...)$, ... and the corresponding columns form bases for R^n, ^nR resp. (as R-modules), called the *standard bases*.

Returning to the case of a general R-module M, we can summarize the relation between M and M^n as follows.

THEOREM 1.2. *Let R be any ring and M a right R-module with endomorphism ring S. Then M^n may be regarded as R_n-module in a natural way, with endomorphism ring S, and there is a lattice isomorphism*

$$\text{Lat}_R(M) \cong \text{Lat}_{R_n}(M^n), \tag{8}$$

where $\text{Lat}_R(M)$ denotes the lattice of R-submodules of M. Moreover, (S, R)-bimodules correspond to (S, R_n)-bimodules under (8).

Proof. The first assertion is just a restatement of (6). To establish (8) recall that M^n consists of row vectors over M; any submodule N of M corresponds to a submodule N^n of M^n and the correspondence

$$N \mapsto N^n \tag{9}$$

is an order-preserving map. Conversely, if P is an R_n-submodule of M^n, then the n projections

$$\pi_i : P \to M \ (i = 1, ..., n)$$

are all equal and associate with P a submodule of M. It is easily seen that the correspondence $P \mapsto \pi_1 P$ is an order-preserving map inverse to (9). Hence (9) is an order isomorphism between lattices, and therefore a lattice-isomorphism. The final assertion follows because the S-action on M and on M^n is unaffected by R. ∎

It is possible to go beyond Theorem 1.2 and obtain an equivalence between the categories of right R-modules and right R_n-modules. We recall that an *equivalence* between two categories $\mathscr{A}$ and $\mathscr{B}$ is given by a pair of functors $F : \mathscr{A} \to \mathscr{B}$, $G : \mathscr{B} \to \mathscr{A}$ such that the compositions (from left to right) $FG : \mathscr{A} \to \mathscr{A}$ and $GF ; \mathscr{B} \to \mathscr{B}$ are naturally equivalent to the identity functor: $FG \simeq 1$, $GF \simeq 1$. For any ring R we denote the category of all right R-modules and all R-homomorphisms between them by $\mathscr{M}_R$. To obtain the category equivalence underlying Theorem 1.2, consider the functors

$$M \mapsto M^n = M \otimes_R R^n \qquad (M \in \mathscr{M}_R) \tag{10}$$

and

$$P \mapsto P^\natural = P \otimes_{R_n} {}^n R \qquad (P \in \mathscr{M}_{R_n}). \tag{11}$$

$P^\natural$ may also be defined as $\text{Hom}_{R_n}(R^n, P)$. It is easily verified that $(M^n)^\natural \cong M$, $(P^\natural)^n \cong P$, and this shows that for any ring R and any $n \geqslant 1$, the categories $\mathscr{M}_R$ and $\mathscr{M}_{R_n}$ are equivalent. Generally, two rings R, S are said to be *Morita-equivalent*: $R \sim S$, if the categories $\mathscr{M}_R$ and $\mathscr{M}_S$ are equivalent, so what we have shown above is that for any ring R and for any integer $n \geqslant 1$, $R \sim R_n$. In general a ring R can be Morita-equivalent to rings not of the form R_n (e.g. when R is itself a matrix ring) but in a special case of importance to us later on, this the only possibility:

THEOREM 1.3. *Let R be a ring over which every finitely generated projective right R-module is free. Then the rings Morita-equivalent to R are precisely the full matrix rings $R_n (n = 1, 2, ...)$.*

Proof. We have seen that $R \sim R_n$. Conversely, let $S \sim R$; then S, being a finitely generated projective S-module (as right S-module) corresponds to a finitely generated projective R-module, because being finitely generated projective is a categorical property. Let S correspond to ^nR in this way, then these modules have isomorphic endomorphism rings, hence by Lemma 1.1 and (7), $S \cong R_n$. ∎

The condition of being Morita-equivalent is invariant under change of sides, i.e. if the ring opposite to R is denoted by R^o, then $R \sim S$ if and only if $R^o \sim S^o$. This is not immediately obvious from the definition and we shall not prove it here in general, but in the case of interest to us it will follow from Proposition 2.6 below.

The equivalence between R and R_n may be used to reduce any categorical question concerning a finitely generated module to a cyclic module. For, given M, generated as right R-module by $u_1, ..., u_n$ say, we apply the functor (10) and pass to the right R_n-module M^n; this is generated by the single element $(u_1, ..., u_n)$. The result may be stated as

THEOREM 1.4. *Any finitely generated R-module M corresponds to a cyclic R_n-module under the category equivalence (10), for a suitable n. In fact it is enough to take n equal to the cardinal of some generating set of M.* ∎

For example, if R is a principal ideal ring, then as is well known (*cf.* Proposition **8.2.2**) any submodule of an n-generator module over R can be generated by n elements. Applying Theorem 1.4 we see that any submodule of a cyclic R_n-module is again cyclic. In particular, R_n is likewise a principal ideal ring. In the other direction, if R_n is a principal ideal ring, any submodule of a cyclic module is cyclic, whence it follows that any submodule of an n-generator R-module can be generated by n elements. This can happen for some $n > 1$ in rings that are not principal ideal rings (*cf.* Webber [70]).

Exercises 0.1

1. If a ring R satisfies right ACC_n (i.e. ACC on n-generator right ideals), show that every finitely generated free right R-module satisfies ACC_n.

2. If R satisfies right ACC_{rs}, show that R_r satisfies right ACC_s.

3. R is a ring in which all finitely generated right ideals are principal; show that the total matrix ring R_n ($n > 1$) has the same property.

4. The ring R is injective, as right module over itself; show that R_n ($n > 1$) has the same property.

5. Let R be a ring with n^2 elements e_{ij} ($i, j = 1, ..., n$) satisfying $e_{ij} e_{kl} = \delta_{jk} e_{il}$, $\Sigma e_{ii} = 1$. Show that $R \cong K_n$, where K is the centralizer of the e_{ij} in R.

6*. Let $R = K_n$ be a total matrix ring and $f: R \to S$ any ring-homomorphism. Show that S is an $n \times n$ total matrix ring: $S = L_n$ and that there is a homomorphism $\phi: K \to L$ which induces f.

0.2 Free and projective modules

Let R be any ring, M an R-module and I a set. The direct power of M with index set I is denoted by M^I, while the direct sum is written $^I M$. This is consistent with our previous notations if we regard the elements of the direct sum and product as consisting of column vectors and row vectors respectively. As an example, $^I R$ has a natural right R-module structure and its endomorphism ring consists of all column finite matrices over R indexed by I (acting on the left); similarly R^I has a natural left R-module structure and its endomorphism ring includes all column finite matrices (this time acting on the right).

With every right R-module M we associate the set of linear functionals on M:

$$M^* = \mathrm{Hom}_R(M, R_R),$$

which has a natural left R-module structure. Similarly, when M is a left R-module, the linear functionals $M^* = \mathrm{Hom}_R(M, {}_R R)$ form a right R-module. In particular, $(^n R)^* \cong R^n$, $(R^n)^* \cong {}^n R$.

A free R-module $F = {}^I R$ is said to have *rank* card(I). When I is infinite, the rank is unique (*cf.* Cohn [65], p. 82), but for finite I this need not be so. We shall say that $^n R$ has *unique rank* if its is not isomorphic to $^m R$ for any $m \neq n$. Using the pairing provided by * we see that $^n R$ has unique rank if and only if R^n has unique rank. If F is a free module of unique rank n, we write $n = rk(F)$.

A ring R is said to have the *invariant basis property* or *invariant basis number* (IBN) if every free R-module has unique rank. Most rings commonly encountered have IBN, but we shall meet examples for which this property fails to hold. A trivial example is the ring consisting of 0 alone. Excluding this case, let us assume that $R \neq 0$ is a ring without IBN; then for some integers $m \neq n$,

$$R^n \cong R^m.$$ (1)

This means that there is a free module with bases of m and n elements respectively. If α, β are the matrices of transformation between these bases, they are $m \times n$ and $n \times m$ matrices respectively, satisfying

$$\alpha\beta = I_m, \qquad \beta\alpha = I_n.$$ (2)

Conversely, the equations (2) define a change from an m-element basis to an n-element basis. Thus the existence of matrices α, β satisfying (2) is necessary and sufficient for (1) to hold.

Let R again be a non-zero ring without IBN. Then there exist positive integers k, h such that

$$R^h \cong R^{h+k}.$$ (3)

The first such pair (h, k) in the lexicographic ordering is called the *type* of the ring R. We observe that for a ring R of type (h, k), $R^m \cong R^n$ if and only if $m = n$ or $m, n \geqslant h$ and $m \equiv n \pmod{k}$.

PROPOSITION 2.1. *Let $f : R \to S$ be a homomorphism between non-zero rings†. If R does not have IBN and its type is (h, k) then S does not have IBN and if its type is (h', k') then $h' \leqslant h, k'|k$.*

Proof. By hypothesis (3) holds, hence there are matrices α, β satisfying (2) with $m = h$, $n = h + k$. Applying f we get such matrices in S, whence it follows that S cannot have IBN and $h \geqslant h', k'|k$. ∎

It is well known that fields (even skew) have IBN; since every non-zero commutative ring has a homomorphism into a field (obtained by taking the quotient by a maximal ideal), it follows that every non-zero commutative ring has IBN. Likewise every ring embeddable in a field (or even one with a homomorphism into a field) has IBN. Further it is not hard to show that every (left or right) Noetherian ring has IBN (*cf.* Ex. 5).

Occasionally we shall need to consider stronger properties of rings than IBN. Let us say that a ring R is *weakly n-finite* if ^nR is not isomorphic to a proper direct summand of itself:

$$^nR \cong {}^nR \oplus K \qquad \text{implies} \quad K = 0.$$ (4)

---

† Here it is important to bear in mind that all our rings have a unit element which is preserved by homomorphisms and passage to subrings.

Clearly weak n-finiteness implies weak n'-finiteness for $n' \leqslant n$. If R satisfies (4) for all, n, it is said to be *weakly finite*. In a weakly n-finite ring, any free module on at most n generators has unique rank, by (4). Hence for non-zero rings, weak finiteness implies IBN; the converse does not hold (*cf.* Cohn [66]).

Condition (4) shows that any surjective endomorphism of ^nR is an automorphism; in fact these conditons are equivalent, for a surjection $\lambda : {}^nR \to {}^nR$ gives rise to an exact sequence

$$0 \to K \to {}^nR \to {}^nR \to 0,$$

which must split, because ^nR is free. If we translate this condition into matrix language, we obtain the following form of weak finiteness:

For any two $n \times n$ matrices α, β over R, $\alpha\beta = I$ implies $\beta\alpha = I$.

Thus a ring is weakly 1-finite precisely when every 1-sided inverse is 2-sided. We also see, on taking determinants, that every commutative ring is weakly finite (even the zero ring), as well as every ring embeddable in a field. A consequence of weak finiteness (actually equivalent to it) is

PROPOSITION 2.2. *Let R be a weakly n-finite ring, then given any sequence of maps*

$$^rR \xrightarrow{\alpha} {}^nR \xrightarrow{\beta} {}^sR \qquad (r + s = n), \tag{5}$$

whose composition is zero, $\beta\alpha = 0$, and such that α has a left inverse and β a right inverse, then there exists an automorphism μ such that the following diagram is commutative

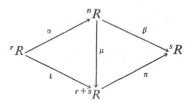

where ι, π are the natural inclusion and projection.

Proof. By hypothesis there exist α', β' such that $\alpha'\alpha = I$, $\beta\beta' = I$. Hence ker β is a direct summand of ^nR, with complement $\beta'\beta(^nR) \cong {}^sR$. Similarly, $\alpha(^rR) \cong {}^rR$ is a direct summand of ker β; calling the complement I, we have $^nR \cong (^rR \oplus I) \oplus {}^sR$. By weak n-finiteness, $I = 0$ and the sequence (5) is split exact. Now a suitable change of basis in ^nR achieves the conclusion. ∎

The content of this Proposition can also be translated into matrix language, and it will be of use to us later in that form. We state it as

PROPOSITION 2.3. *Let R be a weakly n-finite ring and let α, α', β, β' be matrices of orders $n \times r$, $r \times n$, $s \times n$, $n \times s$ respectively, such that $\beta\alpha = 0$, $\alpha'\alpha = I_r$, $\beta\beta' = I_s$, where $r + s = n$. Then there exists $\mu \in \mathbf{GL}_n(R)$ such that*

$$\beta = (I_s\, 0)\mu, \qquad \alpha = \mu^{-1}\begin{pmatrix} 0 \\ I_r \end{pmatrix}. \quad \blacksquare$$

These equations just state that β constitutes the first s rows of μ, while α constitutes the last r columns of μ^{-1}.

Most of the rings we shall meet later have a rather more special property to which we now turn. A ring is said to be *n-projective free* if every n-generator projective right R-module is free, of unique rank. Clearly "n-projective free" implies "n'-projective free" for all $n' \leqslant n$. A ring which is n-projective free for all n is said to be *projective free*.

We observe that every n-projective free ring is weakly n-finite; in particular, every projective free ring is weakly finite and hence has IBN. Of course the converse is false, e.g. not every commutative ring is projective free, and in fact it is a difficult question to decide when a ring is projective free. One of the main uses of this property is to obtain an analogue of Proposition 2.2.

PROPOSITION 2.4. *Let R be an n-projective free ring. Given a commutative triangle*

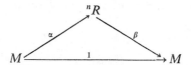

where M is any R-module, there exists an automorphism μ of ^nR such that the following diagram commutes:

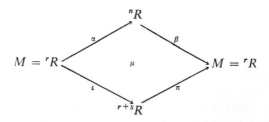

where $n = r + s$ and ι, π are the natural inclusion and projection.

For the proof we observe that the hypothesis shows M to be a direct summand of ^nR, i.e. an n-generator projective and so M is free, of unique rank, r say. Now an appropriate automorphism of ^nR gives the desired diagram. ∎

Again we restate the result in terms of matrices; or rather, the special case in which $M \cong {}^rR$ (the conclusion shows that this is no loss of generality).

PROPOSITION 2.5. *Let R be an n-projective free ring. Given any $n \times r$ matrix α and $r \times n$ matrix β such that $\beta\alpha = I_r$ (whence $r \leqslant n$), there exists $\mu \in \mathbf{GL}_n(R)$ such that*

$$\beta = (I_r, 0)\mu, \qquad \alpha = \mu^{-1}\begin{pmatrix} I_r \\ 0 \end{pmatrix}. \qquad ∎$$

As in Proposition 2.3, these equations relate β, α to the rows of μ and columns of μ^{-1} respectively.

If in Proposition 2.4 we put $\gamma = \alpha\beta$ and restate the result in matrix terms we obtain one half (viz. necessity) of

PROPOSITION 2.6. *A ring R is n-projective free if and only if ^nR has unique rank and each idempotent $n \times n$ matrix γ is conjugate (by an invertible matrix) to a matrix of the form*

$$E_r = \begin{pmatrix} I_r & 0 \\ 0 & 0 \end{pmatrix} \qquad (0 \leqslant r \leqslant n).$$

The proof of the sufficiency is straightforward and may be left to the reader. ∎

Occasionally a generalization of projective free rings is needed. We shall say that a ring R is *projective trivial*, if there exists a projective right R-module P, called the *minimal projective* of R, such that every finitely generated projective right R-module M is of the form ^nP, for an integer n which is uniquely determined by M. Clearly a projective free ring R is projective trivial, with R itself as the minimal projective; but note that the property of being projective trivial, unlike that of being projective free, is a Morita invariant. The precise relationship between these two concepts is elucidated in

PROPOSITION 2.7. *For any ring R, the following properties are equivalent*:

(a) *R is a total matrix ring over a projective free ring,*

(b) *R is Morita equivalent to a projective free ring,*

(c) *R is projective trivial.*

Proof. Clearly (a) ⇒ (b) ⇒ (c). Now assume (c): R is projective trivial, with minimal projective P, say. Since R is finitely generated projective, we have

$$R \cong {}^n P \tag{6}$$

for some positive integer n. Write $T = \text{End}_R(P)$, then by Lemma 1.1, we find on taking endomorphism rings in (6),

$$R \cong T_n.$$

Hence T is again projective trivial, and (6) shows that its minimal projective is T itself. Hence it is projective free, i.e. (a). ∎

From Proposition 2.6 and the left-right symmetry of the condition $R^m \cong R^n$ we deduce that the condition of being n-projective free is left–right symmetric; by Proposition 2.7 we have the

COROLLARY. *The condition of being projective trivial is left–right symmetric.* ∎

We conclude with some results on projective modules, which will be needed later. Let R be any ring, and M a right R-module; we shall say that M is of *type K* if it can be written as a direct sum of (any number of) countably generated modules. Kaplansky has proved the following remarkable result.

THEOREM. 2.8. *Let R be any ring and let M be any R-module of type K. Then any direct summand of M is again of type K.*

Proof. Assume that $M = \Sigma M_i = P \oplus Q$; in order to prove that P is of type K, we first show how to express M as a union of well-ordered ascending sequence, $M = \bigcup S_\alpha$, with the following properties:

(i) *at a limit ordinal, $S_\alpha = \bigcup \{S_\beta | \beta < \alpha\}$,*

(ii) *$S_{\alpha+1}/S_\alpha$ is countably generated,*

(iii) *each S_α is a direct sum of a subfamily of the M_i,*

(iv) *$S_\alpha = P_\alpha \oplus Q_\alpha$, where $P_\alpha = P \cap S_\alpha$, $Q_\alpha = Q \cap S_\alpha$.*

At a limit ordinal the construction is given by (i); it remains to construct $S_{\alpha+1}$ given S_α. Choose some M_j not contained in S_α, and let a countable generating set of M_j be $x_{11}, x_{12}, \dots$. Now split x_{11} into its P- and Q-components. Each of these new elements, in the expression of M as the direct sum of the M_i's, has a non-zero component in only a finite number of the M_i's. The sum of this finite collection of M_i's is again countably generated, by $x_{21}, x_{22}, x_{23}, \dots$, say. Next repeat on x_{12} the treatment just given to x_{11}. The result will be another countable sequence $x_{31}, x_{32}\, x_{33}, \ldots,$ the third row of an infinite matrix we are constructing. We proceed in this fashion, adding enough M_i's to split each x_{mn} into its P- and Q-components, by treating the x's in the order $x_{11}, x_{12}, x_{21}, x_{13}, x_{22}, x_{31}, \ldots$ (along successive diagonals). Let $S_{\alpha+1}$ be the submodule generated by S_α and all the x's. It is clear that (i)–(iv) hold by construction.

To complete the proof we observe that for each α, P_α is a direct summand of S_α which is a direct summand of M. Hence P_α is a direct summand of M, and it follows that it is a direct summand of $P_{\alpha+1}$. Since

$$S_{\alpha+1}/S_\alpha \cong P_{\alpha+1}/P_\alpha \oplus Q_{\alpha+1}/Q_\alpha,$$

$P_{\alpha+1}/P_\alpha$ is countably generated (as image of $S_{\alpha+1}/S_\alpha$). At a limit ordinal, $P_\alpha = \bigcup\{P_\beta | \beta < \alpha\}$, hence $P \cong \Sigma \oplus P_{\alpha+1}/P_\alpha$, so P is of type K. ∎

Any free module, being a direct sum of copies of R, is of type K, hence we have the

COROLLARY. *Every projective module is a direct sum of countably generated modules.* ∎

As an application we prove a result of Bergman on projective modules, which will be used later.

Definition. A ring R is said to be *weakly semihereditary* if, given any finitely generated projective right R-module P, a finite subset A of P and a finite subset B of P^* such that $BA = 0$, there is a direct decomposition $P = P' \oplus P''$ such that $A \subseteq P'$ and $BP' = 0$.

By restating the condition in terms of P^* (and remembering that $P^{**} \cong P$) we see that it is left–right symmetric. Consider now a ring which is *right semihereditary*, i.e. such that every finitely generated right ideal is projective. Let P be a finitely generated projective module with sets A, B satisfying the condition of the above definition. If $B = \{\beta_1, \ldots, \beta_r\}$, consider $\beta_1 : P \to R$; the image is a finitely generated right ideal, hence projective, so P splits over the kernel of β_1. An easy induction shows that $\ker B = \ker \beta_1 \cap \ldots \cap \ker \beta_r$ is a direct summand of P; thus the conclusion of the definition is satisfied. This shows that every right semihereditary ring is weakly semihereditary, and by symmetry the same holds for left semihereditary rings.

In the proof of Bergman's result we shall make use of the following remark. If M is any module and M' a submodule, then any direct summand of M which is contained in M', is also a direct summand of M'. For if $M = P \oplus Q$, where $P \subseteq M'$, then

$$M' = (P \oplus Q) \cap M' = P \oplus (Q \cap M'),$$

by an application of the modular law.

THEOREM. 2.9. *Any projective module over a weakly semihereditary ring is a direct sum of finitely generated modules.*

Proof. Let P be a projective module and A a finite subset; if we can show that A is contained in a finitely generated direct summand of P, we can complete the proof as follows.

By Kaplansky's theorem, any projective module is a direct sum of countably generated modules, so we may assume that our projective module P is countably generated, by $e_1, e_2, \ldots$ say. Suppose that we already have a direct decomposition $P = P_1 \oplus \ldots \oplus P_n \oplus P_n'$ where each P_i is finitely generated and their sum contains $e_1, \ldots, e_n$. By hypothesis we can find a decomposition $P = Q \oplus Q'$, where Q is finitely generated and contains $e_1, \ldots, e_{n+1}$ and a generating set for $P_1, \ldots, P_n$. Then $P_1 \oplus \ldots \oplus P_n$ is contained in Q and it is a direct summand in P, hence it is a direct summand in Q, say $Q = P_1 \oplus \ldots \oplus P_n \oplus P_{n+1}$ and $P = P_1 \oplus \ldots \oplus P_n \oplus P_{n+1} \oplus Q'$. By induction P has a direct summand $P_1 \oplus P_2 \oplus \ldots$ containing $e_1, e_2, \ldots$; but this set generates P, so P has been expressed as a direct sum of finitely generated modules.

It remains to prove the assertion made at the beginning, so let P be projective, say $P \oplus P' = F$ is free, and let A be a finite subset of P. In F consider the projection onto the factor P'; this is a mapping $\alpha : F \to F$ with kernel P. Now take the elements of A and express them in terms of the coordinates in F; only finitely many coordinates will be involved, so A is contained in a finitely generated direct summand F_0 of F. Restrict α to F_0, then $\alpha A = 0$, but $\alpha | F_0$ is equivalent to finitely many functionals (because F_0 is finitely generated), hence by hypothesis, F_0 has a finitely generated direct summand F_1 such that $A \subseteq F_1 \subseteq \ker \alpha = P$. Thus F_1 is a direct summand in F_0, hence in F, hence in P, and $F_1 \supseteq A$. ∎

Exercises 0.2

1. Show that over a ring of type (h, k) every finitely generated module can be generated by $h + k - 1$ elements.

2. Determine all semigroups on one generator. For any ring R show that the isomorphism types of free modules form a semigroup (with respect to direct sum as addition) and hence obtain the classification of rings without IBN.

3. If K is any non-zero ring and I an infinite set, show that $R = \text{End}(^I K)$ does not have IBN and determine its type.

4. Show that a ring of type $(1, k)$ cannot be an integral domain.

5. Show that a ring R with right ACC_n (for some $n \geqslant 1$), is weakly n-finite. Deduce that a ring with right ACC_n for all n is weakly finite.

6. Let R be a ring without IBN and for fixed m, n ($m \neq n$) consider pairs of mutually inverse matrices A, B such that A is $m \times n$ and B is $n \times m$. Show that if A', B' is another such pair, there exists an invertible $m \times m$ matrix P such that $PA = A'$, $BP^{-1} = B'$.

7. Verify that the ring consisting of 0 alone is weakly finite but does not have IBN.

8. Show that the conclusion of Proposition 2.2 is equivalent to weak n-finiteness.

9. Show that a ring R is weakly n-finite if and only if

(1) *every surjective endomorphism of R^n is an automorphism.*

If R is non-zero and has the property described in (1), show that every free homomorphic image of R^n has rank at most n. Deduce that a non-zero weakly finite ring has IBN.

10. Show that weak finiteness is a Morita invariant.

11. Let R be a right semihereditary ring, and let α be any $m \times n$ matrix over R. Show that the right annihilator of α in R_n has the form eR_n, where e is an idempotent $n \times n$ matrix.

12. (Kaplansky [58]). Let M be a countably generated module over a ring R. Assume that each direct summand N of M is such that any $x \in N$ can be embedded in a finitely generated direct summand of N; show that M is a direct sum of finitely generated modules.

13°. Characterize the greatest class of rings closed under homomorphic images, that consists entirely of weakly finite rings.

14. Let R be weakly n-finite and suppose that $P, Q \in R_n$ satisfy $PQ = \begin{pmatrix} I_r & * \\ 0 & I_s \end{pmatrix}$, where $r + s = n$ and $*$ stands for unspecified elements. Then there exists $\mu \in \text{GL}_n(R)$ such that $P\mu$ and $\mu^{-1}Q$ each are of the form $\begin{pmatrix} I_r & * \\ 0 & I_s \end{pmatrix}$.

15. (Bergman [72']). Show that if, in the definition of weakly semihereditary rings, we delete the condition that A resp. B be finite, we obtain a characterization of left resp. right semihereditary rings.

0.3 Local rings

A commutative local ring is generally defined as a ring in which the set of all non-units forms an ideal; equivalently, R/J is a field, where $J = J(R)$ is the Jacobson radical of R. We adopt a corresponding definition in the general case, thus R is said to be a *local ring* if the non-units form a two-sided ideal, i.e. R/J is a field, called the *residue class field* of R.

It will be useful to express this definition in different ways; we shall do this in Proposition 3.2, but first recall the different ways of defining the Jacobson radical (*cf.* Jacobson [64]):

THEOREM 3.1. *Let* R *be any ring and* $a \in R$, *then the following four conditions are equivalent*:

(a) *For every simple right* R-*module* M, $Ma = 0$,

(b) a *belongs to every maximal right ideal of* R,

(c) $1 - ay$ *has a right inverse, for every* $y \in R$,

(d) $1 - xay$ *has an inverse for all* $x, y \in R$,

(a*)–(d*) *the left–right analogues of* (a)–(d).

Proof. (a) $\Rightarrow$ (b). If $\mathfrak{m}$ is a maximal right ideal, then $R/\mathfrak{m}$ is simple; by (a) $(R/\mathfrak{m})a = 0$, i.e. $Ra \subseteq \mathfrak{m}$, so $a \in \mathfrak{m}$.

(b) $\Rightarrow$ (c). Assume that (b) holds, but not (c). Thus for some $y \in R$, $1 - ay$ has no right inverse, i.e. $(1 - ay)R \neq R$. By Zorn's Lemma we can find a maximal right ideal $\mathfrak{m}$ containing $(1 - ay)R$. Thus $1 - ay \in \mathfrak{m}$ and by (b), $a \in \mathfrak{m}$, hence $1 = (1 - ay) + ay \in \mathfrak{m}$, which is a contradiction.

(c) $\Rightarrow$ (a). Let M be simple. Given $u \in M$, if $ua \neq 0$, then $uaR = M$, hence $u = uay$ for some $y \in R$. Thus $u(1 - ay) = 0$, hence by (c), $u = 0$ and so $ua = 0$. Thus $ua = 0$ for all $u \in M$, i.e. $Ma = 0$.

Now (d) $\Rightarrow$ (c) trivially; we complete the proof by showing that (a) + (c) $\Rightarrow$ (d). By (a), $Ma = 0$ for any simple M, hence $Mxa \subseteq Ma = 0$ for any x, i.e. xa satisfies (a) and hence (c). Thus $1 - xay$ has a right inverse $1 - b$ say:

$$(1 - xay)(1 - b) = 1, \tag{1}$$

i.e. $b = xay(b - 1)$, hence $Mb = 0$ for all simple M, so b satisfies (a) and hence (c), i.e. $1 - b$ has a right inverse $1 - c$ say:

$$(1 - b)(1 - c) = 1. \tag{2}$$

By (1) and (2), $1 - xay = (1 - xay)(1 - b)(1 - c) = 1 - c$. Inserting this value in (2), we find

$$(1 - b)(1 - xay) = 1,$$

and this equation, together with (1) shows that $1 - xay$ has an inverse, namely $1 - b$.

This proves the equivalence of (a)–(d); the symmetry follows because (d) is left–right symmetric. ∎

The set of all $a \in R$ satisfying the conditions of this theorem is called the *Jacobson radical* of R and is denoted by $\mathbf{J}(R)$. By (b) it is characterized as the intersection of all maximal right ideals, or also (by (b*)) as the intersection of all maximal left ideals. Hence it is a two-sided ideal of R.

PROPOSITION 3.2. *For any ring R the following conditions are equivalent*:

(a) *R is a local ring, i.e. $R/\mathbf{J}(R)$ is a (skew) field*,

(b) *the non-units of R form an ideal*,

(c) *for any $a \in R$, either a or $1 - a$ has a right inverse*.

Proof. (a) $\Rightarrow$ (b). If R/J is a field, J is the unique maximal ideal and hence consists of non-units. Given $u \notin J$, we have $uv \equiv vu \equiv 1 \pmod{J}$ for some $v \in R$, hence $uv = 1 + n$ ($n \in J$) is a unit, therefore so is u. Thus J is the set of all non-units.

(b) $\Rightarrow$ (c) is clear, and to prove (c) $\Rightarrow$ (a), let $u \in R$, $u \notin J$. If u has no right inverse, neither does ux, for any $x \in R$, hence $1 - ux$ always has a right inverse, hence $u \in J$, a contradiction. Thus u has a right inverse, say $uv = 1$; clearly $v \notin J$, so v again has a right inverse, w say. Now $w = uvw = u$, thus $uv = vu = 1$ and u is a unit, i.e. (a) holds. ∎

Here (a) or (b) is the usual form of the definition of a local ring, while (c) is the easiest to verify.

The proof shows that a local ring is weakly 1-finite; taking $J = 0$, we obtain the useful (and well known)

COROLLARY. *A non-zero ring in which every non-zero element has a right inverse, must be a field.* ∎

Sometimes a more general notion of local ring is needed. Let us call R a *matrix local ring* if $R/\mathbf{J}(R)$ is simple Artinian. By Wedderburn's theorem this means that $R/J \cong K_t$, where K is a field and $t \geqslant 1$. When $t = 1$, we are back in the case of a local ring; by contrast this is sometimes called a *scalar local ring*, but generally we omit the qualifier, so 'local ring' will as usual mean 'scalar local ring'.

Later we shall need the fact that all local rings are projective free. Since any local ring has a field as homomorphic image, it clearly has IBN (Proposition 2.1), so by Proposition 2.6 we need only show that every idempotent matrix over a local ring is conjugate to E_r, for some r.

Let R be a local ring, then $R/J \cong K$ for some field K; we shall write $a \mapsto \bar{a}$ for the canonical homomorphism $R \to R/J$. Given any idempotent $n \times n$ matrix $\gamma = (\gamma_{ij})$ over R, the image $\bar{\gamma} = (\bar{\gamma}_{ij})$ is an idempotent matrix over K.

Hence there exist $\alpha, \beta \in R_n$ such that

$$\alpha\beta \equiv \beta\alpha \equiv I \ (\text{mod } J_n), \qquad \beta\gamma\alpha \equiv E_r \ (\text{mod } J_n),$$

where $0 \leqslant r \leqslant n$. Now it is well known (and easily verified, *cf.* Jacobson [64], p. 11) that J_n is the Jacobson radical of R_n. Hence if $\alpha\beta = I + \lambda$, say, where $\lambda \in J_n$, then $I + \lambda$ is invertible, and putting $\beta' = \beta(I + \lambda)^{-1}$ we see that α has a right inverse β' in R_n. By symmetry it has a left inverse and so is invertible. Thus $\gamma' = \alpha^{-1}\gamma\alpha$ is an idempotent matrix such that $\bar{\gamma}' = E_r$, and it follows that $\gamma' = E_r + \lambda'$, where $\lambda' \in J_n$.

Now $\mu = (2E_r - I) + \lambda'$ is congruent (mod J_n) to a matrix with 1 or -1 on the main diagonal and zeros elsewhere, hence it is invertible and

$$\gamma'\mu = \gamma'\left(E_r - (I - \gamma')\right) = \gamma'E_r$$
$$= \left(\gamma' - (I - E_r)\right)E_r = \mu E_r.$$

Hence $\mu^{-1}\gamma'\mu = E_r$, as claimed. Thus we have proved

THEOREM 3.3. *Every local ring is projective free.* ∎

In fact it can be shown that every projective ideal of a local ring is free (Ex. 11), but we shall not need this stronger result.

Exercises 0.3

1. Let R be a ring and J its Jacobson radical. For any right R-module M, if $MJ = M$, then any quotient N of M satisfies $NJ = N$. Show further, that every finitely generated non-zero module has a simple quotient, and use Theorem 3.1 to deduce Nakayama's lemma: If M is a finitely generated right R-module such that $MJ = M$, then $M = 0$.

2. Use Nakayama's lemma to prove Theorem 3.3.

3. Let R be a ring with Jacobson radical J. By considering the kernel of the homomorphism $R_n \to (R/J)_n$, induced by the natural homomorphism $R \to R/J$, show that $\mathbf{J}(R_n) = J_n$.

4. Show that if R is a matrix local ring, then so is R_n.

5. Show that any matrix local ring is weakly finite.

6. Show that a weakly semihereditary local ring is an integral domain.

7. Let K be a field and R a subring such that for any $x \in K$, either x or x^{-1} lies in R. If a is a non-unit in R, show that $a(a - 1)^{-1} \in R$; deduce that R is a local ring.

8*. A ring R is such that for any $x \in R$, either x has a left inverse or $1 - x$ has a right inverse. Prove that R is weakly 1-finite and hence that R is a local ring.

9. Let $\mathbf{H}$ be the field of real quaternions, $\mathbf{H}(x)$ the field of rational functions in a commuting indeterminate, and R the subring of fractions with denominator

congruent to 1 (mod $x^2 + 1$). Verify that $R/(x^2 + 1) R \cong C_2$, and hence that R is a matrix local ring which is not a matrix ring over a scalar local ring.

10*. Show that any Artinian matrix local ring is a total matrix ring over a scalar local ring.

11*. (Kaplansky [58]). Let P be a projective module over a local ring. Show that any element of P can be embedded in a free direct summand of P; deduce that *every* projective module over a local ring is free.

0.4 Eigenrings and centralizers

Let R be any ring and $\mathfrak{a}$ a right ideal in R. The set

$$\mathbf{I}(\mathfrak{a}) = \{x \in R | x\mathfrak{a} \subseteq \mathfrak{a}\}$$

is called the *idealizer* of $\mathfrak{a}$ in R. It is a subring of R containing $\mathfrak{a}$ as an ideal and may be characterized as the largest subring of R in which $\mathfrak{a}$ is a two-sided ideal. The quotient ring

$$\mathbf{E}(\mathfrak{a}) = \mathbf{I}(\mathfrak{a})/\mathfrak{a}$$

is called the *eigenring* of $\mathfrak{a}$. In the commutative case, $\mathbf{I}(\mathfrak{a}) = R$ and the eigenring is just the quotient ring $R/\mathfrak{a}$; more generally this holds whenever $\mathfrak{a}$ is a two-sided ideal in R. Its significance in the general case is illustrated by

PROPOSITION 4.1. *If $\mathfrak{a}$ is a right ideal in a ring R, then*

$$\mathbf{E}(\mathfrak{a}) \cong \operatorname{End}_R(R/\mathfrak{a}). \tag{1}$$

Proof. Any $a \in \mathbf{I}(\mathfrak{a})$ defines an R-endomorphism α_a of $R/\mathfrak{a}$ by the rule

$$\alpha_a : \bar{x} \to \overline{ax}, \tag{2}$$

where $\bar{x}$ is the class of x (mod $\mathfrak{a}$). By the definition of $\mathbf{I}(\mathfrak{a})$ this is a well-defined map of $R/\mathfrak{a}$ which is easily seen to be an R-endomorphism. Thus we have a ring homomorphism

$$\alpha : \mathbf{I}(\mathfrak{a}) \to \operatorname{End}_R(R/\mathfrak{a}) \tag{3}$$

given by $a \to \alpha_a$. Clearly $\alpha_a = 0$ if and only if $a \in \mathfrak{a}$, therefore $\ker \alpha = \mathfrak{a}$. To complete the proof it remains to show that α is surjective. Let $\lambda \in \operatorname{End}_R(R/\mathfrak{a})$, say $\lambda \bar{1} = \bar{a}$, then for any $x \in R$,

$$\lambda \bar{x} = (\lambda \bar{1})x = \bar{a}x = \overline{ax} = \alpha_a \bar{x},$$

hence $\lambda = \alpha_a$ and so (3) is an isomorphism. ∎

Similarly, if $\mathfrak{a}, \mathfrak{b}$ are any right ideals of R, then each R-homomorphism $R/\mathfrak{a} \to R/\mathfrak{b}$ is completely specified by an element $c \in R$ such that $c\mathfrak{a} \subseteq \mathfrak{b}$; if we put

$$I(\mathfrak{a}, \mathfrak{b}) = \{x \in R | x\mathfrak{a} \subseteq \mathfrak{b}\},$$

then it can be shown, as in the proof of Proposition 4.1, that

$$\text{Hom}_R(R/\mathfrak{a}, R/\mathfrak{b}) \cong \mathbf{I}(\mathfrak{a}, \mathfrak{b})/\mathfrak{b}, \tag{4}$$

regarded as isomorphism of $(\mathbf{E}(\mathfrak{b}), \mathbf{E}(\mathfrak{a}))$-bimodules.

The eigenring of an element (i.e. of the right ideal generated by it) is closely related to its centralizer, and to some extent both may be treated by the same method, by the device of adjoining an indeterminate. The basic result is

THEOREM 4.2. *Let R be a ring and $S = R[t]$ the ring obtained by adjoining a commuting indeterminate to R. Given $a,\ b \in R$, write*

$$C = \mathbf{C}(a, b) = \{x \in R | xa = bx\},$$
$$I = \mathbf{I}((t - a)S,\ (t - b)S)) = \{s \in S | s(t - a) \in (t - b)S\}.$$

Then

$$I = C \oplus (t - b)S, \tag{5}$$

and hence

$$\mathbf{C}(a, b) \cong \text{Hom}_S(S/(t - a)S,\ S/(t - b)S). \tag{6}$$

Proof. Any $f \in S$ admits the decomposition

$$f_t = f_b + (t - b)h \qquad (h \in S), \tag{7}$$

where for any $f = f_t = \Sigma t^i c_i$ we write $f_b = \Sigma b^i c_i$. This follows on division by the monic polynomial $t - b$, and the decomposition (7) is clearly unique. To establish (5) it only remains to show that $f_t \in I$ if and only if $f_b \in C$. If $f_b \in C$, then $f_t \in I$ by (7) because $C \subseteq I$. Conversely, let $f_t \in I$, then since $(t - b)h \in I$, it follows that $f_b \in I$, i.e.

$$f_b(t - a) = (t - b)g$$

for some $g \in S$. By comparing degrees in t we see that $g \in R$, and equating coefficients, we find $g = f_b$, $f_b a = b f_b$, whence $f_b \in C$. Now (6) follows from (5) on dividing by $(t - b)S$ and applying the isomorphism (4). ■

By putting $b = a$ we can express the centralizer of a as an eigenring:

COROLLARY. *The centralizer of an element $a \in R$ is isomorphic to the eigenring of $t - a$ in the polynomial ring $R[t]$.* ■

The following result is well known in the special case of matrix rings over a field, where it is used to obtain the canonical form of a matrix.

PROPOSITION 4.3. *Let R be any ring and t a commuting indeterminate. Then two elements a, b of R are conjugate under $\mathbf{U}(R)$ if and only if $t - a$ and $t - b$ satisfy a comaximal relation*

$$f(t - a) = (t - b)g \tag{8}$$

in $R[t]$. Moreover, in any such comaximal relation (8) f and g may be taken to lie in $\mathbf{U}(R)$.

Proof. If a, b are conjugate, say $ua = bu(u \in \mathbf{U}(R))$, then clearly $u(t - a) = (t - b)u$ is a comaximal relation. Conversely, assume a comaximal relation (8). By subtracting an expression $(t - b)h(t - a)$ from both sides, we get an equation

$$u(t - a) = (t - b)v, \tag{9}$$

where $u = f - (t - b)h$, $v = g - h(t - a)$. Here we may choose h so that u has degree 0 in t, i.e. $u \in R$. Then on comparing degrees in (9) we find that $v \in R$, while a comparison of highest terms shows that $v = u$ and so

$$ua = bu. \tag{10}$$

Further, since $u = f \pmod{(t - b) R[t]}$, u and $t - b$ are still right comaximal, say

$$up + (t - b)q = 1 \qquad (p, q \in R[t]). \tag{11}$$

By subtracting an appropriate right multiple of (9), we can reduce (11) to the case where p has degree 0. Then $q = 0$, by comparing degrees, and (11) shows p to be a right inverse to u. By the symmetry of (10), u also has a left inverse and so is a unit. Now (10) shows a and b to be conjugate, as asserted. ∎

Exercises 0.4

1. In any ring R, if $ab' = ba'$, show that $a'b$ lies in the idealizer of $Rb'b$ and that of $a'aR$.

2. Let R be a ring and t a commuting indeterminate. Given $a, b \in R^*$, show that the elements $a^n b$ ($n = 0, 1, 2, \ldots$) are right linearly dependent over R if and only if $a - t$ and b are right commensurable in $R[t]$.

3. Use Proposition 4.3 to show that two $n \times n$ matrices A and B over a field k are similar if and only if $tI - A$, $tI - B$ are associated in $k_n[t]$.

4. Let R be a ring and t a commuting indeterminate. If $t - a$ and $h = h(t)$ satisfy a comaximal relation

$$f . (t - a) = h . g,$$

show that g can be taken to lie in R, but not in general f.

5*. (Robson [72]). (a) Let R be a ring, $\mathfrak{a}$ a right ideal and A its idealizer. Show that any simple right R-module M is either simple as right A-module or a homomorphic image of $R/\mathfrak{a}$.

(b) Let $\mathfrak{a} = \mathfrak{m}_1 \cap \ldots \cap \mathfrak{m}_k$, where the $\mathfrak{m}_i$ are maximal right ideals and $B = \{b \in R \mid b\mathfrak{a} \subseteq \mathfrak{m}_1\}$. Show that $B/\mathfrak{m}_1$ is a simple right A-module.

(c) With the notation as before, let M be a simple right R-module. Then M is simple as right A-module, unless for some i, $M \cong R/\mathfrak{m}_i$ and $R\mathfrak{a} \nsubseteq \mathfrak{m}_i$. In that case M has a unique composition series

$$R \supset A + \mathfrak{m}_i \supset \mathfrak{m}_i.$$

0.5 Groups and rings of fractions

Let S be a semigroup; there always exists a group $\mathbf{G}(S)$ with a homomorphism $\alpha : S \to \mathbf{G}(S)$ such that any homomorphism of S into a group can be uniquely factored by α (i.e. the accompanying triangle commutes).

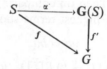

$\mathbf{G}(S)$ is unique up to isomorphism and is called the *universal group* of S; it can be obtained e.g. by writing down a presentation of S as semigroup, in terms of generators and defining relations, and treating this now as a group presentation.

More generally, if T is a subsemigroup of S, we can define the *universal T-inverting semigroup* on S as a semigroup S_T with a homomorphism $\alpha : S \to S_T$ which maps each element of T to an invertible element of S_T, briefly, α is *T-inverting*, and any T-inverting homomorphism of S can be uniquely factored by α. We shall be mainly concerned with the question: when is $\alpha : S \to S_T$ an injection? When this is the case, S_T is called a *semigroup of fractions* for S. Clearly a necessary condition is that S should admit cancellation by T:

$$st = s't \quad \text{or} \quad ts = ts' \quad \text{implies} \quad s = s' \qquad (s, s' \in S, \ t \in T). \tag{1}$$

In the commutative case this is also sufficient, but in general it is only the first of an infinite sequence of conditions which are necessary and sufficient for S to be embeddable in S_T. These are the Malcev conditions (*cf.* e.g. Cohn [65]).

We shall not discuss the Malcev conditions in more detail, but limit ourselves to some special cases, where the embedding is more easily proved directly. The most important of these is the case where, on identifying S

with its image in S_T, every element of S_T can be written as $st^{-1}(s \in S,\ t \in T)$. In particular, $t^{-1}s$ must have this form, say $t^{-1}s = s_1 t_1^{-1}$ and on multiplying up we find $st_1 = ts_1$. Let us define a *right denominator set* in a semigroup S as a subsemigroup T satisfying (1) and

$$sT \cap tS \neq \varnothing \qquad \text{for all} \qquad s \in S,\ t \in T. \tag{2}$$

PROPOSITION 5.1. *Let S be a semigroup and T a right denominator set in S, then the natural mapping $S \to S_T$ into the universal T-inverting semigroup is an embedding and every element of S_T has the form $st^{-1}(s \in S,\ t \in T)$.*

We sketch the proof. Define an equivalence on $S \times T$ by the rule: $(s_1, t_1) \sim (s_2, t_2)$ whenever there exist $u_1, u_2 \in S$ such that $t_1 u_2 = t_2 u_1 \in T$, $s_1 u_2 = s_2 u_1$. This is easily seen to be an equivalence. Now define a multiplication on the equivalence classes as follows: Given $(s_1, t_1),\ (s_2, t_2)$, we can find $u_1 \in T,\ u_2 \in S$ such that $t_1 u_2 = s_2 u_1$. Then $(s_1, t_1) \sim (s_1 u_2, t_1 u_2)$ $\sim (s_1 u_2, s_2 u_1)$ and now $(s_1, t_1) . (s_2, t_2) \sim (s_1 u_2, s_2 u_1) . (s_2 u_1, t_2 u_1) \sim (s_1 u_2, t_2 u_1)$. It is easily checked that this multiplication depends only on the equivalence classes of the factors and is associative, so that we have a homomorphism $s \mapsto (s, 1)$. This is injective, since $(s_1, 1) \sim (s_2, 1)$ means $s_1 u_2 = s_2 u_1$ for some $u_1 = u_2 \in T$, whence $s_1 = s_2$. Moreover, $(t, 1)$ has the inverse $(1, t)$, so that we can write $(s, t) = (s, 1) . (1, t) = st^{-1}$, on identifying S with its image. ∎

An element c of a semigroup S is said to be *right large* if $cS \cap aS \neq \varnothing$ for all $a \in S$. The set L of right large elements is always a subsemigroup. For let $a, b \in L$ and take $c \in S$. Then there exist $x, y \in S$ such that $ax = cy$ and there exist $u, v \in S$ such that $bu = xv$, hence $abu = axv = cyv$. If moreover, S admits cancellation by L, then L is a right denominator set. In particular this holds if S is a cancellation semigroup, so we have the

COROLLARY 1. *In a cancellation semigroup S the set L of right large elements is a right denominator set and hence $S \to S_L$ is an embedding.* ∎

In particular, if every element of S is right large, we obtain a sufficient condition for S to have a group of fractions, i.e. to be embeddable in a group:

COROLLARY 2. *Let S be a semigroup and $\mathbf{G}(S)$ its universal group. Then the natural mapping $S \to \mathbf{G}(S)$ is an embedding and $\mathbf{G}(S) = SS^{-1}$ if and only if S is a cancellation semigroup and every element of S is right large.*

The sufficiency of the conditions follows from Proposition 5.1 and the necessity from the remarks preceding that proposition. ∎

There is another sufficient condition which we shall need later, due to Doss [48], who proves its sufficiency by verifying the Malcev conditions. Since that verification is itself non-trivial, we shall sketch a direct proof, which at the same time provides us with a normal form for the elements of $\mathbf{G}(S)$.

THEOREM 5.2. *Let S be a cancellation semigroup and L the set of its right large elements. If $aS \cap bS \neq \varnothing$ implies $ab' = ba'$ with either a' or b' in L, then S is embeddable in a group.*

Proof (sketch). By Proposition 5.1, Corollary 1, S is embeddable in S_L. Writing again S for S_L we now have a cancellation semigroup satisfying the condition

$$aS \cap bS \neq \varnothing \quad \text{implies} \quad aS \subseteq bS \quad \text{or} \quad bS \subseteq aS. \tag{3}$$

Any cancellation semigroup satisfying (3) is said to be *rigid*. Thus we must show that every rigid semigroup is embeddable in a group.

Let us put U for the group of units of S and write $a \sim b$ if $a = ubv$ for some $u, v \in U$. For each $a \in S$ we define two subsemigroups of U by the equations

$$U_1(a) = \{u \in U | ua \in aU\},$$

$$U_{-1}(a) = \{v \in U | av \in Ua\}.$$

Clearly if $a \sim b$, $U_1(a) \cong U_1(b)$ and $U_1(v) = U_{-1}(v) = U$ for any $v \in U$. Now let $T_1(a)$, $T_{-1}(a)$ be left transversals (i.e. complete sets of left coset representatives) of $U_1(a)$, $U_{-1}(a)$ respectively in U, with 1 represented by itself, and let A be a complete set of representatives of the $\sim$-classes in S, with U represented by 1. Further, define $\mathbf{G}(S)$ as permutation group on the set of expressions

$$t_1 a_1^{\varepsilon_1} t_2 a_2^{\varepsilon_2} \dots t_r a_r^{\varepsilon_r} u \tag{4}$$

where $t_i \in T_{\varepsilon_i}(a_i)$, $a_i \in A$, $\varepsilon_i = \pm 1$ $(i = 1, \dots, r)$, $u \in U$, subject to the condition that if $t_i = 1$ and $a_{i-1} = a_i$, then $\varepsilon_{i-1} + \varepsilon_i \neq 0$ $(i = 2, \dots, r)$. It is a routine (though tedious) exercise to verify that the resulting permutation group contains S as semigroup. Moreover, the elements of $\mathbf{G}(S)$ have a unique normal form (4). ∎

The constructions given in this section can be applied to an integral domain R, to embed R^* in a group. Nevertheless R need not be embeddable in a field. But if R is an integral domain and R^* a right denominator set— so that we can embed R^* in its universal group $\mathbf{G}(R^*)$—it turns out that we can extend the addition on R^* on $K = \mathbf{G}(R^*) \cup \{0\}$ in such a way as to obtain a field containing R as subring. More generally, we can form a ring of fractions with respect to a right denominator set.

Thus let R be any ring and T a right denominator set. By Proposition 5.1 there is an embedding $R \to R_T$ of R into the universal T-inverting semigroup. To define addition in R_T we observe that any two elements of T have a common right multiple in T: given $t_1, t_2 \in T$, by hypothesis there exist $s \in S$, $t \in T$ such that $t_1 t = t_2 s$, and this lies in T. Hence any two elements of R_T may be brought to a common denominator: $s_1 t^{-1}$ and $s_2 t^{-1}$ say. Now their sum is defined by the equation

$$s_1 t^{-1} + s_2 t^{-1} = (s_1 + s_2)t^{-1}.$$

It is easily checked that this is independent of the form in which the elements of R_T are written, and that R_T forms a ring with respect to this addition, with R as subring. We summarize this as

THEOREM 5.3. *Let R be any ring and T a right denominator set, i.e. T is a subsemigroup such that for any $r \in R$ and $t \in T$, $rT \cap tR \neq \varnothing$, and $rt = 0$ or $tr = 0$ implies $r = 0$. Then the universal T-inverting semigroup R_T has a ring structure such that the natural mapping $R \to R_T$ becomes a ring monomorphism.* ■

The ring R_T is called the *ring of fractions* or the *localization* of R by T.

The case where R^* is a right denominator set is of particular importance. It is easily seen that this is so precisely when R is an integral domain satisfying the following condition (known as the *right Ore condition*)

$$aR \cap bR \neq 0 \qquad (a, b \in R^*). \tag{5}$$

An integral domain satisfying (5) is called a *right Ore ring*. Thus Theorem 5.3 yields the

COROLLARY. *Any right Ore ring R can be embedded in a field K; in fact $K = R_{R^*}$ is the universal R^*-inverting semigroup of R.* ■

The field K obtained here is also called the *field of right fractions* of R. By symmetry, every left Ore ring, defined in an analogous manner, can be embedded in a field of left fractions. In particular, every commutative integral domain is a left and right Ore ring.

An integral domain R that is not a right Ore ring must contain two non-zero elements a, b that are right incommensurable: $aR \cap bR = 0$. It follows that the right ideal $aR + bR$ is a free right R-module of rank 2. Moreover, the elements $a^n b$ $(n = 0, 1, 2, \ldots)$ are right linearly independent;

for if $\Sigma a^i b c_i = 0$, then by cancelling as many factors a as possible on the left we can write this as

$$bc_0 + abc_1 + \dots a^r bc_r = 0 \qquad (c_0 \neq 0),$$

hence $bc_0 \in aR \cap bR$, a contradiction. This proves

PROPOSITION 5.4. *An integral domain that is not a right Ore ring contains free right ideals of any finite or countable rank.* ∎

In particular we note the

COROLLARY. *Any right Noetherian integral domain is a right Ore ring.* ∎

Examples of non-Ore domains are free associative algebras on two or more generators (soon to be formally defined).

Let R, A, B be any rings, $\alpha : R \to A$, $\beta : R \to B$ two homomorphisms and M an (A, B)-bimodule. Then an (α, β)-*derivation* from R to M is a mapping $\delta : R \to M$ which is additive and satisfies

$$(xy)^\delta = x^\alpha y^\delta + x^\delta y^\beta. \qquad (6)$$

In particular, if $A = R$ and $\alpha = 1$, we speak of a (right) β-*derivation*. Putting $x = y = 1$ in (6) and observing that $1^\alpha = 1^\beta = 1$, we see that any (α, β)-derivation satisfies

$$1^\delta = 0. \qquad (7)$$

As an example, let $R = A = B = k(t)$ be the field of rational functions in t over some field k, and let f' be the usual derivative of f, then on taking $\alpha = \beta = 1$, we obtain the familiar formula

$$(fg)' = fg' + f'g$$

as special case of (6).

With any (A, B)-bimodule M we can associate the ring $(A, B; M)$ consisting of all matrices

$$\begin{pmatrix} a & m \\ 0 & b \end{pmatrix} \qquad (a \in A, \ b \in B, \ m \in M)$$

with the usual matrix addition and multiplication. The (A, B)-bimodule property just ensures that we get a ring in this way:

$$\begin{pmatrix} a & m \\ 0 & b \end{pmatrix} \begin{pmatrix} a' & m' \\ 0 & b' \end{pmatrix} = \begin{pmatrix} aa' & am' + mb' \\ 0 & bb' \end{pmatrix}.$$

Given maps $\alpha : R \to A$, $\beta : R \to B$, $\delta : R \to M$, we can define a map $(\alpha, \beta ; \delta) : R \to (A, B ; M)$ by the rule

$$x \to \begin{pmatrix} x^\alpha & x^\delta \\ 0 & x^\beta \end{pmatrix}$$

and it is easily verified that this is a ring-homomorphism if and only if α, β are homomorphisms and δ is an (α, β)-derivation. This alternative method of defining derivations is often useful.

Keeping the same notation, take $m \in M$ and define a mapping $\delta_m : R \to M$ by the rule

$$\delta_m : x \mapsto x^\alpha m - m x^\beta.$$

This mapping is easily verified to be an (α, β)-derivation; it is called the *inner* (α, β)-derivation induced by m. A derivation not of this form is called *outer*.

Exercises 0.5

1. Let S be a cancellation semigroup; show that any central subsemigroup is a right (and left) denominator set.

2. Let S be a semigroup, T a right denominator set and S_T its group of fractions; show that any finite set of fractions can be brought to a common denominator.

3. Let S be a semigroup and T a subsemigroup such that (i) for any $s, s' \in S$, $t \in T$, $ts = ts'$ implies $st' = s't'$ for some $t' \in T$, (ii) for any $s \in S$, $t \in T$, $sT \cap tS \neq \emptyset$. If S_T is the universal T-inverting semigroup and $f : S \to S_T$ the natural mapping, show that every element of S_T can be written $sf(tf)^{-1}$ ($s \in S$, $t \in T$). Determine the kernel of f, i.e. the subset of elements (s, s') of $S \times S$ such that $sf = s'f$.

4. Show that the direct limit of any system of rigid semigroups is again rigid. Give a proof of Theorem 5.2 in case S is *conical* (i.e. $uv = 1$ implies $u = v = 1$) as well as rigid.

5. Let R be a ring and T a right denominator set. If R is (right or left) Ore, Noetherian or Artinian, show that R_T has the same property.

6. Let R be a right Ore ring and K its field of fractions. If $A \in R_n$ is a left non-zerodivisor show that it is a left non-zerodivisor in K_n and hence is invertible with an inverse of the form Bd^{-1}, $B \in R_n$, $d \in R^*$. Deduce that every right zerodivisor in R_n is a left zerodivisor. Does the reverse inclusion hold generally?
Show that the set of left non-zerodivisors in R_n is a right denominator set.

7. Let R be any ring; show that any R^*-inverting homomorphism into a field must be injective.

8. Let R be an integral domain. Show that any Ore subring of R is contained in a maximal Ore subring.

9. Show that any right denominator set in an integral domain R consists of right large elements. Deduce that if R^* is a right denominator set, R satisfies the right Ore condition.

10. Show that every left factor of a right large element is again right large. In an integral domain is the same true of every right factor?

11. Let R be a right Ore ring with right ACC_1 and $\mathfrak{a}$ an ideal of R which is principal as left ideal. If $R/\mathfrak{a}$ is an integral domain, show that it is again a right Ore ring.

12. Let R be an integral domain which is not a right Ore ring and let $n \geqslant 1$. Show that ^nR can be embedded in R as a right ideal, and if ^nR does not have unique rank, show that ^nR contains a strictly descending chain of direct summands that are free of rank n. Deduce that if an integral domain R satisfies right ACC_n then R^n has unique rank (as left or right R-module).

13. Let R, R' be rings and T, T' right denominator sets in R, R' respectively. Show that any homomorphism $\alpha: R \to R'$ which maps T into T' extends in just one way to a homomorphism of the rings of fractions. Deduce that any α-derivation on R can also be extended to R_T.

14. Given any derivation δ of a ring R, show that $K = \ker \delta$ is a subring of R. Show that any unit of R which lies in K is a unit in K, and deduce that the kernel of δ acting on a local ring is itself a local ring.

15. Prove Leibniz's formula for derivations:

$$(ab)\, \delta^n = \Sigma \binom{n}{i} (a\delta^i)\, (b\delta^{n-i}).$$

More generally, if δ is an α-derivation, show that

$$(ab)\, \delta^n = \Sigma a\delta^i . bf_i^n(\alpha, \delta),$$

where $f_i^n(\alpha, \delta)$ is the coefficient of t^i in the formal expansion of $(t\alpha + \delta)^n$.

16. If δ is a derivation on an integral domain of characteristic p, show that δ^p is again a derivation.

17. If δ is a nilpotent derivation of exponent r on an integral domain K (i.e. $\delta^r = 0$, $\delta^{r-1} \neq 0$), and $r > 1$, show that K has prime characteristic p, and $r = p^t$. (Hint: Apply δ^r to ab, where $b\delta \neq 0$, $b\delta^2 = 0$, and use Leibniz's formula to show that $p|r$. Now repeat the argument with δ replaced by δ^p.)

0.6 Modules over Ore rings

Many general results on modules over commutative integral domains hold more generally *either* for right modules over right Ore rings *or* for right modules over left Ore rings. The results of this section are mainly intended to supplement the results proved later for more special classes of rings, but some will find application later.

Let R be a ring and let M be a right R-module. An element $x \in M$ is called a *torsion element* if $xa = 0$ for some $a \in R^*$. When R is a right Ore ring the set tM of all torsion elements of M is a submodule, the *torsion* *submodule* of M. To prove that tM is a submodule, let $xa = 0$ and $a, b \in R^*$. If $ab' = ba' \neq 0$, then $xba' = xab' = 0$, hence $xb \in tM$ for all $b \in R^*$; the closure under addition is proved similarly. If $tM = 0$ we say that M is *torsion free*, and if $tM = M$, we call M a *torsion module*. Clearly for any module M, tM is a torsion module and M/tM is torsion free; moreover, these two properties serve to determine tM.

The process of forming fractions can be applied to modules as well as rings:

PROPOSITION 6.1. *Let R be a right Ore ring, K its field of fractions, and M a right-module. Then $M \otimes K$ can be described as the set of formal products xb^{-1} ($x \in M$, $b \in R^*$) subject to the relations $xb^{-1} = x'b'^{-1}$ if and only if there exist u, $v \in R^*$ with $xu = x'v$, $bu = b'v$.*

Proof. Any element of $M \otimes K$ has the form $x = \Sigma x_i \otimes a_i b_i^{-1}$. If b is a common right multiple for the $b_i : b_i c_i = b$, then

$$x = \Sigma x_i \otimes a_i c_i b^{-1} = (\Sigma x_i a_i c_i)b^{-1}.$$

Thus every element of $M \otimes K$ has the form xb^{-1}. Given $p = xb^{-1}$ and $p' = x'b'^{-1}$, there exist u, $v \in R^*$ such that $bu = b'v = c$, and we can write $p = xuc^{-1}$, $p' = x'vc^{-1}$. Clearly $p = p'$ if and only if $pc = p'c$, i.e $xu = x'v$. ∎

The proposition shows that $xb^{-1} = 0$ for some $b \in R^*$ if and only if $xu = 0$ for some $u \in R^*$, i.e. precisely if x is a torsion element. Hence we have the

COROLLARY. *Let R be a right Ore ring and K its field of fractions. For any right R-module M, the kernel of the canonical mapping*

$$M \to M \otimes K \tag{1}$$

is the torsion submodule tM of M. In particular, (1) is an embedding precisely when M is torsion free. ∎

In order to discuss the structure of K as R-module we shall need the notion of a locally free module; it is convenient to define this quite generally. Let R be any ring; a right R-module M is said to be *locally free* if every finite subset of M is contained in a free submodule of M. This free submodule may always be taken to be of finite rank, for the given subset requires only finitely many of the basis elements to express it and we may discard the rest. In dealing with tensor products we shall need the *independence property of the tensor product*, which may be expressed as follows:

LEMMA 6.2. *Let R be any ring, M a right R-module and N a left R-module which moreover is locally free. Given $a_1, \ldots, a_r \in M$ and $b_1, \ldots, b_r \in N$ if the a's are right linearly independent over R and*

$$\Sigma a_i \otimes b_i = 0 \qquad (2)$$

in $M \otimes N$, then $b_i = 0$ $(i = 1, \ldots, r)$.

Proof. Let N' be the submodule of N generated by $b_1, \ldots, b_r$. In $M \otimes N'$ it may well happen that $\Sigma a_i \otimes b_i \neq 0$, but by adjoining a finite number of elements to N' we reach a finitely generated submodule N^* of N such that (2) holds in $M \otimes N^*$. Since N is locally free, we may assume N^* to be free, by enlarging it further, if necessary. Let (u_λ) be a left R-basis for N^* and put $b_i = \Sigma \beta_{i\lambda} u_\lambda$, then (2) in $M \otimes N^*$ shows that

$$\Sigma a_i \beta_{i\lambda} \otimes u_\lambda = \Sigma a_i \otimes \beta_{i\lambda} u_\lambda = 0.$$

Since the u_λ form a basis of N^*, $\Sigma a_i \beta_{i\lambda} = 0$ for all λ, but the a_i are linearly independent, hence $\beta_{i\lambda} = 0$ and this shows that $b_i = 0$. ∎

We note that the lemma is false without the proviso on N, by the usual counter-example for tensor products: take $R = \mathbf{Z}$, $M = \mathbf{Z}$ and $N = \mathbf{Z}/2\mathbf{Z}$. If M, N are generated by e, f respectively, then $2e$ is independent in M and $2e \otimes f = e \otimes 2f = 0$, although $f \neq 0$.

Let us return to a right Ore ring R with its field of fractions K. The latter can be expressed as

$$K = RR^{*-1} = \lim_{\rightarrow} \{Ra^{-1} | a \in R^*\}.$$

Each Ra^{-1} is a free left R-module; it follows that K is locally free as left R-module and hence, by Lemma 6.2, if a family of elements in a right R-module is linearly independent, so is its image in $M \otimes K$. Therefore the dimension of $M \otimes K$ as vector space over K equals the cardinality of any maximal linearly independent subset of M. This number is an invariant of M which we shall denote by $rk\, M$ and call the *rank* of M. In particular, $rk\, M = 0$ if and only if M is a torsion module. On free modules the rank clearly agrees with our previous definition of rank, and since tensoring with K preserves exactness, we have

PROPOSITION 6.3. *If $0 \rightarrow M' \rightarrow M \rightarrow M'' \rightarrow 0$ is an exact sequence of right R-modules over a right Ore ring R, then*

$$rk\, M = rk\, M' + rk\, M''.$$

In particular if N is a submodule or a homomorphic image of M, then $rk\, N \leqslant rk\, M$. ∎

The last assertion, relating to homomorphic images, holds (under an appropriate definition of rank) for a large class of rings, including all that can be embedded in fields, and hence most of the rings considered later. But apart from this none of the other assertions hold with 'left' in place of 'right' Ore ring. Thus let R be any left Ore ring (or indeed any integral domain) that is not a right Ore ring and let x, y be two right linearly independent elements of R, then R contains the right ideal $xR + yR$ which is isomorphic to ^2R; this shows that the first part of Proposition 6.3 cannot be extended to such rings. For an example showing that the Corollary of Proposition 6.1 does not extend, see Exercise 5.1.9. However, we shall see later that the Corollary remains true when R is a left Bezout domain (cf. Exercise 1.1.11).

The following property of left Ore rings is not shared by right Ore rings in general (cf. Exercise 3).

PROPOSITION 6.4. *Let R be a left Ore ring and K its field of fractions. Then any right K-module, considered as right R-module, is locally free.*

Proof. Let M be a finitely generated R-submodule of a K-vector space V, which may without loss of generality be taken to be ^nK, for some finite n. We can choose a common left denominator $c \in R^*$ for the components of the finite generating set of M; then $M \subseteq c^{-1}R^n$ and the latter is a free R-module. ∎

Combining this result with Proposition 6.1, Corollary we obtain the

COROLLARY. *If R is a left and right Ore ring, then any finitely generated torsion free R-module is embeddable in a free R-module.* ∎

Exercises 0.6

1. Let K be a field, F a subfield of its centre and E a commutative field containing F. Show that $K \otimes_F E$ is an Ore ring, provided that it is an integral domain.

2. Let E be a commutative field, F a subfield such that E/F is algebraic, and A an F-algebra which is a right Ore ring with field of fractions K. If $A \otimes E$ is an integral domain, show that it is a right Ore ring with field of fractions $K \otimes E$ (all tensor products are understood to be taken over F).

3. Let R be a right but not left Ore ring and K its field of fractions. Show that K, as a right R-module, has rank 1 but is not locally free.

4. (Gentile [60]). Let R be a right Ore ring. If every finitely generated torsion free right R-module can be embedded in a free right R-module, show that R is also left Ore. Note that this is a converse to Proposition 6.4, Corollary; investigate the truth of other possible converses (cf. Ex. 1.4.4).

5. Let R be a right Ore ring, K its field of fractions and $\mathfrak{a}$ any non-zero right ideal. Show that $\mathfrak{a} \otimes_R K \cong K$ (as right R-modules). Deduce that for any left or right Ore ring R with field of fractions K, $K \otimes_R K \cong K$. (This is equivalent to the assertion that the embedding $R \to K$ is an *epimorphism* in the category of rings; it actually holds for any ring embeddable in a field, *cf.* Exercise 7.2.6).

6. Let R be a right Ore ring and K its field of fractions; verify that K is the injective hull of R.

7. (Bergman [67]). Let R be a left Ore ring and K its field of fractions. Prove that the following conditions on a finitely generated right R-module M are equivalent:

(a) the canonical mapping $M \to M \otimes K$ is an embedding,
(b) M is embeddable in a K-space (regarded as right R-module),
(c) M is embeddable in a free right R-module,
(d) $\text{Hom}_R (M, R)$ distinguishes elements of M,
(e) $\text{Hom}_R (M, K)$ distinguishes elements of M.

8^0. Find the relations between (a)–(e) of Exercise 7 when (i) K is a field and R a subring generating K, as field, (ii) K is any ring and R a subring. Find conditions on the finitely generated R-module M for (a)–(e) to be equivalent.

0.7 Free associative algebras

Much of the subsequent discussion will centre on free algebras and related rings; in this section we shall describe them briefly and list some of their simpler properties.

Free algebras are most easily defined by the universal mapping property: Given a commutative ring K and a set X, there exists a K-algebra $A = K\langle X \rangle$ and a mapping $i : X \to A$ such that any mapping $\theta : X \to B$ into another K-algebra B can be factored uniquely by i. Thus there exists a unique homomorphism $\theta' : A \to B$ such that the accompanying diagram

commutes. This algebra A, clearly unique up to isomorphism, is easily proved to exist by general considerations (*cf.* Cohn [65], p.170). However, its main use stems from the fact that its elements have a rather simple normal form, which is easily described and can also be used to prove its existence.

Consider first the free semigroup S_X on a set X. Its elements can be expressed as products

$$x_1 x_2 \ldots x_n \qquad (x \in X, \; n \geqslant 0),$$

including the empty product to represent 1. Sometimes it is more convenient to take X indexed: $X = (x_i)$, then the general element of S_X takes the form

$$x_I = x_{i_1} x_{i_2} \ldots x_{i_n},$$

where $I = (i_1, \ldots, i_n)$ runs over all finite sequences of suffixes (including the empty sequence). Now the free associative algebra $K\langle X \rangle$ on X over K may be described as the semigroup algebra of S_X. Its elements are uniquely expressible in the form

$$\Sigma x_I a_I \qquad (a_I \in K, \text{ almost all } 0). \tag{1}$$

We remark that since the elements of X do not commute, $K\langle X \rangle$ is non-commutative, unless X consists of a single element x. In that case $K\langle x \rangle$ becomes just the polynomial ring $K[x]$ in a single variable over K. One of our main objectives, in later chapters, will be to give a detailed description of $K\langle X \rangle$ in which each property of $K[x]$ has its counterpart.

We first check that $K\langle X \rangle$ indeed satisfies the characteristic property of free algebras. Given any K-algebra B and any mapping $\theta : x_i \mapsto b_i$ of X into B, consider the mapping

$$\theta' : \Sigma x_I a_I \mapsto \Sigma b_I a_I \qquad (b_I = b_{i_1} \ldots b_{i_n});$$

this is a well defined mapping by the uniqueness of the normal form (1). It is easily seen to be a homomorphism and it extends the given mapping θ. Moreover, θ' is determined by its values on a generating set, hence for each θ there is just one θ', as claimed.

Thus any mapping $X \to A$ extends to a unique homomorphism $K\langle X \rangle \to A$, and similarly, given an (A, B)-bimodule M, any mapping $X \to (A, B; M)$ extends to a unique homomorphism, i.e. given homomorphisms α, β of $K\langle X \rangle$ into A, B respectively, any mapping $X \to M$ extends to a unique (α, β)-derivation.

In general we cannot recover K from $K\langle X \rangle$, but when K is a field (the case that will mainly occupy us later), then K is determined by $K\langle X \rangle$ as the set of units, together with zero. The generating set X is also not generally determined, but at least its cardinality is unique, and is called the *rank* of $K\langle X \rangle$. To prove that the rank is well-defined, let A be any K-algebra and $\alpha : K\langle X \rangle \to A$ any homomorphism; the (α, α)-derivations from $K\langle X \rangle$ to A form a K-module D say, and by the universal mapping property for derivations we have an isomorphism of K-modules

$$D \cong A^X. \tag{2}$$

Now choose A to be a free K-module of rank $r > 1$ (e.g. $A = K_2$) and observe that K has IBN. Choose α in any way, then on comparing ranks in (2), if X is finite with d elements, we find the rank of D to be r^d, and this determines d uniquely. Of course when X is infinite, then every generating set of $K\langle X\rangle$ is infinite, and all have the same cardinality, by Proposition II.5.5 of Cohn [65].

The product $x_1x_2 \ldots x_n$ will be called a *monomial* of $K\langle X\rangle$ and $n = l(x_1x_2 \ldots x_n)$ its *length*; for a general element (1) of $K\langle X\rangle$ we define its *degree* as

$$d(f) = \max \{l(x_I)|a_I \neq 0\},$$

and its *order* as

$$o(f) = \min \{l(x_I)|a_I \neq 0\}.$$

The degree (and the order) so defined depend of course on the choice of the free generating set X; more precisely, if Y is another free generating set, we have a bijection

$$\alpha : X \to Y, \tag{3}$$

which extends to a unique automorphism of $K\langle X\rangle$. Thus the different (ordered) free generating sets of $K\langle X\rangle$ correspond to the automorphisms. The automorphism α given by (3) preserves the degree if and only if the expression of each element of Y in terms of X is linear, and it preserves the order if and only if the expression of each element of Y in terms of X is of order 1. The ideal of $K\langle X\rangle$ generated by X is called the *augmentation ideal*; X and Y have the same augmentation ideal if and only if the expression of each element of Y in terms of X has zero constant term. In that case the corresponding automorphism α is said to be *augmentation preserving*.

In later work the degree of a non-commutative polynomial (1) will play an essential role. We shall quite generally define a *degree-function* in a ring R as a function which assigns to each $a \in R$ a degree $d(a)$ such that

D.1. For $a \in R^*$, $d(a) \geqslant 0$, while $d(0) = -\infty$,

D.2. $d(a - b) \leqslant \max \{d(a), d(b)\}$,

D.3. $d(ab) = d(a) + d(b)$.

By D.3, $d(1) = 0$. By D.2 we have, as for valuations,

$$d(a) = d(-a), \tag{4}$$

and

$$d(a + b) \leqslant \max \{d(a), d(b)\}, \tag{5}$$

with equality holding whenever $d(a) \neq d(b)$ ("every triangle is isosceles"). By D.1 and D.3, R^* is closed under multiplication, i.e. every non-zero ring with a degree function is necessarily an integral domain.

For example, the degree defined earlier on $K\langle X \rangle$ is a degree-function in this sense, provided that K is an integral domain. In particular, this shows that the free algebra over a field is an integral domain.

Exercises 0.7

1. Let R be a free k-algebra and $\mathfrak{a}$ the augmentation ideal. Show that $\bigcap \mathfrak{a}^n = 0$; find an ideal not contained in $\mathfrak{a}$ with the same property.

2. Let R be the group algebra (over a commutative field k) of a free group of rank t. Show how to compute t in terms of R.

3. Given a commutative field k and two disjoint sets X and Y, the *mixed* free k-algebra on X, Y is defined as the k-algebra $R = k\langle X, Y, Y^{-1}\rangle$ generated by X, Y and the inverses of elements of Y which is universal for Y-inverting mappings of $X \cup Y$ into k-algebras. Prove that the numbers of invertible and of non-invertible free generators in a mixed free algebra are independent of the choice of free generating set.

4. Let $R = k\langle x_1, \ldots, x_q \rangle$ be the free associative algebra and

$$D = \Sigma x_i \left(\frac{\partial}{\partial x_i} \right)$$

the derivation which is the identity on the x_i. Show that for any element $a = \Sigma a_n$ of R, where a_n is homogeneous of degree n, $a^D = \Sigma n a_n$. (This generalizes Euler's theorem on homogeneous functions).

5. (Jategaonkar [69″], Koševoi [70]). Let R be an integral domain which is not a right Ore ring. If $xR \cap yR = 0$, show that x and y generate a free algebra (over $\mathbf{Z}$ or $\mathbf{GF}(p)$); deduce that an integral domain contains a free algebra on two free generators unless it is a left and right Ore ring.
    Given a left but not right Ore ring, obtain an embedding of a free algebra of rank 2 in a field.

6. In $k\langle x, y \rangle$ show that the elements xy^r ($r = 0, 1, \ldots$) form a free generating set, and deduce that the free algebra of countable rank can be embedded in the free algebra of rank 2.

7. With every ring R we can associate another ring R^{ab}, the ring R made abelian, which is obtained by dividing R by the ideal generated by all commutators. Thus the natural mapping $R \to R^{ab}$ is universal for homomorphisms of R into commutative rings.
    Given any ring R, denote by S the subring generated by the kernel of the natural mapping $R \to R^{ab}$. Given $a, b \in R$, write $c = [a, b] = ab - ba$ and establish the identity

$$c[ac, bc] = ca[c, bc] + cb[ac, c] + c^4.$$

Deduce that S^{ab} is nilpotent.
    If R is any free algebra, verify that R^{ab} is the corresponding polynomial ring; hence show that in this case S is not free.

0.8 Skew polynomial rings

There is one type of ring that is especially useful in providing examples and counter-examples at the simplest level. It generalizes the usual polynomial ring in one indeterminate $F[x]$ over a field F, but differs from it in that F need not be commutative or commute with x.

Let R be any ring and S a ring containing R as subring, as well as an element x such that every element of the ring A generated by R and x is uniquely expressible in the form

$$f = a_0 + xa_1 + \dots + x^n a_n \qquad (a_i \in R). \tag{1}$$

Moreover, the function

$$d(f) = \max \{i | a_i \neq 0\} \tag{2}$$

is assumed to be a degree function on A, in the sense explained in 0.6. Of the properties D.1–3 listed there, only D.3 represents a restriction on the base ring R; it entails that R is an integral domain and moreover, for any $a \in R$, there exist $a^\alpha, a^\delta \in R$ such that

$$ax = xa^\alpha + a^\delta \qquad (a \in R). \tag{3}$$

In the first place we note that a^α, a^δ are uniquely determined by a, and $a^\alpha = 0$ if and only if $a = 0$. Next we have, by (3), $(a + b)x = x(a + b)^\alpha + (a + b)^\delta$, $ax + bx = xa^\alpha + a^\delta + xb^\alpha + b^\delta$, whence

$$(a + b)^\alpha = a^\alpha + b^\alpha, \qquad (a + b)^\delta = a^\delta + b^\delta. \tag{4}$$

Thus α, δ are additive mappings of R. Similarly, by comparing the expressions for $a(bx)$ and $(ab)x$ we find

$$(ab)^\alpha = a^\alpha b^\alpha, \qquad (ab)^\delta = a^\delta b^\alpha + ab^\delta. \tag{5}$$

By putting $a = b = 1$ and observing that R is an integral domain, we see that

$$1^\alpha = 1, \qquad 1^\delta = 0. \tag{6}$$

Hence α is an endomorphism of R; in fact it is a monomorphism by the remark following (3). Now (4) and (5) show that δ is just an α-derivation of R. The law (3), with the uniqueness of (1), suffices to determine the multiplication in A in terms of R, α and δ. For by the distributive law we need only know $x^m a \cdot x^n b$ and from (3) we see that

$$x^m a \cdot x^n b = x^{m+1} a^\alpha x^{n-1} b + x^m a^\delta x^{n-1} b.$$

By induction on n this defines $x^m ax^n b$ in all cases. Thus A is completely fixed when R, α, δ are given. We shall write

$$A = R[x; \alpha, \delta] \tag{7}$$

and call A the *skew polynomial ring* in x over R determined by α, δ. When $\delta = 0$, we shall write $R[x; \alpha]$ instead of $R[x; \alpha, 0]$.

We now show that conversely, any injective endomorphism α of an integral domain and any α-derivation δ may be used to define a skew polynomial ring $R[x; \alpha, \delta]$ by the rule (3).

Consider the direct power $M = R^N$ as right R-module and define an additive group endomorphism of M by

$$x : (a_i) \mapsto (a_i^\delta + a_{i-1}^\alpha) \qquad (a_{-1} = 0). \tag{8}$$

If we identify R with its image in End (M) by right multiplication then the action of the endomorphism x defined by (8) satisfies the rule

$$
\begin{aligned}
(c_i) \, ax = (c_i a) \, x &= ((c_i a)^\delta + (c_{i-1} a)^\alpha) \\
&= c_i^\delta a^\alpha + c_i a^\delta + c_{i-1}^\alpha a^\alpha \\
&= (c_i^\delta + c_{i-1}^\alpha) \, a^\alpha + (c_i) \, a^\delta \\
&= (c_i) \, (x \cdot a^\alpha + a^\delta).
\end{aligned}
$$

Hence (3) holds, and moreover every element of the subring of End (M) generated by R and x can be brought to the form (1). This form is unique, because when $a_0 + xa_1 + \ldots + x^n a_n$ is applied to $(1, 0, 0, \ldots)$ it produces $(a_0, a_1, \ldots, a_n, 0, 0, \ldots)$. Moreover, the function $d(f)$ defined by (2) is easily seen to be a degree-function (using the fact that R is an integral domain and α a monomorphism). Thus we have proved

THEOREM 8.1. *Let R be an integral domain with an injective endomorphism α and an α-derivation δ, then there is a skew polynomial ring $R[x; \alpha, \delta]$, and every skew polynomial ring arises in this way.* ∎

PROPOSITION 8.2. *Any skew polynomial ring $A = R[x; \alpha, \delta]$ is an integral domain. If R is right Noetherian, so is A.*

Proof. That A is an integral domain follows already from the existence of a degree function. Now assume that R is right Noetherian; the proof that A is right Noetherian follows as in the commutative case (Hilbert basis theorem): If $\mathfrak{a}$ is a right ideal of A, let $\mathfrak{a}_0 = \Sigma_1^k c_i R$ be the ideal of its leading coefficients and suppose that

$$f_i = x^{n_i} c_i + \ldots \in \mathfrak{a}.$$

Writing $n = \max \{n_i\}$, we can reduce every element of $\mathfrak{a}$ to a polynomial of degree less than n and a linear combination of $f_1, \ldots, f_k$. For each degree $i < n$ there is a finite basis B_i for the polynomials of degree at most i in $\mathfrak{a}$ and the union of all the B_i and $f_1, \ldots, f_k$ forms a finite basis for $\mathfrak{a}$. ∎

This result is not left–right symmetric because in (3) the coefficients were written on the right. One therefore sometimes introduces the *left* skew polynomial ring, in which the commutation rule instead of (3) is

$$xa = a^\alpha x + a^\delta. \tag{9}$$

In general $R[x; \alpha, \delta]$ will not be a left skew polynomial ring; but when α is an automorphism of R, with inverse β say, then on replacing a by a^β in (3) and rearranging terms we obtain

$$xa = a^\beta x - a^{\beta\delta}. \tag{10}$$

Thus we have

PROPOSITION 8.3. *The ring $R[x; \alpha, \delta]$ is a left skew polynomial ring provided that α is an automorphism.* ∎

It follows that when R is left Noetherian and α is an automorphism, then $R[x; \alpha, \delta]$ is also left Noetherian; without the hypothesis on α this need not hold (*cf.* Exercise 2).

From the Corollary to Proposition 5.4 we see that any skew polynomial ring $K[x; \alpha, \delta]$ over a field K is a right Ore ring and hence has a field of fractions, which we shall write $K(x; \alpha, \delta)$. More generally, let R be a right Ore ring with field of fractions K. If α is any injective endomorphism of R and δ an α-derivation, they can be extended to K (*cf.* Ex. 5.13) and we have the inclusions

$$R[x; \alpha, \delta] \subseteq K[x; \alpha, \delta] \subseteq K(x; \alpha, \delta).$$

Any element u of $K(x; \alpha, \delta)$ has the form fg^{-1}, where $f, g \in K[x; \alpha, \delta]$. On bringing the coefficients of f and g to a common right denominator we can write them as $f = f_1 c^{-1}$, $g = g_1 c^{-1}$, where $f_1, g_1 \in R[x; \alpha, \delta]$, $c \in R^*$. Hence $u = fg^{-1} = f_1 g_1^{-1}$ and we have proved

PROPOSITION 8.4. *A skew polynomial ring over a right Ore ring is again a right Ore ring.* ∎

We now give some examples of skew polynomial rings, both as illustration and for later use.

1. The *complex-skew* polynomial ring $\mathbf{C}[x; \bar{\ }]$ consists of polynomials with complex coefficients and with commutation rule

$$ax = x\bar{a}, \qquad \text{where } \bar{a} \text{ is the complex conjugate of } a.$$

We observe that the centre of this ring is $\mathbf{R}[x^2]$, the ring of real polynomials in x^2. The residue-class ring mod $x^2 + 1$ is the field of real quaternions.

2. Let $K = F(t)$ be the field of rational functions in an indeterminate t over a commutative field F. The usual derivative $f \mapsto f'$ defines a derivation on K, and this gives rise to a skew polynomial ring $R = K[x; 1, ']$, the ring of differential operators.

3. Let F be a commutative field of characteristic $p \neq 0$ and E/F a separable extension of degree p, say $E = F(\xi)$, where $\xi^p - \xi \in F$. The mapping $\alpha : f(\xi) \mapsto f(\xi + 1)$ defines an automorphism of order p and we have the skew polynomial ring $E[\eta; \alpha]$.

4. Let k be any commutative field and denote by $\mathbf{W}_1(k)$ the algebra on x, y over k with the defining relation $xy - yx = 1$. This ring $\mathbf{W}_1(k)$, called the *Weyl algebra* on x, y over k, may also be defined as skew polynomial ring $A[y; 1, ']$, where $A = k[x]$ and $'$ is the derivation with respect to x. The example in No. 2 above is obtained by localizing at the set of all monic polynomials in x over k, and the example in No. 3 by putting $\xi = xy$, $\eta = y$ and then localizing at the set of all monic polynomials in ξ over k.

The Weyl algebra is useful as an example of a finitely generated infinite-dimensional algebra which in characteristic 0 is simple. For in any non-zero ideal we can pick an element $f(x, y)$ with the least possible x-degree. The ideal still contains $df/dx = fy - yf$; this is of lower degree and so must be 0. Therefore $f = f(y)$ is a polynomial in y alone. If its y-degree is minimal, then $df/dy = xf - fx = 0$, hence $f = c \in k^*$; so the ideal contains a non-zero element of the ground field, and so must be the whole ring. This shows $\mathbf{W}_1(k)$ to be simple.

Besides the polynomial ring we shall also need the ring of formal power series. Taking first the case of a zero derivation, we can describe the formal power series ring $A = R[[x; \alpha]]$ over a ring R as the set of all infinite series

$$f = a_0 + xa_1 + x^2a_2 + \dots \tag{11}$$

with componentwise addition, and multiplication based on the commutation rule $ax = xa^\alpha$. With each power series f we associate its *order* $o(f)$, defined as the suffix of the first non-zero coefficient in (11). This satisfies the conditions for an order function, analogous to D.1–3:

O.1 For $a \in R^*$, $o(a) \in \mathbf{N}$ while $o(0) = \infty$,

O.2. $o(a - b) \geqslant \min \{o(a), o(b)\}$,

O.3. $o(ab) = o(a) + o(b)$.

As before this implies that A is an integral domain, and from (3) we see that $o(f)$ will not be an order function unless $\delta = 0$. However we may now allow higher powers; the most general commutation formula in a formal power series ring is of the form

$$ax = xa^{\delta_0} + x^2 a^{\delta_1} + x^3 a^{\delta_2} + \dots \tag{12}$$

where $\delta_0, \delta_1, \delta_2, \dots$ is a sequence of maps of the ground ring R into itself. As before we can show that they are additive and satisfy $1^{\delta_0} = 1$, $1^{\delta_i} = 0$ for $i > 0$. Moreover δ_0 must be an endomorphism of R, while the δ's satisfy

$$(ab)^{\delta_n} = \Sigma a^{\Delta_i^n} b^{\delta_i}, \qquad (n > 0), \tag{13}$$

where Δ_i^n is the coefficient of t^{n+1} in $(\Sigma t^{k+1} \delta_k)^{i+1}$. Such a sequence $(\delta_0, \delta_1, \dots)$, with $\delta_0 = \alpha$, is sometimes called a *higher α-derivation*; we shall not need a special notation for the ring defined by such a higher derivation.

As an example of a ring with commutation formula (12) let us take a skew polynomial ring $K[y; \alpha, \delta]$ where K is a field. In the field of fractions $K(y; \alpha, \delta)$ take the subring generated by K and y^{-1}. Writing $x = y^{-1}$, we have $ax^{-1} = x^{-1} a^\alpha + a^\delta$, hence

$$xa = a^\alpha x + xa^\delta x$$
$$= a^\alpha x + a^{\delta\alpha} x^2 + xa^{\delta\alpha} x$$
$$= \dots$$

hence in the limit,

$$xa = a^\alpha x + a^{\delta\alpha} x^2 + a^{\delta^2\alpha} x^3 + \dots, \tag{14}$$

and this is indeed of the form (12), with $\delta_n = \delta^n\alpha$, except for a change of sides. In particular, if δ is nilpotent, say $\delta^{r+1} = 0$, then (14) reduces to a polynomial formula

$$xa = a^\alpha x + a^{\delta\alpha} x^2 + \dots + a^{\delta^r\alpha} x^{r+1}. \tag{15}$$

Of course not every higher α-derivation is of the special form $\delta_n = \delta^n\alpha$, but if in (12), $\delta_i = 0$ for $i > r$, then (with another mild restriction) we do indeed have $\delta_n = \delta^n\alpha$.

THEOREM 8.5. *Let A be the ring of polynomials in an indeterminate x with coefficients in a field K, with the normal form*

$$f = a_0 + a_1 x + \dots + a_n x^n \qquad (a_i \in K), \tag{16}$$

such that $o(f) = \min \{i \,|\, a_i \neq 0\}$ is an order function, and

$$xa = a^\alpha x + a^{\delta_1} x^2 + \dots + a^{\delta_r} x^{r+1}. \tag{17}$$

Assume further that (i) *r is independent of a,* (ii) α *is an automorphism of K and* (iii) $\alpha, \delta_1, \ldots, \delta_r$ *are right linearly independent over K, in the sense that*

$$a^{\alpha} b_1 + a^{\delta_1} b_2 + \ldots a^{\delta_r} b_{r+1} = 0 \quad \text{implies} \quad b_1 = b_2 = \ldots = b_{r+1} = 0.$$

Then A is obtained from a skew polynomial ring $B = K[y; \alpha, \delta]$, where δ is a nilpotent α-derivation satisfying $\delta^{r+1} = 0$ by adjoining $x = y^{-1}$ and taking polynomials in x. Conversely, every skew polynomial ring with a nilpotent α-derivation δ leads to a ring satisfying (17), *with $\delta_i = \delta^i \alpha$.*

Proof. The converse has already been established. To prove the direct part, we have from (17), by induction on n,

$$x^n a = a_{nn} x^n + a_{nn+1} x^{n+1} + \ldots + a_{nk} x^k, \tag{18}$$

where k depends on n but not on a. We shall write (18) as

$$x^n a = \Sigma a_{ni} x^i, \tag{19}$$

where the summation is over all i and $a_{ni} = 0$ for $i < n$ or $i > k$. Clearly $a \mapsto (a_{ni})$ is a matrix representation of K over itself. From (17) we find, by induction on n,

$$a_{nn} = a^{\alpha^n}. \tag{20}$$

For $n = 1$ we have from (19),

$$x(ab) = \Sigma (ab)_{1i} x^i,$$
$$(xa)b = \Sigma a_{1i} x^i b = \Sigma a_{1i} b_{ij} x^j.$$

Hence for $j > r + 1$,

$$\Sigma a_{1i} b_{ij} = 0 \quad (j = r + 2, \ldots).$$

Now $a_{11} = a^{\alpha}$, $a_{1i} = a^{\delta_{i-1}}$ and all these elements are right linearly independent over K, by hypothesis, so

$$b_{ij} = 0 \quad \text{for} \quad i = 1, \ldots, r+1; \quad j = r+2, \ldots.$$

Thus (18) takes the form

$$x^i a = a_{ii} x^i + a_{ii+1} + a_{ii+1} x^{i+1} + \ldots + a_{ir+1} x^{r+1}. \tag{21}$$

In particular, taking $i = r + 1$ and remembering (20), we find

$$x^{r+1} a = a^{\alpha^{r+1}} x^{r+1}. \tag{22}$$

Similarly, for $i = r$, (21) becomes

$$x^r a = a^{\alpha^r} x^r + a_{rr+1} x^{r+1}.$$

Let us put $\alpha^{-1} = \beta$ and write a^y for a_{rr+1}, then using (22) we may write this relation formally as

$$x^{-1}a = x^r a^{\beta^{r+1}} x^{-r-1} = a^\beta x^{-1} + a^{\beta^{r+1}y}.$$

If we define δ by setting $a^\delta = -a^{\beta^r y}$, this relation takes on the form

$$x^{-1}a = a^\beta x^{-1} - a^{\beta\delta}. \tag{23}$$

A comparison with (10) shows that we have

$$ax^{-1} = x^{-1}a^\alpha + a^\delta.$$

Thus A is obtained from the skew polynomial ring $k[y; \alpha, \delta]$ by taking the subring generated by K and $x = y^{-1}$. To find the relation between δ and the δ_i in (17), apply y^{-1} to (17) and use (23):

$$a = x^{-1}(a^\alpha x + a^{\delta_1} x^2 + \dots + a^{\delta_r} x^{r+1})$$
$$= a + a^{\delta_1 \beta} x + \dots + a^{\delta_r \beta} x^r - a^\delta x - a^{\delta_1 \beta\delta} x^2 - \dots - a^{\delta_r \beta\delta} x^{r+1}.$$

Equating coefficients, we find $\delta_1\beta = \delta$, $\delta_i\beta = \delta_{i-1}\beta\delta$, $\delta_r\beta\delta = 0$, hence by induction, $\delta_i\beta = \delta^i$ or

$$\delta_i = \delta^i\alpha, \quad (i = 1, \dots, r), \quad \delta^{r+1} = 0 \quad \blacksquare$$

Exercises 0.8

1. Let $R = K[x; \alpha, \delta]$ be a skew polynomial ring. Show that K may be regarded as a right R-module be letting each $a \in K$ correspond to right multiplication by a and letting x correspond to the action of δ. When is this representation faithful?

2. Let $R = K[x; \alpha, \delta]$ be a skew polynomial ring where K is an integral domain and α an endomorphism such that $R^\alpha a \cap R^\alpha = 0$ for some $a \in R^*$. Show that R is not left Ore. If K is a field, show that R is left Ore if and only if α is an automorphism.

3. (Ore [32]). Let R be a skew polynomial ring over a field K and let f, g be polynomials of degrees m, n respectively. Denote by K_0 the centralizer in K of f, and assume that the dimension of K as right K_0-space is r. Show that $\operatorname{Hom}_R (R/fR, R/gR)$ is a right K_0-space of dimension at most rm (use formula (4) of 0.4).
Deduce that the eigenring of any non-zero element centralizing K is finite-dimensional over K.

4. Let K be a field with centre C and let $R = K[x; \alpha, \delta]$ be a skew polynomial ring. Show (i) if $\alpha | C \neq 1$, the variable can be changed to make $\delta = 0$, (ii) if $\alpha | C = 1$ and $\delta | C \neq 0$, the variable can be changed to make $\alpha = 1$. Deduce that for a suitable choice of variable either $\delta = 0$ or $\alpha = 1$ or C is the centre of R.

5. Let $R = k\langle x, y \rangle$; show that the elements $[yx^{[r]}] = [\dots[y, x], \dots], x] \, (r = 0, 1, \dots)$ form a free generating set. Denoting the subalgebra generated by these elements by S and the derivation on S defined by $a \mapsto [a, x]$ by δ, show that $R \cong S[x; 1, \delta]$.

6^0. To what extent are the conditions (i)–(iii) in Theorem 8.5 necessary?

Notes and comments on Chapter 0

Much of this material is part of the folklore and the citations given below are probably far from exhaustive. The formulation of Morita theory in 0.1 can easily be extracted from the general account (*cf.* e.g. Bass [68], and for the form in which it appears here, Cohn [67]). It is all that is needed for projective trivial rings; nearly all the rings occurring in the sequel fall into this class.

The first thorough discussion of IBN occurs in Leavitt [57]; for weak finiteness see Cockcroft–Swan [61]. A recent account of both (as well as an intermediate property) can be found in Cohn [66]. Theorem 2.8 is due to Kaplansky [58], who uses it to prove (*inter alia*) a commutative form of Theorem 2.9: Over a commutative semihereditary ring every projective module is a direct sum of finitely generated ideals. This result was generalized to right modules over a *right* semihereditary ring by Albrecht [61] and to right modules over a *left* semihereditary ring by Bass [64]. Both results are included in Theorem 2.9, which is taken from Bergman [72'].

The concepts of idealizer and eigenring seem to have been first used by Ore [32] in his papers on the formal theory of differential equations. Proposition 4.1 is due to Fitting [35]; special cases of 4.2–3 are well known (see e.g. Amitsur [58]), they are stated in this form by Cohn [70].

Since Ore's original construction (Theorem 5.3, Corollary, Ore [31]) there have been innumerable papers dealing with extensions, analogues for semigroups etc. For a fairly complete survey see Elizarov [69]. Proposition 5.4, Corollary is due to Goldie [58] (the customary proof actually gives the proposition). The discussion in 0.6 is taken from Bergman [67]; the Corollary of Proposition 6.4 occurs in Gentile [60].

Skew polynomial rings were first studied systematically by Ore [33]. Proposition 8.2, the analogue of the Hilbert basis theorem, is well known; Proposition 8.4 is due to Curtis [52]. Rings with the commutation rule (12) have been studied by Smits [68'], to whom Theorem 8.5 is due.

1. Firs, Semifirs and n-firs

Nearly all the rings discussed in this book have the property that all their (one-sided) ideals are free as left resp. right modules, of unique rank. This leads to the notion of *free ideal ring*, or *fir*, which forms the subject of this chapter. Firs may be regarded as generalizing the notion of a principal ideal domain.

If one only demands that all finitely generated right ideals, or all n-generator right ideals, shall be free of unique rank, one obtains the wider classes of *semifirs* and *n-firs* respectively. These notions are automatically left–right symmetric and arise naturally if one postulates that certain dependence relations should be trivializable. A number of equivalent conditions, and the relation to the commutative case, is discussed in **1.1**, while firs and their basic properties occupy **1.2**. In **1.3** we discuss a special case which arises when we insist on 'trivializing' our relations by subgroups of the general linear group. This will be of importance later when we come to meet firs (such as free associative algebras) in which every invertible matrix is a product of elementary and diagonal matrices.

From the homological point of view, firs are rather trivial. They are hereditary rings in which every projective is free (of unique rank). We can describe not merely firs, but also matrix rings over firs in this way, and in fact give a categorically invariant description of these classes. This is done in **1.4**, with a description of flat modules over semifirs. In **1.5** we establish the link with weak Bezout rings (they are just the 2-firs) and consider what happens when fractions are introduced. **1.6** is devoted to the notion of inertia, useful for studying extensions of firs.

1.1 Dependence relations

We begin with a study of conditions under which every relation of the form

$$x \cdot y = x_1 y_1 + \ldots + x_n y_n = 0 \tag{1}$$

is in some sense trivial. To be precise, let us call the relation (1) in a ring R *trivial* if for each $i = 1, \ldots, n$ either $x_i = 0$ or $y_i = 0$. Of course every non-zero ring R has non-trivial relations (1), e.g. $x = (a, ab), y = (bc, -c)^T$ for any

$a, b, c \in R$. Let us say that the n-term relation (1) is *trivialized* by the invertible matrix μ if the relation $x\mu \cdot \mu^{-1} y = 0$ is trivial. We shall be concerned with rings in which every relation (1) can be trivialized in this way. Some equivalent conditions for this to happen are given in

THEOREM 1.1. *Let R be a non-zero ring and n a positive integer. Then the following conditions are equivalent:*

(a) *every m-term relation $\Sigma x_i y_i = 0$ ($m \leqslant n$) holding in R can be trivialized by an invertible $m \times m$ matrix,*

(b) *given elements $x_1, \ldots, x_m \in R$ ($m \leqslant n$) which are right linearly dependent, there exist $m \times m$ matrices μ, v over R such that $\mu v = I$ and the vector $(x_1, \ldots, x_m) \mu$ has at least one zero component,*

(c) *any right ideal of R generated by $m \leqslant n$ right linearly dependent elements has a family of fewer than m generators,*

(d) *any right ideal of R on at most n generators is free, of unique rank,*

(e) *if F is a free right R-module, any submodule of F generated by at most n elements is free, and R, ^2R, $\ldots$, ^nR all have unique rank,*

(f) *if ϕ is a mapping of ^mR ($m \leqslant n$) into a free right R-module, then there exists an integer $r \leqslant m$ and isomorphisms μ, v such that the following diagram is commutative;*

$$\begin{array}{ccccc} \ker \phi & \longrightarrow & {}^mR & \overset{\phi}{\longrightarrow} & \operatorname{im} \phi \\ \downarrow{\scriptstyle \mu} & & \downarrow & & \downarrow{\scriptstyle v} \\ {}^rR & \longrightarrow & {}^rR \oplus {}^sR & \longrightarrow & {}^sR \end{array}$$

where $r + s = m$ and the lower maps are the natural inclusion and projection.

(a*)–(f*). *The left–right analogues of (a)–(f).*

Proof. We shall show (a) $\Rightarrow \ldots \Rightarrow$ (f) $\Rightarrow$ (a). Since (a) is left–right symmetric, the conditions must then be equivalent to their left–right analogues. The implication (a) $\Rightarrow$ (b) is evident.

(b) $\Rightarrow$ (c). Let $\mathfrak{a}$ be a right ideal generated by $x_1, \ldots, x_m$ and suppose that the x_i are right linearly dependent. By (b) there exist $\mu, v \in R_m$ such that $\mu v = I$ and if $x' = x\mu$, then $x_1' = 0$. Clearly $x_1', \ldots x_m' \in \mathfrak{a}$ and since $x = x\mu v = x'v$, $\mathfrak{a}$ is generated by $x_2', \ldots, x_m'$.

(c) $\Rightarrow$ (d). If $\mathfrak{a}$ is a right ideal on at most n generators, (c) allows us to reduce the number of generators until we get a linearly independent generating set; so $\mathfrak{a}$ will be free on at most n generators. Now let $m \leqslant n$ be the least integer such that $\mathfrak{a}$ is free of rank m and assume that $\mathfrak{a}$ does not have unique rank, say $\mathfrak{a} \cong {}^mR \cong {}^{m+k}R$ ($k > 0$). Then $\mathfrak{a}$ has a surjective endomorphism with kernel $^k R$. If we take a set of m generators of $\mathfrak{a}$, their images will again generate $\mathfrak{a}$ but will be linearly dependent, because the mapping has non-zero kernel. Hence by the previous argument $\mathfrak{a}$ will be free of rank $m' < m$, contradicting our choice of m.

(d) $\Rightarrow$ (e). In proving the first assertion we may assume F to be of finite rank, since our submodule, being finitely generated, can only involve finitely many components. So let $F \cong {}^k R$ and use induction on k. For $k = 0$ there is nothing to prove; so we may assume $k > 0$ and take a submodule H of $^k R$ on at most n generators. Write $p : {}^k R \to R$ for the projection on the last component. By (d) the right ideal $p(H)$ of R is free, hence the exact sequence

$$0 \to \ker (p|H) \to H \to p(H) \to 0$$

splits. Therefore $\ker (p|H)$ is a direct summand of H, hence a homomorphic image of H and so generated by at most n elements. It lies in ^{k-1}R and so is free by the induction hypothesis. Hence $H = \ker (p|H) \oplus p(H)$ is free.

Next observe that R is an integral domain. For if $a \in R$, then $aR \cong R/\mathfrak{n}$, where $\mathfrak{n}$ is the right annihilator of a in R. Since $n \geqslant 1$, aR is free and we have $R = \mathfrak{n} \oplus \mathfrak{a}$, where $\mathfrak{a} \cong aR$. This shows $\mathfrak{n}$ to be a homomorphic image of R, hence principal and so also free. By the uniqueness of the rank of R, either $\mathfrak{n} = 0$ or $\mathfrak{a} = 0$, i.e. a is either a left non-zerodivisor or zero, whence R is an integral domain. Now if R is a right Ore ring, it clearly has invariant basis number. Otherwise it contains right ideals isomorphic to $^k R$ for all k (by Proposition 0.5.4) and by (d) those with $k \leqslant n$ have unique rank.

(e) $\Rightarrow$ (f). Given ϕ as in (f), (e) tells us that im ϕ is free, say im $\phi \cong {}^s R$. Then the exact sequence

$$0 \to \ker \phi \to {}^m R \to \text{im } \phi \to 0$$

splits, whence $\ker \phi$ is a direct summand of $^m R$, hence also free, say $\ker \phi \cong {}^r R$. Thus our exact sequence is isomorphic to the direct sum decomposition

$$0 \to {}^r R \to {}^{r+s} R \to {}^s R \to 0,$$

and since $^m R$ has unique rank, $r + s = m$.

(f) $\Rightarrow$ (a). In the relation given by (a) let us consider $y = (y_i)$ as an element of $^m R$ and regard $x = (x_i)$ as a linear functional $\phi : {}^m R \to R$ with $y \in \ker \phi$. If we take μ as in (f) and represent it by an $m \times m$ matrix, we see that the last $m - r$ components of μy must be 0, while $\phi \mu^{-1}$, since it annihilates $^r R$, must have its first r components 0. Written as a row vector, $\phi \mu^{-1}$ is $x \mu^{-1}$; so we have shown that $x \mu^{-1} \cdot \mu y = 0$ is a trivial relation. ∎

A ring satisfying the equivalent conditions of Theorem 1.1 is called an *n-fir*. The reason for the name will become clear when we come to discuss the case of a *fir* (= free ideal ring, *cf.* **1.**2). Of the above defining conditions we shall find (a) and (f) the most useful tools, while (b) is the easiest to verify. Conditions (c) and (d) are weaker in appearance than the others since they do not explicitly tell us anything about the modules ^mR.

Clearly an n-fir is also an n'-fir, for all $n' < n$. Consequently a theorem of the form "If R is an n-fir, then for all $m \leqslant n$, R has the property $P(m)$" is equivalent to the statement "Any n-fir has the property $P(n)$". We shall often state results in the latter simpler form, with the understanding that $P(m)$ holds for $m < n$ because R is also an m-fir.

Note that a 1-fir is simply an integral domain. At the other extreme we have a ring which is an n-fir for all n; such a ring is called a *semifir*. From (d) of Theorem 1.1 we obtain the following characterization of n-firs and semifirs:

COROLLARY 1. *A ring R is an n-fir if and only if every right ideal on at most n generators is free of unique rank. Further, R is a semifir if and only if R has invariant basis number and every finitely generated right ideal is free.* ∎

Clearly an n-fir either has invariant basis number or is of type (h, k) where $h > n$ (*cf.* **0.**2). Hence by (f) we obtain

COROLLARY 2. *Any n-fir is weakly n-finite and a semifir is weakly finite.* ∎

We shall often need the following dimension formula:

PROPOSITION 1.2. *Let R be a $2n$-fir and A, B any n-generator submodules of a free R-module. Then the exact sequence*

$$0 \to A \cap B \to A \oplus B \to A + B \to 0$$

splits and

$$rk(A + B) + rk(A \cap B) = rk\,A + rk\,B.$$

Proof. The mapping $(a, b) \mapsto a - b$ maps $A \oplus B$ onto $A + B$, with kernel $A \cap B$. Now apply Theorem 1.1 (f). ∎

Let us see what becomes of our definitions in the commutative case. More generally, let us take a right Ore ring R; in R any two elements are right linearly dependent, therefore any free right ideal has rank at most 1. Hence if a right Ore ring is a 2-fir, condition (c) of Theorem 1.1 shows that any two elements generate a principal right ideal. An integral domain with this property is called a *right Bezout domain*; and it is not hard to show the truth of

PROPOSITION 1.3. *For any ring R the following conditions are equivalent*:

(a) *R is an integral domain in which any two elements generate a principal right ideal (i.e. R is a right Bezout domain)*,

(b) *R is an integral domain in which every finitely generated right ideal is principal*,

(c) *R is a 2-fir and a right Ore ring.* ∎

In particular, a commutative 2-fir is just an integral domain in which all finitely generated ideals are principal, i.e. a *Bezout* domain. A comparison of (b) and (c) yields the

COROLLARY. *In the commutative case every 2-fir is a semifir.* ∎

Analogous remarks apply to left Bezout domains (defined correspondingly). Thus for right or left Ore rings our chain of conditions from 2-fir to semifir collapses to a single condition. By contrast there are, in the general case, for each n, n-firs that are not $(n + 1)$-firs, as examples in Cohn [66, 69″] show.

The following property of submodules of free modules over a right Bezout domain is often useful.

PROPOSITION 1.4. *Let R be a right Bezout domain*; *then any finitely generated submodule of ^nR is free of rank at most n*.

Proof. Let M be any finitely generated submodule of $F = {}^nR$ and denote the projection on the last factor by ε. Then $\varepsilon(M)$ is a finitely generated right ideal of R, and hence principal, i.e. free of rank 0 or 1. Thus $M = \ker \varepsilon \oplus F_1$, where F_1 is free of rank at most 1, and $\ker \varepsilon$ is a finitely generated submodule of ^{n-1}R. Now the result follows by induction on n. ∎

When R is a principal right ideal domain, the hypothesis of finite generation is not needed, and we obtain the

COROLLARY. *Let R be a principal right ideal domain*; *then any submodule of ^nR is free of rank at most n*. ∎

Exercises 1.1

1. For each $n \geqslant 1$, determine which of the following are n-firs: k, $k[x]$, $k[x, y]$ (k a commutative field), $\mathbf{Z}$, $\mathbf{Z}[x]$.

2*. For which n is it true that every subring of an n-fir is again an n-fir? Give an example of a non-commutative integral domain which cannot be embedded in a semifir.

3. For any $n \geqslant 1$, show that the direct limit of n-firs is an n-fir. Deduce that the direct limit of semifirs is a semifir.

4. Let R be an n-fir and $X \subseteq R^n$, $Y \subseteq {}^n R$ such that $x \cdot y = 0$ for all $x \in X$, $y \in Y$. Show that there exists $\mu \in \mathbf{GL}_n(R)$ which simultaneously trivializes all these relations.

5. Let R be an n-fir and S a subring which is also a homomorphic image of R (i.e. S is a *retract* of R). Show that S is again an n-fir.

6*. (G. Bergman) Let R be an n-fir and G a finite group of R-module automorphisms of ${}^n R$. Show that the set of fixed elements is a free submodule.

7. Let R be an n-fir and $x_1 y_1 + \ldots + x_n y_n = 0$ a relation in R. Show that $rk(\Sigma x_i \, R) + rk(\Sigma R y_i) \leqslant n$.

8. Let R be the ring of polynomials in an indeterminate x with rational coefficients and integral constant term. Show that R is a Bezout domain but not a principal ideal domain.

9^0. Is the inverse limit of a system of semifirs necessarily a semifir? The same problem for n-firs (for a fixed n).

10^0. A ring R is said to be a *pseudosemifir*, if every finitely generated right ideal is free, but R does not have IBN. Is the notion so defined left–right symmetric?

If a pseudosemifir is of type (h, k), show that (i) R is a $(h - 1)$-fir, and (ii) every finitely generated module can be generated by $h + k - 1$ elements.

11. Let R be a left Bezout domain. Show that a finitely generated torsion free right R-module is free.

1.2   Firs and α-firs

A *free right ideal ring*, or *right fir* for short is a ring in which all right ideals are free of unique rank, as right R-modules. By Theorem 1.1, Corollary 1, such a ring is certainly a semifir and hence has invariant basis number. In particular it is always an integral domain. *Left firs* are defined correspondingly. More generally, let α be an infinite cardinal, then by a *right α-fir* we understand a ring in which all right ideals on α or fewer generators are free, of unique rank, and similarly for left α-firs. Clearly this generalizes the definition given in 1.1, but unlike the latter, the above definition turns out not to be left–right symmetric (*cf.* the examples in 2.9 and 8.6).

Our first aim is to prove that in a 2-sided fir, submodules of free modules are free. This follows from the more general

PROPOSITION 2.1. *If all right ideals in a ring R are free, then all submodules of a free right R-module are free.*

Proof. Let $M = {}^I R$ be the given free right R-module; well-order I and write M_i for the submodule spanned by the basis-elements $\leqslant i$. Given any submodule N of M, consider the projection of $N \cap M_i$ on the ith component. This is a right ideal of R, hence free. Choose a basis for this right ideal and take a set of inverse images in $N \cap M_i$, B_i say. Then $\bigcup B_i$ is easily seen to be a basis for M. ∎

In the particular case of firs we have

COROLLARY 1. *Any submodule of a free right module over a right fir is again free.* ∎

COROLLARY 2. *Let α be an infinite cardinal and R a ring in which all right ideals generated by at most α elements are free. Then in any free right R-module, all submodules generated by at most α elements are free.*

This will follow from Proposition 2.1 once we show that in the above proof, if N is generated by $\leqslant \alpha$ elements, then for each i, the submodule $N \cap M_i$ is also generated by $\leqslant \alpha$ elements. In proving this fact we may, by an induction on α, assume that in free R-modules, all submodules generated by $< \alpha$ elements are free.

Let N be generated by $\{a_\beta | \beta < \alpha\}$, where α has been identified with the first ordinal of cardinal α. For each $\beta < \alpha$, let N_β be the submodule of N generated by $\{a_\gamma | \gamma \leqslant \beta\}$. Further, write $M = M_i \oplus M_i'$ where as before M_i is spanned by the basis-elements $\leqslant i$ and M_i' is spanned by those that are $> i$. Now fix i and consider the projection of N into M_i'. Since N_β has $< \alpha$ elements the image of N_β under this map is free, so N_β splits over the kernel $N_\beta \cap M_i$ of this projection. Hence $N_\beta \cap M_i$ is generated by card (β) elements, and it follows that $N \cap M_i = \bigcup (N_\beta \cap M_i)$ is a union of α submodules each generated by $\leqslant \alpha$ elements, and hence is generated by $\leqslant \alpha$ elements as asserted. ∎

In the presence of the Ore condition the definition of a fir can again be simplified. Thus a right Ore ring cannot have a right ideal which is free on more than one generator, hence we get

PROPOSITION 2.2. *A right fir satisfies the right Ore condition if and only if it is a principal right ideal domain. In particular, a commutative fir is just a principal ideal domain.* ∎

Since a right Noetherian domain is right Ore, the result just proved shows that a fir is not Noetherian except in the rather special case of a principal ideal domain. Nevertheless, there is a chain condition which is satisfied by all firs, namely the ACC on n-generator right ideals. We shall say that a module M satisfies ACC_n if the ACC holds for the set of submodules of M which can be generated by at most n elements. If a ring R satisfies ACC_n as right R-module, we shall express this by saying that R satisfies *right* ACC_n; the term 'left ACC_n' is defined analogously.

THEOREM 2.3. *A right $\aleph_0$-fir satisfies right ACC_n for all n.*

Proof. By passing to the $n \times n$ matrix ring we are, by Theorem 0.1.2, reduced to proving that a total matrix ring R over a right $\aleph_0$-fir satisfies right ACC_1. Consider an ascending chain of principal right ideals in R:

$$a_1 R \subseteq a_2 R \subseteq \dots . \tag{1}$$

If the union is finitely generated, it can be generated by a finite subset of $a_1, a_2, \dots$, and hence must equal some $a_i R$. In the contrary case the union is a direct sum of countably many copies of the minimal projective, and hence is free, with basis (c_λ) say. Let $c_1 = \Sigma_1^r a_i u_i = a_r u$, where we may assume that u is a non-unit (by increasing r if necessary). Now $a_r = \Sigma c_\lambda v_\lambda$ and hence

$$c_1 = a_r u = \Sigma c_\lambda v_\lambda u.$$

Comparing coefficients, we find that $v_1 u = 1$, and since R, being equivalent to a semifir, is weakly finite (*cf.* 0.2), it follows that $u v_1 = 1$, so u is a unit, which is a contradiction. Therefore the sequence (1) must terminate. ■

This result applies in particular to right firs. We also note

COROLLARY 1. *Any free right module over a right $\aleph_0$-fir has ACC_n for all n.*

This is obvious for a right Ore ring; otherwise any ascending chain of finitely generated submodules of a free module is contained in a free module of countable rank, and this is isomorphic to a right ideal of the ring, so the result follows from Theorem 2.3. ■

Let R be a left and right $\aleph_0$-fir and take any $c \in R^*$. Then $cR \neq R$ if and only if c is a non-unit, and in that case there is a maximal principal right ideal $p_1 R$ such that $cR \subseteq p_1 R \subset R$. This means that $c = p_1 c_1$ and p_1 is an atom. Repeating the procedure on c_1 we see that unless it is a unit, we can write $c_1 = p_2 c_2$, where p_2 is an atom. Continuing in this fashion we get a strictly ascending sequence of principal left ideals

$$Rc \subset Rc_1 \subset Rc_2 \subset \dots$$

which must terminate. Therefore every non-zero element of R is either a unit or a product of atoms. A ring with this property is said to be *atomic*, and what we have proved can be stated as

COROLLARY 2. *Any left and right $\aleph_0$-fir is atomic.* ∎

Theorem 2.3 can also be used to factorize matrices over $\aleph_0$-firs provided they have no zero-divisors as factors. This will be done in Chapter 5 in a more general context, where this rather cumbersome condition on the factors is expressed in a different form.

Finally we consider the effect of maximum conditions in the Ore case.

PROPOSITION 2.4. *A ring is a principal right ideal domain if and only if it is a right Bezout domain and satisfies right* ACC_1.

Proof. Clearly a principal right ideal domain is right Bezout and satisfies right ACC_1. Conversely, if R is right Bezout and satisfies right ACC_1, let $\mathfrak{a}$ be a right ideal of R and take a maximal principal right ideal cR contained in $\mathfrak{a}$. If $cR \subset \mathfrak{a}$, let $d \in \mathfrak{a}$, $d \notin cR$, then $cR \subset cR + dR = c_1 R \subseteq \mathfrak{a}$, contradicting the maximality of cR, hence $cR = \mathfrak{a}$, i.e. every right ideal is principal. ∎

Exercises 1.2

1. Give a direct proof that every principal right ideal domain is a right fir.

2*. Show that an $\aleph_0$-fir which is also a right Ore ring is a principal right ideal domain.

3. A non-zero ring without IBN, in which all right ideals are free, is called a right *pseudofir*. Show that a pseudofir cannot be weakly finite. Which results of this section carry over to pseudofirs?

4. Show that any integral domain with left and right ACC_1 is atomic.

5. By expressing $k[x, y]$ as an intersection of principal ideal domains in $k(x, y)$, show that an intersection of firs need not be a fir. What can be said about the intersection of two firs?

6. If α is an infinite cardinal, show that any right α-fir that is not a right fir has right ideals that are not projective (Hint: Use Theorem 0.2.9).

7. In a two-sided fir R, let $\mathfrak{a}$ be a two-sided ideal and $\mathfrak{b}$ a right ideal. Examine the possible relations between $rk(\mathfrak{a} + \mathfrak{b})$ and $rk(\mathfrak{b})$.

8^0. Investigate join-irreducible (or meet-irreducible) right ideals in firs.

9^0. (Robson) Find a principal right ideal domain which is left Ore (and hence left Bezout) but not atomic.

10^0. If all maximal right ideals in a semifir R are finitely generated, is R necessarily a right fir?

11^0. If R is a two-sided fir, and not a left or right Ore ring, is the Jacobson radical of R necessarily zero?

12. Show that a semifir is a right fir if and only if it is right hereditary.

1.3 Strong G_n-rings

In **1.1** we achieved the triviality of dependence relations by using invertible matrices. Such matrices are most tractable when they can be expressed as products of elementary and diagonal matrices, as is the case over a field, or over the ring **Z** of integers. However, over a general ring not every invertible matrix is a product of elementary and diagonal matrices. and it is reasonable to consider relations that are trivialized by products of elementary matrices or by matrices in some subgroup $G_n(R)$ of $\mathbf{GL}_n(R)$.

Let R be a ring and n a positive integer or the symbol ∞; we shall call a family of subgroups

$$G_1(R) \subseteq \mathbf{GL}_1(R), \dots, G_n(R) \subseteq \mathbf{GL}_n(R)$$

(with no last term if $n = \infty$) *coherent* if for all m with $1 \leqslant m < n$, G_{m+1} contains all the $m + 1$ images of G_m obtained by letting it act on all but one component of ^{m+1}R, leaving that component fixed (or equivalently: obtained by inserting a row and column intersecting in 1, with zeros elsewhere, in each matrix of G_m).

Examples of infinite coherent families:

1. The general linear group $\{\mathbf{GL}_n(R)\}$, $n = 1, 2, \dots$

2. $\{\mathbf{GE}_n(R)\}$, where $\mathbf{GE}_n(R)$ is generated by all elementary matrices (matrices differing from the unit matrix in only one off-diagonal place) and diagonal matrices.

3. $\{\mathbf{E}_n(R)\}$, where $\mathbf{E}_n(R)$ is generated by all elementary matrices.

4. $\{\mathbf{O}_n(R)\}$, the orthogonal matrices over a commutative ring R.

5. $\{\mathbf{Tr}_n(R)\}$, the upper unitriangular matrices over R, i.e. matrices having 1's down the main diagonal, 0's below and arbitrary elements above. This type of matrix will be of considerable importance to us later on.

In order to study rings in which every relation can be trivialized by some member of a coherent family of groups $\{G_n\}$ we prove a theorem analogous to Theorem 1.1.

THEOREM 3.1. *Let R be a ring, n a positive integer or ∞ and $\{G_m(R)\}$ $(m = 1, \dots, n)$ a coherent family of matrix groups. Then the following conditions on R are equivalent:*

(a) *every m-term relation $\Sigma x_i y_i = 0$ $(m \leqslant n)$ holding in R can be trivialized by a member of $G_m(R)$,*

(b) *if $x_1, \dots, x_m \in R$ are right linearly dependent $(m \leqslant n)$, there exists $\mu \in G_m(R)$ such that $(x_1, \dots, x_m)\mu$ has at least one zero component,*

(c) *given* $x_1, \ldots, x_m \in R$ $(m \leqslant n)$ *there exists* $\mu \in G_m(R)$ *such that the non-zero components of* $(x_1, \ldots, x_m)\mu$ *are right linearly independent,*

(d) *if* $x_1, \ldots, x_m$ $(m \leqslant n)$ *are elements of a free right R-module M, there exists* $\mu \in G_m(R)$ *such that the non-zero terms of* $(x_1, \ldots, x_m)\mu$ *are right linearly independent,*

(e) *if* ϕ *is a mapping of* ^mR $(m \leqslant n)$ *into a free right R-module, then there exist an integer* $r \leqslant m$, *and* $\mu \in G_m(R)$ *such that the following diagram is commutative:*

$$\begin{array}{ccccc} \ker\phi & \longrightarrow & {}^mR & \overset{\phi}{\longrightarrow} & \operatorname{im}\phi \\ \downarrow{\scriptstyle\mu'} & & \downarrow{\scriptstyle\mu} & & \downarrow{\scriptstyle\mu''} \\ {}^rR & \longrightarrow & {}^rR \oplus {}^sR & \longrightarrow & {}^sR \end{array}$$

where $r + s = m$, *the lower maps are the natural inclusion and projection and* μ', μ'' *are the mappings induced by* μ.

(a*)–(e*). *The left–right analogues of* (a)–(e).

Proof. Again (a) is left–right symmetric, so we need only prove (a) $\Rightarrow \ldots \Rightarrow$ (e) $\Rightarrow$ (a). Of these, (a) $\Rightarrow$ (b) is immediate and (b) $\Rightarrow$ (c) follows by an obvious induction, using the coherence assumption.

(c) $\Rightarrow$ (d). Since $x_1, \ldots, x_m$ can only involve finitely many components of M, we may take M to be of finite rank, say $M = {}^kR$. We use induction on k, the case $k = 0$ being vacuous.

Let u be the projection of M onto ^{k-1}R by its first $k - 1$ components and v the projection on the last component. By induction hypothesis there exists $\mu \in G_m$ such that the non-zero terms of $(u(x_1), \ldots, u(x_m))\mu$ are right linearly independent. Let the zero terms be the i_1th, ..., i_hth, then by (c) we can find $v \in G_h$ such that $(v(x_{i_1}), \ldots, v(x_{i_h}))v$ has all non-zero terms right linearly independent. Denote by v' the image of v in G_m that makes it act on the i_1th, ..., i_hth components only. Then it is easy to see that the non-zero terms of $(x_1, \ldots, x_m)\mu v'$ are right linearly independent.

(d) $\Rightarrow$ (e). Let $e_1, \ldots, e_m$ be the standard basis of ^mR and let $\mu \in G_m(R)$ be chosen by (d) so that all non-zero terms of $(\phi(e_1), \ldots, \phi(x_m))\mu$ are linearly independent. This can be written $(\phi(e_1\mu), \ldots, \phi(e_m\mu))$; if the non-zero components are the i_1th, ..., i_sth, then $(\phi(x_{i_1}\mu), \ldots, \phi(x_{i_s}\mu))$ forms a basis of $\phi(^mR)$, giving us a natural isomorphism $\mu'' : \phi(^mR) \cong {}^sR$. The remaining terms $e_j\mu$ will form a basis of $\ker\phi$ and so we get the desired commutative diagram. Finally (e) $\Rightarrow$ (a) follows as in Theorem 1.1. ∎

A ring satisfying the above equivalent conditions will be called a *strong* $\{G_m\}_{m=1,\ldots,n}$*-ring.* This will be abbreviated to "strong G_n-ring" if n is finite, and "strong G-ring" if $n = \infty$. Thus for example, a strong GL_n-ring is just an n-fir, and a strong GL-ring is a semifir. On the other hand, the Euclidean algorithm shows that $\mathbf{Z}$, and more generally any Euclidean domain, is a strong GE-ring. We shall meet many other examples of strong GE-rings in the next chapter.

The general relation between n-firs and strong G_n-rings is given by the following result:

PROPOSITION 3.2. *Let R be a ring, n a positive integer or ∞ and let $\{G_m\}$ $(m \leqslant n)$ be a coherent family of matrix groups. Further, denote by U_m the set of $m \times m$ matrices of the form $\delta\tau\pi$, where δ is an invertible diagonal matrix, τ an upper unitriangular matrix and π a permutation matrix. Then R is a strong G_n-ring if and only if R is an n-fir and*

$$\mathbf{GL}_m(R) = U_m G_m(R) \qquad (1 \leqslant m \leqslant n). \tag{1}$$

Proof. $\Rightarrow$. Assume that R is a strong G_n-ring, then clearly R will be an n-fir. Now let $\alpha \in \mathbf{GL}_m(R)$; for $i = 1, \ldots, m$ define $f_i : {}^m R \to {}^i R$ as the map α followed by projection onto the first i components. Writing $K_i = \ker f_i$ we see that $K_i = \alpha^{-1}({}^i 0 \oplus {}^{m-i}R)$, so K_i is free of rank $m - i$. Applying Theorem 3.1(e) to f_1 and noting that here $r = m - 1$, we see that there is a $\mu_1 \in G_m(R)$ such that $\mu_1 K_1$ is the set of vectors in ${}^m R$ for which some component, say the i_1th, is zero.

Now $\mu_1 K_2$ is the kernel of a linear mapping on $\mu_1 K_1$ and has dimension $m - 2$. If we apply the same reasoning as above—first deleting the i_1th component and working in ${}^{m-1}R$, and then using the coherence of $\{G_m\}$ to bring the matrix so found from G_{m-1} to $G_m(R)$—we obtain $\mu_2 \in G_m(R)$ such that $\mu_2 \mu_1 K_2$ is the set of elements of ${}^m R$ with two components, say the i_1th and i_2th, equal to zero, while $\mu_2 \mu_1 K_1$ is still the set with vanishing i_1th component. Continuing this process, we get $\mu = \mu_{m-1} \cdots \mu_2 \mu_1 \in G_m(R)$ such that μK_j is the set of all vectors with some j components zero $(j = 1, \ldots, m)$. Thus for a suitable permutation π of the coordinates $\pi\mu K_j = {}^j 0 \oplus {}^{m-j}R$. Hence $\mu\alpha^{-1}$ preserves each of the spaces ${}^j 0 \oplus {}^{m-j}R$. Adjusting by an appropriate $\tau \in \mathbf{Tr}_m(R)$ and then by an appropriate diagonal matrix δ, we ensure that $\delta\tau\pi\mu\alpha^{-1} = I$, whence $\alpha \in U_m G_m(R)$, i.e. (1) holds.

$\Leftarrow$. We shall verify condition (e) of Theorem 3.1. Given ϕ as in (e), we can find $\mu \in \mathbf{GL}_m(R)$, as in Theorem 1.1(f), and by (1) write it in the form $\mu = \delta\tau\pi\mu'$, $\mu' \in G_m(R)$. Now it is easy to verify that $\mu' = \pi^{-1}\tau^{-1}\delta^{-1}\mu$ will still have property (e). ∎

This Proposition shows that any strong G_n-ring $(n \geqslant 1)$ is an integral domain.

Note that we always have

$$U_m \subseteq \mathbf{GE}_m(R). \tag{2}$$

For triangular and diagonal matrices this is clear, while for permutation matrices it follows from the equation

$$\begin{pmatrix} 0 & 1 \\ 1 & 0 \end{pmatrix} = \begin{pmatrix} 1 & 0 \\ 0 & -1 \end{pmatrix} \begin{pmatrix} 1 & 1 \\ 0 & 1 \end{pmatrix} \begin{pmatrix} 1 & 0 \\ -1 & 1 \end{pmatrix} \begin{pmatrix} 1 & 1 \\ 0 & 1 \end{pmatrix},$$

expressing a transposition as an element of $\mathbf{GE}_2(R)$.

Let us agree to call R a G_n-*ring* if $G_n(R) = \mathbf{GL}_n(R)$ and a G-*ring* if it is a G_n-ring for all n; then our conclusion can be stated as

COROLLARY 1. *Let* $\{G_n\}$ *be a coherent family such that* $G_m(R) \supseteq \mathbf{GE}_m(R)$ $(m = 1, \ldots, n)$; *then* R *is a strong* G_n-*ring if and only if* R *is an* n-*fir and a* G_m-*ring for all* $m \leqslant n$. *Thus* R *is a strong* G-*ring if and only if it is a semifir and a* G-*ring.* ∎

In Proposition 3.2, $\mathbf{GL}_m(R)$ may itself be replaced by a member of a coherent family $\{H_m\}$; this leads to the following generalization, proved in exactly the same way as Proposition 3.2 itself:

COROLLARY 2. *Let* U_m *be as in Proposition* 3.2 *and let* $\{G_m(R)\}$, $\{H_m(R)\}$ $(m \leqslant n)$ *be two coherent families such that* $G_m(R) \subseteq H_m(R)$. *Then* R *is a strong* G_n-*ring if and only if* R *is a strong* H_n-*ring and*

$$H_m(R) \subseteq U_m G_m(R) \qquad (1 \leqslant m \leqslant n). \ ∎ \tag{3}$$

When $H_m(R) \supseteq \mathbf{GE}_m(R)$, (3) becomes an equality and we have

COROLLARY 3. *If* $\{G_m(R)\}$, $\{H_m(R)\}$ $(m \leqslant n)$ *are two coherent families such that* $H_m(R) \supseteq G_m(R) . \mathbf{GE}_m(R)$, *then* R *is a strong* G_n-*ring if and only if* R *is a strong* H_n-*ring and* $G_m(R) = H_m(R)$ $(m \leqslant n)$. ∎

Exercises 1.3

1. Show that the subgroup of $\mathbf{GL}_n(R)$ generated by the set U_n defined in Proposition 3.2 is $\mathbf{GE}_n(R)$.

2. Show that a strong $\mathbf{Tr}_n$-ring is the same as an integral domain for $n = 1$ and is the same as a field for $n > 1$.

3. Let R be a (commutative) discrete valuation ring and $G_n(R)$ the group of invertible matrices whose entries below the main diagonal lie in the maximal ideal. Show that R is a strong G-ring.

Let R be a local ring. What can be said about the form of R when R is a strong G-ring?

4*. If $O_n(R)$ denotes the orthogonal group (when R is commutative) show that the field of real numbers is a strong O-ring. For which other fields is this true?

5. (Silvester) Let R be a ring and $a, b \in R$. Show that $\mathrm{diag}\,(1, 1 + ab)$ is associated to $\mathrm{diag}\,(1 + ba, 1)$ in R_2. Deduce that $1 + ab$ is a unit in R if and only if $1 + ba$ is.

By evaluating $b(1 + ab)^{-1}\,a$, give a direct proof of this fact.

1.4 Homological properties of firs and semifirs

From the point of view of category theory, firs and semifirs are rather simple objects. On the one hand, these rings are hereditary and semi-hereditary respectively, while on the other hand they are projective-free. This makes it particularly easy to describe the Morita class containing a given fir or semifir (cf. **0.2**).

THEOREM 4.1. *For any ring R the following are equivalent*:

(a) *R is a total matrix ring over a semifir*,

(b) *R is Morita equivalent to a semifir*,

(c) *R is right semihereditary and projective-trivial*,

(c*) *the left–right analogue of* (c).

Proof. Since (a) is left–right symmetric, we need only prove (a) $\Rightarrow$ (b) $\Rightarrow$ (c) $\Rightarrow$ (a). Of these, (a) $\Rightarrow$ (b) is immediate and (b) $\Rightarrow$ (c) follows because being right hereditary and projective-trivial are categorical properties, possessed by semifirs.

Now let R be a ring satisfying (c), by Proposition **0.2.7**, R is a total matrix ring over a projective-free ring, say $R = T_n$, where T is projective-free and like R right semihereditary. But this means that every finitely generated right ideal of R is free, of unique rank, i.e. R is a semifir. ∎

The equivalence of (c) and (c*) is of interest in view of the fact that examples are known of right but not left semihereditary rings (Chase [61]).

In the same way we get a Morita-invariant description of firs.

THEOREM 4.2. *For any ring R the following are equivalent*:

(a) *R is a total matrix ring over a right fir*,

(b) *R is Morita-equivalent to a right fir*,

(c) *R is right hereditary and projective-trivial*.

Only (c) $\Rightarrow$ (a) requires proof. If R satisfies (c), then by Theorem 4.1, R is a total matrix ring over a semifir, T say. Now T is like R, right hereditary so any right ideal is projective, and by Theorem **0.2.9**, a direct sum of finitely generated right ideals, i.e. free. Hence R is a right fir. ∎

It is also possible to give an explicit description of flat modules over semifirs. We use the following criterion for flatness in semifirs.

LEMMA 4.3. *Let R be a semifir. Then a right R-module U is flat if and only if, for any finite left linearly independent family $a_1, \ldots, a_n$ in R and any $u_1, \ldots, u_n \in U$,*

$$\Sigma u_i a_i = 0 \quad \text{implies} \quad u_i = 0 \ (i = 1, \ldots, n). \tag{1}$$

Proof. Let $\lambda : R^n \to R$ be the linear map defined by $(x_i) \mapsto \Sigma x_i a_i$. Since the a_i are left linearly independent over R, the sequence

$$0 \to R^n \xrightarrow{\lambda} R$$

is exact, and (1) states that the induced sequence

$$0 \to U \otimes R^n \cong U^n \to U \otimes R = U \tag{2}$$

is exact. Writing $C = \operatorname{coker} \lambda$, we see that C is a finitely related cyclic R-module, and all such modules arise in this way, because R is a semifir. Now the exactness of (1) means that $\operatorname{Tor}_R^1(U, C) = 0$. It follows from the properties of Tor that this holds for all modules C, whence U is flat. Conversely, when U is flat, $\operatorname{Tor}_R^1(U, C) = 0$, hence (2) is then exact and (1) holds. Thus (1) is necessary and sufficient for U to be flat. ∎

THEOREM 4.4. *Let R be a semifir, and U a right R-module. Then U is flat if and only if every finitely generated submodule of U is free, i.e. if and only if U is locally free.*

Proof. Clearly U is flat whenever all its finitely generated submodules are flat, and this is so whenever the latter are free. Conversely, let U be flat and take a finitely generated submodule U'. Let U' be generated by $u_1, \ldots, u_n$, where n is chosen minimal. If U' is not free on the u's, then there is a non-trivial relation $\Sigma u_i a_i = 0$. By Lemma 4.3, the a_i are left linearly dependent over R, hence there is an invertible transformation $a_h' = \Sigma p_{hi} a_i$, $a_i = \Sigma q_{ij} a_j'$, where the non-zero a_j' are left linearly independent. Now U' is also generated by the elements $u_j' = \Sigma u_i q_{ij}$ and $\Sigma u_i' a_i' = 0$. But not all the a_i' vanish, say $a_1' \neq 0$, therefore $u_1' = 0$. It follows that U' can be generated by $u_2', \ldots, u_n'$ and this contradicts the choice of n; thus U' is free, as asserted. ∎

Over a left Bezout domain every finitely generated torsion free module is free (Exercise 1.11), therefore we obtain the

COROLLARY. *Let R be a left Bezout domain, then a right R-module is flat if and only if it is torsion free.* ∎

Exercises 1.4

1. Show that any weakly semihereditary local ring is a semifir (and hence left and right semihereditary).

2. Let R be a semifir and S a ring such that $R_n \cong S_n$. Show that $R \cong S$. Does this still hold if R is merely assumed to be an n-fir?

3. Show that the Corollary of Theorem 4.4 fails for semifirs that are not left Bezout.

4. Let R be a left Bezout domain and K its field of fractions. Show that for any torsion free right R-module M, the mapping $M \to M \otimes K$ is an embedding.

5. Given any ring R and modules U_R, $_R V$, $\mathrm{Tor}_1(U, V)$ may be defined as the abelian group generated by the symbols (u, A, v) where u is a row in U, v a column in V and A a matrix over R such that $uA = Av = 0$, subject to the relations

$$(u, A, v) + (u', A, v) = (u + u', A, v),$$
$$(u, A, v) + (u, A, v') = (u, A, v + v'),$$
$$(uA, B, v) = (u, AB, v), (u, A, Bv) = (u, AB, v),$$
$$(u, A, v) + (u, A', v) = (u, A + A', v),$$

whenever the left-hand sides are defined (*cf.* MacLane [63], p. 150). What simplification is possible if R is (i) a principal ideal domain, (ii) a fir?

6^0. Investigate conditions on a ring R which imply that all flat modules are locally free.

1.5 Further properties of semifirs

Of the chain of n-fir conditions, the case $n = 1$ is really too general to be of interest; thus a 1-fir is just an integral domain. On the other hand, 2-firs form an important class, e.g. in the commutative case they already comprise all semifirs.

By Theorem 1.1(c) a 2-fir can be characterized as an integral domain in which any two right linearly dependent elements generate a principal right ideal: the case $m = 1$ of (c) asserts that R is an integral domain and $m = 2$ is the above ideal condition. We reformulate this definition as

THEOREM 5.1. *A 2-fir R may be characterized by either of the following two equivalent conditions*:

(i) *R is an integral domain in which any two principal right ideals with non-zero intersections have a sum which is principal,*

(i*) *the left–right analogue of* (i). ∎

By analogy with Bezout domains (in which *any* two principal right ideals have a principal sum) such a ring is also called a *weak Bezout ring*.

However, we shall usually prefer the shorter name 2-fir, which designates the same class.

Let R be a 2-fir and consider two principal right ideals aR, bR with non-zero intersection. If we apply Proposition 1.2 to the sequence

$$0 \to aR \cap bR \to aR \oplus bR \to aR + bR \to 0,$$

we find that $aR \cap bR$ and $aR + bR$ are both free, of positive ranks adding up to 2, hence each is of rank 1, i.e. principal. This proves the

COROLLARY. *In a 2-fir the intersection of two principal right (or left) ideals is again principal.* ∎

By combining Theorem 5.1 with this corollary we obtain

THEOREM 5.2. *An integral domain R is a 2-fir if and only if, for every $c \neq 0$, the principal right ideals containing c form a sublattice $\mathbf{L}(cR, R)$ of the lattice $\mathrm{Lat}_R(R)$ of all right ideals. Hence in a 2-fir this lattice $\mathbf{L}(cR, R)$ is modular.* ∎

In a commutative 2-fir (i.e. a Bezout domain) this lattice is a sublattice of the lattice of all fractional principal ideals, a lattice-ordered group and hence distributive (*cf.* Birkhoff [67]). In Chapter **4** we shall investigate for which 2-firs $\mathbf{L}(cR, R)$ is distributive.

We have seen (in **0.5**) that any right Ore ring has a field of fractions. This depended essentially on the fact that each pair of non-zero elements has a non-zero common right multiple (i.e. is right commensurable). In a principal ideal domain any two non-zero elements have a *least* common right multiple, and this enables one to get a normal form for the elements of the field of fractions. It is convenient to formulate this property more generally for 2-firs. We begin with a remark which assures us of a supply of right commensurable elements.

LEMMA 5.3. *In a 2-fir R, any two right comaximal elements have a least common right multiple.*

Proof. Let a, b be a right comaximal in R, say

$$ad' - bc' = 1. \tag{1}$$

If a or b is a unit, there is nothing to prove, so assume a, b to be non-units. By (1), they, as well as c', d' must then be non-zero. Now $ad'b = b(c'b + 1)$ is a non-zero common right multiple, hence $aR \cap bR \neq 0$, and so $aR \cap bR = mR$, as asserted. ∎

THEOREM 5.4. *Let a_0, b_0 be left commensurable elements in a 2-fir R, and assume that R is embeddable in a ring S in which b_0 has an inverse. Then the element $a_0 b_0^{-1}$ of S can be written in the form $a' b'^{-1}$, where a', b' are left comaximal, and if*

$$a' b'^{-1} = a'' b''^{-1}, \qquad (2)$$

then $a'' = a'u$, $b'' = b'u$ for some $u \in R$. In particular, if a_0, b_0 are left comaximal, they are right associated to a', b' respectively.

Proof. By hypothesis $Ra_0 \cap Rb_0 \neq 0$, so let $ab_0 = ba_0$ be a LCLM of a_0, b_0 and let

$$ab' = ba' \qquad (3)$$

be a LCRM of a, b. Then a', b' are left comaximal and $a_0 = a'd, b_0 = b'd$ for some $d \in R$. Hence b' is invertible and $ba_0 b_0^{-1} = a = ba'b'^{-1}$, so $a_0 b_0^{-1} = a'b'^{-1}$. Moreover, if (2) holds, then $b^{-1}a = a''b''^{-1}$, hence $ab'' = ba''$, and so by (3), $a'' = a'u$, $b'' = b'u$ as asserted. ∎

We next show that the *n*-fir property is preserved by localization.

PROPOSITION 5.5. *Let R be an n-fir and let T be a right denominator set in R. Then the universal T-inverting ring R_T is again an n-fir.*

Proof. Let $a_1, \ldots, a_m$ $(m \leqslant n)$ be any elements of R_T and denote by $\mathfrak{a}$ the right ideal generated by them. Finding a common denominator, we can write them as $a_i = u_i t$, where $u_i \in R$, $t \in T$. Clearly $u_1, \ldots, u_m$ again generate $\mathfrak{a}$. If they are right linearly dependent over R_T, say $\Sigma u_i b_i = 0$, where the $b_i \in R_T$ are not all 0, then by bringing the b's to a common denominator and multiplying up, we get a relation $\Sigma u_i v_i = 0$ $(v_i \in R)$ and here $v_i = 0$ if and only if $b_i = 0$. Hence $u_1, \ldots, u_m$ are right linearly dependent over R, so $\Sigma u_i R$ can be generated by fewer than m elements, and a fortiori, so can $\mathfrak{a} = \Sigma u_i R_T$. This shows that R_T is an *n*-fir. ∎

When R is a left or right Bezout domain, we can embed R in a field of fractions. Later (in Chapter 7) we shall see how to embed any semifir in a field; but for the moment we observe that if R is a semifir, the semigroup R^* can be embedded in a group. This depends on the following theorem of A. A. Klein [69]. The proof consists in taking the Malcev conditions for the embeddability of a semigroup, expressing them in matrix form and so showing that they are satisfied; it will not be given here.

Klein's theorem. *Let R be an integral domain such that every nilpotent $n \times n$ matrix μ over R satisfies $\mu^n = 0$. Then R^* is embeddable in a group.* ∎

Let R be a semifir, and μ a nilpotent $n \times n$ matrix over R. Then μ defines a descending chain of submodules of ^nR:

$$V_0 \supseteq V_1 \supseteq V_2 \supseteq \dots \supseteq V_k \supseteq V_{k+1} = 0, \tag{4}$$

where $V_0 = {}^nR$, $V_i = \mu V_{i-1}$ and the chain ends in 0 because μ is nilpotent. Each V_i is finitely generated, as homomorphic image of V_0, and as submodule of V_0 is free. Moreover, V_i is a homomorphic image of V_{i-1} and hence $rk\,(V_i) \leqslant rk\,(V_{i-1})$, with equality if and only if $V_i = V_{i-1}$. If for some i, $V_i = V_{i-1}$, then $V_i = V_{i+1} = \dots = 0$. Hence the ranks in (4) decrease strictly until we reach 0 and since $rk\,(V_0) = n$, it follows that 0 is reached after at most n steps, i.e. $\mu^n = 0$.

Applying Klein's theorem, we therefore obtain

THEOREM 5.6. *Let R be a semifir, then the multiplicative semigroup of non-zero elements of R is embeddable in a group.* ∎

Exercises 1.5

1. Show that in a 2-fir, any finite set of pairwise right commensurable elements generates a principal right ideal.

2. Let R be a ring; if the polynomial ring $R[x]$ is a 2-fir, show that R must be a field.

3. Let R be a 2-fir with a right large non-unit c. Show that the localization R_c is a right Bezout domain.

4*. (Klein [69]) Show that any ring satisfying Klein's nilpotence condition is weakly finite.

5^0. Is the semigroup of non-zero elements of a 2-fir always embeddable in a group? (For atomic 2-firs the answer is 'yes', see Cohn [71]).

1.6 Inert extensions

When we come to discuss extensions of rings, we shall be particularly interested in those extensions that preserve factorizations. This leads to the notion of an n-inert extension, which turns out to be closely related to the notion of an n-fir, as we shall see below.

In what follows we shall be dealing with row and column vectors from a ring. We shall generally denote the components of such a vector by a latin suffix, thus a has the components a_i and a_λ the components $a_{\lambda i}$. The precise range will be indicated in brackets, when it is not clear from the context. We also continue to use the notation ab for the product of a row a and a column b, i.e. $\Sigma a_i b_i$.

Let R be a ring and X a subring of R. Given $a \in R^n$, $b \in {}^nR$, then the product $ab = \Sigma a_i b_i$ is said to lie *trivially* in X, if for each $i = 1, ..., n$, either a_i and b_i lie in X or $a_i = 0$ or $b_i = 0$.

Definition. The subring X of R is said to be *n-inert* in R if for any families (a_λ) of rows in R^n and (b_μ) of columns in ${}^nR$, such that $a_\lambda b_\mu \in X$ for all λ, μ, there exists $P \in \mathbf{GL}_n(R)$ such that on writing $a_\lambda' = a_\lambda P$, $b_\mu' = P^{-1}b_\mu$, each product $a_\lambda' b_\mu'$ lies trivially in X. If X is n-inert in R for all $n \geqslant 1$, we say that X is *totally inert* in R.

For example, allowing for the moment subrings not containing 1, an n-fir may be described as a ring in which the zero-element is v-inert for $v = 1, ..., n$. More generally we have the following relation between these definitions.

PROPOSITION 6.1. *Let R be a ring and X a totally inert subring of R, then every finitely generated right ideal of R is the direct sum of a free right ideal and a right ideal generated by a subset of X. Similarly for left ideals.*

Proof. Let $\mathfrak{a}$ be a finitely generated right ideal of R, and let n be the least integer such that $\mathfrak{a}$ can be generated by n elements, $e_1, ..., e_n$ say. Let $b_\lambda \in {}^nR$ be the family of all columns satisfying $(e_1, ..., e_n)b_\lambda \in X$. By total inertia we can apply an invertible matrix so that all these products lie trivially in X. Keeping the same notation, we find that the $e_1, ..., e_n$ still generate $\mathfrak{a}$, and by the minimality of n no e_i can vanish. So we may assume that $e_1, ..., e_r \in X$ while $b_{\lambda j} = 0$ for $j > r$. Let $\mathfrak{a}_1$ be generated by $e_1, ..., e_r$ and $\mathfrak{a}_2$ by $e_{r+1}, ..., e_n$, then

$$\mathfrak{a} = \mathfrak{a}_1 + \mathfrak{a}_2, \tag{1}$$

and for any $b_1, ..., b_n \in R$ such that $\Sigma e_i b_i = 0$ we must have $b_{r+1} = ... = b_n = 0$, hence the sum in (1) is direct and $\mathfrak{a}_2$ is free on $e_{r+1}, ..., e_n$. This proves the assertion; the result for left ideals follows by symmetry. ∎

Let us examine more closely the situation which arises in a totally inert subring. Suppose we have two families $a_\lambda \in R^n$ and $b_\mu \in {}^nR$, such that each product $a_\lambda b_\mu$ lies trivially in X. We can divide the range $\{1, 2, ..., n\}$ into 3 parts N', N'' and $N_0 : i \in N'$ if for some a_λ the ith coordinate is not in X, $i \in N''$ if for some b_μ the ith coordinate is not in X and $i \in N_0$ in all other cases. Then it is clear that $b_{\mu i} = 0$ for $i \in N'$, $a_{\lambda i} = 0$ for $i \in N''$ and $a_{\lambda i}, b_{\mu i} \in X$ for $i \in N_0$. Thus taking the families finite for simplicity, if the

a_λ form the rows of a matrix A and the b_μ form the columns of a matrix B, then for a suitable ordering of the columns of A and rows of B,

$$A = (A' \ 0 \ A_0) \qquad B = \begin{pmatrix} 0 \\ B'' \\ B_0 \end{pmatrix} \tag{2}$$

where A_0, B_0 have coefficients in X. Now the definition of n-inertia shows the truth of

PROPOSITION 6.2. *Let R be a ring and X an n-inert subring. Given an $r \times n$ matrix A and an $n \times s$ matrix B such that all elements of AB lie in X, there exists $P \in \mathbf{GL}_n(R)$ such that AP, $P^{-1}B$ have the respective forms shown in* (2). ∎

To give an example, any field k (even skew) is totally inert in the polynomial ring $k[x]$. For, given any families of row and column vectors a_λ, b_μ over $k[x]$ such that all $a_\lambda b_\mu \in k$, we may assume that $a_\lambda \neq 0$ for some λ. Fix such a λ, say $\lambda = 0$, and apply an invertible matrix which brings a_0 to the form $(d, 0, ..., 0)$, where d is the highest common factor of the coefficients of a_0 in $k[x]$. If the transformed columns are $b_\mu' = (b_{\mu 1}', ..., b_{\mu n}')^T$, then $db_{\mu 1}' \in k$ and $d \neq 0$, hence *either* $b_{\mu 1}' = 0$ for all μ, *or* $d, b_{\mu 1}' \in k$ and $a_{\lambda 1}' \in k$ for all λ. In either case $\Sigma_2^n a_{\lambda i}' b_{\mu i}' \in k$ and an induction on n shows that k is totally inert in $k[x]$. Other examples will be encountered in Chapters 2 and 5.

Exercises 1.6

1. Prove the transitivity of inertia: If R is n-inert in S and S is n-inert in T, then R is n-inert in T.

2. Given an integral domain R and a commuting indeterminate t, show that R is 1-inert in $R[t]$.

3. Let K be a field, and R a subring which is a unique factorization domain. Show that R is 1-inert in K.

4. Let R be a unique factorization domain and K its field of fractions. Show that $R[t]$ is 1-inert in $K[t]$ (this is essentially Gauss's lemma).

5. If R is 1-inert in S, show that any atom in R remains an atom in S.

6. Prove the following stronger form of Proposition 6.1. If R is a ring with a totally inert subring X, then any finitely generated right ideal $\mathfrak{a}$ is the direct sum of a right ideal $\mathfrak{a}_1$ generated by a subset of X and a right ideal $\mathfrak{a}_2$ which is free mod $\mathfrak{a}_1$ and whose rank is completely determined by $\mathfrak{a}$.

Notes and comments on Chapter 1

The basic references for firs and semifirs are Cohn [64'] and, for the presentation used here, Bergman [67]. n-firs arose out of the construction in Cohn [66] and a treatment planned as a joint paper by Bergman and the author has been absorbed in this book (this is the reference [1] of Cohn [69″]) α-firs were first defined in Bergman [67], where Theorem 2.3 was proved, generalizing the fact (proved in Cohn [67]) that any right fir has right ACC_n. The fact that modules over firs satisfy ACC_n can be viewed as a result in universal algebra (cf. B. Baumslag and G. Baumslag [71]).

The treatment of strong G_n-rings is taken from Bergman [67] and generalizes earlier results on strong GE_n-rings in Cohn [66, 69]. Theorem 4.1 and 4.2 first occur in Cohn [66″], the remainder of 1.4 is new (for an application of Theorem 4.4, cf. Jensen [69]).

Weak Bezout rings were defined and studied in Cohn [63]; the present weaker form of their definition is due (independently) to Bergman [67], Bowtell [67] and Williams [68]. This observation appears here as Theorem 5.1, Corollary. Commutative Bezout domains are studied by Jaffard [60] and Bourbaki [65]; the name is intended to convey that any two coprime elements a, b satisfy the 'Bezout identity' $au - bv = 1$.

The miscellaneous facts in 1.5 are for the most part well known; Theorem 5.6 appears in Klein [69]. The notion of inertia described in 1.6 is the natural generalization of the commutative concept introduced in Cohn [68]; of course the underlying idea already occurs in Gauss's lemma and the notion of 'inertial' primes in algebraic number theory.

2. The weak algorithm

Just as the Euclidean algorithm singles out some important classes of principal ideal domains, so there are some special classes of firs that may be described by a 'weak' algorithm. This generalizes the Euclidean algorithm, to which it reduces in the commutative case.

We begin in **2**.1 by recalling the division algorithm. Corresponding to n-firs we have the n-term weak algorithm defined (for any filtered rings) in **2**.2. To enable us to study this algorithm efficiently we look at graded rings in **2**.3; it turns out that the n-term weak algorithm on a filtered ring can be described entirely in terms of the associated graded ring. The weak algorithm can be used to characterize free algebras (**2**.4), and in **2**.5 we give Bergman's classification of all rings with a weak algorithm. In **2**.6–7 we turn to a closer study of the 2-term weak algorithm, using non-commutative continuants, and obtain a presentation of the $\mathbf{GL}_2$ for such rings.

The non-commuting power series rings can be characterized in analogous fashion by an 'inverse weak algorithm'. This is in many ways simpler than the weak algorithm; in fact the commutative analogue is so simple that it has never been thought worthy of study. However, in the general case it provides some very explicit information on the relations in free power series rings, which are explored in **2**.8. The final section **2**.9 deals with algorithms with transfinite range; this is useful for constructing one-sided counter-examples.

2.1 The division algorithm

We recall briefly the division algorithm used in deriving the familiar Euclidean algorithm. Let R be an integral domain and ϕ a function on R^* with non-negative integer values such that

$$\phi(ab) \geqslant \phi(a) \quad \text{for all} \quad a, b \in R^*. \tag{1}$$

The ring R is said to satisfy the *division algorithm* relative to the function ϕ if

DA. *For any* $a, b \in R$, $b \neq 0$, *there exist* $q, r \in R$ *such that*

$$a = bq + r, \qquad \phi(r) < \phi(b). \tag{2}$$

Here the domain of ϕ is extended to R by putting $\phi(0) = -\infty$. Strictly speaking this should be called the *right* division algorithm, since it is not left–right symmetric, but we shall usually omit the qualifying adjective.

We shall find it convenient to replace this condition DA by the following apparently weaker condition A, which is however equivalent to it:

A. *For any* $a, b \in R^*$ *such that* $\phi(a) \geqslant \phi(b)$, *there exists* $c \in R$ *such that*

$$\phi(a - bc) < \phi(a). \tag{3}$$

If DA holds, A follows by taking $c = q$. Conversely, assume A; given $a, b \in R^*$, choose $q \in R$ such that $\phi(a - bq)$ takes its least value. If $\phi(a - bq) \geqslant \phi(b)$, then by A, there exists $c \in R$ such that $\phi(a - bq - bc) < \phi(a - bq)$, which contradicts the choice of q, hence $\phi(a - bq) < \phi(b)$ and we obtain DA on writing $r = a - bq$. Note that this equivalence of A to DA does not depend on (1).

We note that any condition such as DA or A is relative to a value function ϕ, satisfying (1). But in fact it is not necessary to assume that (1) holds. If R satisfies an algorithm relative to any positive integer valued function, it satisfies the division algorithm relative to a function satisfying (1). To prove this assertion, let θ be any function on R^*, taking values in N and such that A holds relative to θ. Let us put $a + bR = \{a + br \,|\, r \in R\}$ for short, and for any subset S of R^*, define the *derived set*

$$S' = \{s \in S \,|\, a + sR \subseteq S \quad \text{for some} \quad a \in R\}.$$

Now define S_n recursively by putting

$$S_0 = R^*, \qquad S_{n+1} = S_n'. \tag{4}$$

We assert that

$$\theta(x) \geqslant n \quad \text{for all} \quad x \in S_n. \tag{5}$$

For $n = 0$ this holds by definition. Let $n > 0$ and take $x \in S_n$ then by the induction hypothesis, there exists $a \in R$ such that $\theta(a + xq) \geqslant n - 1$ for all $q \in R$. But by DA there exists $q \in R$ such that $\theta(a + xq) < \theta(x)$, hence $\theta(x) > n - 1$, i.e. $\theta(x) \geqslant n$ as asserted.

We observe that the definition of S_n was quite independent of any function θ. Thus the S_n can be defined in any ring R, and they form a descending sequence of sets there:

$$R^* = S_0 \supseteq S_1 \supseteq \dots . \tag{6}$$

When a function θ exists on R, satisfying A (or equivalently DA), we claim that

$$\bigcap_n S_n = \varnothing. \tag{7}$$

For take $x \in R^*$ and choose $n > \theta(x)$, then by what has been shown, $x \notin S_n$ and (7) follows. By (7) we can define a function ϕ on R^* by

$$\phi(x) = \max \{n | x \in S_n\}. \tag{8}$$

This definition states that $\phi(x) \geq n$ if and only if $x \in S_n$. Taking $n = \phi(x)$ we see from (5) that

$$\phi(x) \leq \theta(x). \tag{9}$$

For any $a, b, c \in R$, $c + abR \subseteq c + aR$, hence $ab \in S_n$ implies $a \in S_n$, i.e. $\phi(ab) \geq \phi(a)$. In particular, $\phi(1) \leq \phi(b)$ for all $b \in R^*$, a fact which can also be verified directly. Summing up, we have

THEOREM 1.1. *Let R be a ring which satisfies the division algorithm with respect to some function assuming non-negative integer values on R^*. Then there is a function ϕ on R^*, defined by (4) and (8), such that*

(i) *$\phi(a)$ is a non-negative integer for $a \in R^*$, $\phi(0) = -\infty$,*

(ii) *$\phi(1) = 0$,*

(iii) *$\phi(ab) \geq \phi(a)$,*

and R satisfies the division algorithm relative to ϕ. Moreover, ϕ is the 'smallest' function for which the algorithm holds, in the sense that

$$\phi(a) \leq \theta(a) \quad \text{for all} \quad a \in R,$$

for any function θ satisfying the division algorithm. ∎

Since $\phi(a)$ forms a bound on the number of steps needed to apply the Euclidean algorithm to a (cf. **2.7** below), the function ϕ described in Theorem 1.1 leads to the fastest algorithm, and the theorem may be expressed by saying that when a division algorithm exists in R, then there is a uniquely determined fastest algorithm. Such a ring is called *Euclidean*.

The sets S_n in (4) depend only on R, therefore we obtain the

COROLLARY. *A ring R satisfies the division algorithm (relative to some function) if and only if the sets S_n defined in (4) have empty intersection.* ∎

The principal applications of the division algorithm are to two classes of rings:

(i) *rings of algebraic integers,*

(ii) *polynomial rings over fields.*

In (i) the role of the function ϕ is played by the norm; in (ii) the degree of the polynomial is used for ϕ. These two functions show rather different behaviour and since we shall mainly be concerned with generalizations of (ii) we shall consider rings satisfying the division algorithm relative to a *degree-function*, as defined in 0.7, in a little more detail.

Let R be any ring with a degree function d, then any unit of R has degree 0, for if $ab = 1$, then $d(a) + d(b) = d(ab) = d(1) = 0$, hence $d(a) = d(b) = 0$. In particular, if R is a field, every non-zero element has degree 0 and the division algorithm holds trivially. Further, the ring of polynomials in a single indeterminate over a field: $k[x]$, satisfies the division algorithm relative to the usual degree function. More generally, if α is any endomorphism of k and δ an α-derivation, then the skew polynomial ring $k[x; \alpha, \delta]$ satisfies the division algorithm. This is most easily verified in the form A: Let $a = a_0 + xa_1 + \ldots + x^n a_n$, $b = b_0 + xb_1 + \ldots + x^m b_m$ $(a_n, b_m \neq 0, n \geqslant m)$, then $d(a - b \cdot b_m^{-1} x^{n-m} a_n) < n = d(a)$. We note that $k[x; \alpha, \delta]$ does *not* satisfy the left division algorithm, unless α is an automorphism of k.

We now show that the examples just given exhaust the rings with a right division algorithm relative to a degree function.

THEOREM 1.2. *Let R be a ring with a right division algorithm relative to a degree function. Then the elements of non-positive degree form a field k, and R is either k or of the form $k[x; \alpha, \delta]$.*

Conversely, every ring of the form k or $k[x; \alpha, \delta]$, with any degree function for which $d(x)$ is positive, satisfies the right division algorithm.

Proof. Suppose that R is a ring with a degree function satisfying condition A, and write $k = \{a \in R | d(a) \leqslant 0\}$. By the properties of the degree function, k is a subring of R. Given $a \in k^*$, we have $d(a) = 0$, hence there exists $b \in R$ such that $d(ab - 1) < d(a) = 0$, so $ab = 1$, and $d(b) = 0$, i.e. $b \in k^*$. Thus every non-zero element of k has a right inverse in k, whence k is a field. If R has no elements of positive degree, then $R = k$ and the result follows. Otherwise we take an element x say, of least positive degree in R and assert that every element of R is of the form

$$\Sigma_0^n x^i a_i \qquad (a_i \in k, \quad n \geqslant 0). \tag{10}$$

For if this were not so, there would be elements not of this form; let b be an element of least degree which is not of the form (10). By DA there exists $q \in R$ such that $d(b - xq) < d(x)$. By the definition of x it follows that $b - xq \in k$. Thus for some $a \in k$ we have

$$b = xq + a, \tag{11}$$

and $d(q) < d(x) + d(q) = d(xq) \leqslant \max\{d(b), d(a)\} = d(b)$. Therefore, by the choice of b, q must be of the form $q = \Sigma x^i a_i \ (a_i \in k)$. Inserting this expression in (11) we get

$$b = \Sigma x^{i+1} a_i + a,$$

which contradicts the assumption that b is not of the form (10). Moreover, the form (10) for any element of R is unique, for otherwise we should have a relation

$$x^n c_n + \ldots + c_0 = 0,$$

say, with $c_n \neq 0$. Hence

$$d(x^n) = d(x^n c_n) \leqslant \max\{d(x^i c_i) | i = 0, 1, \ldots, n-1\},$$

i.e. $nd(x) \leqslant (n-1)d(x)$, in contradiction with the assumption $d(x) > 0$. Finally, we have for any $a \in k$,

$$ax = xa_1 + a_2,$$

where $d(a_2) < d(x)$ and $d(a_1) \leqslant d(ax) - d(x) = 0$, hence $a_1, a_2 \in k$. By the uniqueness of the form (10) it follows (as in **0.8**) that $a \mapsto a_1$ is an endomorphism, α say, of k and $a \mapsto a_2$ is an α-derivation δ, thus $R = k[x; \alpha, \delta]$ as asserted.

To prove the converse, consider a skew polynomial ring $k[x; \alpha, \delta]$ with a degree function d. We have already seen that the elements of k^* must have degree 0, and by hypothesis, $d(x) = \lambda > 0$, hence the degrees $d(x^n) = n\lambda$ are all different for different n. From the properties of the degree functions (**0.7**) it follows that for $a_n \neq 0$,

$$d(x^n a_n + \ldots + a_0) = \max\{d(x^i a_i) | i = 0, 1, \ldots, n\} = n\lambda.$$

Thus all degrees are multiples of λ. We may therefore divide the degrees by λ and so obtain the usual degree function on $k[x; \alpha, \delta]$. As we saw earlier, the right division algorithm holds for this degree function. ∎

Exercises 2.1

1. Show that the ring of integral quaternions over the rationals satisfies the Euclidean algorithm relative to the norm function. Does this still hold for the ring of quaternions with integer coefficients? (Recall that every principal ideal domain is integrally closed in its field of fractions.)

2. (Sanov [67]). Let R be a commutative Euclidean domain relative to a function ϕ, and on R_n define $|A| = \phi(\det A)$. Show that for any $A, B \in R_n$, with $|A| \neq 0$, there exist $P, Q \in R_n$ such that

$$B = AQ + P, \qquad 0 < |P| < |A| \quad \text{or} \quad P = 0.$$

Use this result to obtain a reduction to triangular form for matrices over R.

3^0. (Samuel). Determine all rings with a division algorithm in which the number of atomic factors of an element serves as value function.

4^0. Investigate rings with a division algorithm of the form DA relative to some function ϕ, such that q, r are uniquely determined by a and b.

5. (Hasse [28]). Show that a commutative integral domain R is a principal ideal domain if and only if there is an integer valued function ϕ on R^* such that (i) $a \mid b$ implies $\phi(a) < \phi(b)$ with equality if and only if $aR = bR$, (ii) if neither of a, b divides the other, then there exist $p, q, r \in R$ such that $pa + qb = r$, $\phi(r) < \min \{\phi(a), \phi(b)\}$.

6^0. Generalize Exercise 5 to a characterization of Bezout domains.

7^0. Investigate rings with a positive *real* valued function satisfying the conditions (i)–(ii) of Exercise 5.

2.2 The n-term weak algorithm

Any generalization of the Euclidean algorithm will necessarily depend on the form of the value function. We shall not make the most general choice, but take our function to be a *filtration*. This means that we have a map v from R^* to $\mathbf{N}$ with the properties:

V.1. $v(x) \geqslant 0$ for all $x \neq 0$ (by convention $v(0) = -\infty$),

V.2. $v(x - y) \leqslant \max \{v(x), v(y)\}$,

V.3. $v(xy) \leqslant v(x) + v(y)$,

V.4. $v(1) = 0$.

Such a function is also called a *pseudo-valuation*. If equality holds in V.3, we have a degree function as defined in **0.7** (this is an exponential valuation with the sign changed). Even in the general case we shall call $v(x)$ the *degree* of x.

Given any filtration, let us write R_n for the set of elements of degree at most n; then the R_n are subgroups of the additive group of R such that

(a) $0 = R_{-\infty} \subseteq R_0 \subseteq R_1 \subseteq \dots$,

(b) $\bigcup R_n = R$,

(c) $R_i R_j \subseteq R_{i+j}$,

(d) $1 \in R_0$.

Conversely, any series of subgroups R_n of the additive group of R satisfying (a)–(d) leads to a filtration v, with $v(a)$ defined as $\min \{n \mid a \in R_n\}$, as is easily seen. We remark that every ring has the *trivial* filtration

$$v(x) = \begin{cases} 0 & \text{if } x \neq 0, \\ -\infty & \text{if } x = 0. \end{cases}$$

Let R be any filtered ring, with filtration v; we shall say that a family (a_i) $(i \in I)$ of elements of R is *right dependent relative to the filtration v*, or *right v-dependent*, if there exist elements $b_i \in R$, almost all 0, such that

$$v(\Sigma a_i b_i) < \max_i \{v(a_i) + v(b_i)\},$$

or if some $a_i = 0$. Otherwise the family (a_i) is *right v-independent*.

We note that any right v-independent family will be right linearly independent over R. Of course the converse will not necessarily hold. In fact, linear dependence is just the special case of v-dependence obtained by taking the trivial filtration for v.

We shall say that an element a of R is *right v-dependent* on a family (a_i) if $a = 0$ or if there exist $b_i \in R$, almost all 0, such that

$$v(a - \Sigma a_i b_i) < v(a), \quad \text{while} \quad v(a_i) + v(b_i) \leqslant v(a) \quad \text{for all} \quad i.$$

In the contrary case a is said to be *right v-independent* of (a_i). Note that dependence on a family (a_i) is unaffected by adjoining or removing 0 from the family.

The corresponding conditions for *left v-dependence* are defined analogously.

We note that if a is right v-dependent on a set B, then it is right v-dependent on the set of members of B of degree at most $v(a)$.

If a is right v-dependent on a set B and each element of B is right v-dependent on a set C, then a is right v-dependent on C.

If one member of a family is right v-dependent on the rest, the family will be right v-dependent, but the converse is not generally true. In fact, the converse constitutes the "weak algorithm", as expressed in the following

Definition. A ring R with a filtration v is said to satisfy the *n-term weak algorithm* with respect to v, if given any right v-dependent family $a_1, ..., a_m$ $(m \leqslant n)$ with $v(a_1) \leqslant ... \leqslant v(a_m)$, some a_i is right v-dependent on $a_1, ..., a_{i-1}$. If R satisfies the n-term weak algorithm for all n, we say that R satisfies the *weak algorithm* for v.

Strictly speaking the notion here defined should be described as the weak *right* algorithm, but we shall see in **2.3** that the property of satisfying the (n-term) weak algorithm for a given filtration is left–right symmetric.

Note that for $n' \leqslant n$, the n-term weak algorithm entails the n'-term weak algorithm. Thus in any filtered ring R we shall be able to prove more about R, the larger n is. Let us define the *dependence number of R relative to the filtration v*, written $\lambda_v(R)$, as the least integer n for which the n-term weak algorithm fails, or $+ \infty$ if no such integer exists. Thus $\lambda_v(R) = \infty$ means that R satisfies the weak algorithm for v; $\lambda_v(R) > 1$ means that the 1-term weak algorithm holds, which is the case if and only if $v(ab) = v(a) + v(b)$ for all $a, b \in R$, i.e. v is a degree function. In particular, such a ring will be an integral domain.

In order to have an invariant associated with a given ring (independently of the filtration) we define, for any ring R, the *dependence number* $\lambda(R)$ as the supremum of the $\lambda_v(R)$ as v runs over all filtrations of R. Since every ring has the trivial filtration, this set is never empty, so $\lambda(R)$ is always defined. It is an invariant, associated with any ring, necessarily a positive integer or ∞. We note that $\lambda(R) = \infty$ means that there exists, for every n, a filtration v such that $\lambda_v(R) > n$. Such a ring may or may not actually satisfy a weak algorithm, i.e. there may or may not exist a filtration v for which $\lambda_v(R) = \infty$; examples of both possibilities exist (*cf.* Bergman [67]).

As in the case of λ_v, the condition $\lambda(R) > 1$ has a simple interpretation:

PROPOSITION 2.1. *A ring R is an integral domain if and only if $\lambda(R) > 1$.*

Proof. The sufficiency follows by choosing v with $\lambda_v(R) > 1$. Conversely, in an integral domain, $\lambda_v(R) > 1$ holds for the trivial filtration v. ∎

In view of this result, the notion of dependence as defined here is of no interest for rings other than integral domains. In fact we shall almost exclusively be occupied with integral domains in this book.

To illustrate the notion of dependence, let us consider the commutative case, or a little more generally, the case of Ore rings. In a right Ore ring, any set of more than one element is clearly right v-dependent. Therefore if $\lambda_v(R) > 2$, any element of R is right v-dependent on any element of lower degree, i.e. given $a, b \in R$, if $v(a) \geqslant v(b)$, then there exists $c \in R$ such that

$$v(a - bc) < v(a).$$

This is precisely the classical division algorithm, in the form A. Conversely, if the classical division algorithm holds in R, any element of R is right v-dependent on any non-zero element of lower degree and this in turn shows that $\lambda_v(R) = \infty$. These results may be summed up as

PROPOSITION 2.2. *If R is a filtered right Ore ring, then there are exactly three possibilities:*

    (i) $\lambda_v(R) = 1$: v *is not a degree function,*

    (ii) $\lambda_v(R) = 2$: v *is a degree function but the division algorithm does not hold,*

    (iii) $\lambda_v(R) = \infty$: v *is a degree function and the division algorithm holds in R.* ∎

In contrast to this result, for non-commutative rings λ can have any positive integral value (Cohn [66], Bergman [67]). All this is of course in strict parallel with the n-fir condition (*cf.* **1**.1).

We now come to the connexion between the weak algorithm and semifirs. First we need a general result on filtered rings:

LEMMA 2.3. *Let R be a filtered ring. Then any n-tuple of elements of R can be reduced by a member of* $\mathbf{GE}_n(R)$ *acting on the right, to an n-tuple each of whose non-zero terms is v-independent of the remaining terms.*

Proof. If a non-zero term of an n-tuple is right v-dependent on the rest, then by adding to it an appropriate right linear combination of the remaining terms, its degree can be reduced, the degrees of the other terms being unaffected. This clearly corresponds to right action by a member of $\mathbf{GE}_n(R)$. We repeat the process until no such terms remain. It must terminate since the set of values assumed by v is well-ordered: $-\infty, 0, 1, \ldots$ and hence so is the set of degrees $\{v(a_1), \ldots, v(a_n)\}$ in the lexicographic ordering. ∎

Now let R be a ring satisfying the n-term weak algorithm relative to a filtration v. Given any m-tuple $(m \leqslant n)$, whose non-zero elements are right v-dependent, by the n-term weak algorithm, one of the non-zero terms is right v-dependent on the rest. By the lemma we can apply a member of $\mathbf{GE}_m(R)$ so as to obtain an m-tuple whose non-zero elements are right v-independent and hence right linearly independent. This establishes

THEOREM 2.4. *Let R be a ring such that $\lambda(R) > n$. Then R is a strong GE_n-ring, and in particular an n-fir.* ∎

In any filtered ring R, the set $R_0 = \{a \in R | v(a) \leqslant 0\}$ is clearly a subring; if moreover, $\lambda_v(R) > 2$, then for any $a \in R_0$ the pair $(1, a)$ is right v-dependent, whence if $a \neq 0$, 1 is right v-dependent on a, i.e. there exists $b \in R_0$ such that $v(1 - ab) < v(1) = 0$. Thus $ab = 1$, so a is invertible. Hence R_0 is a field whenever R satisfies 2-term weak algorithm for v. For this reason, in considering filtered rings, we usually confine our attention to k-rings (k a field), where every element of degree 0 lies in k.

Let us consider more closely the structure of an ideal in a filtered k-ring R, where $k = R_0$, the set of elements of non-positive degree, is a field. Given a right ideal $\mathfrak{a}$ in such a ring, we shall call a set $B \subseteq \mathfrak{a}$ a *weak v-basis* for $\mathfrak{a}$ if (i) all elements of $\mathfrak{a}$ are right v-dependent on B, and (ii) no element of B is right v-dependent on the rest of B.

It is clear that such a basis generates $\mathfrak{a}$ as right ideal. But it need not be a minimal generating set and it need not be v-independent.

Every right ideal $\mathfrak{a}$ of R has a weak v-basis, which may be constructed as follows: For any integer $t \geqslant 0$, $\mathfrak{a}_t = \mathfrak{a} \cap R_t$ is clearly a right k-space; moreover, the set $\mathfrak{a}_t'$ of all elements of $\mathfrak{a}_t$ that are right v-dependent on $\mathfrak{a}_{t-1}$ is also a right k-space. For evidently $\mathfrak{a}_t'$ is closed under right multiplication by elements of k; closure under addition is clear if the sum has degree t, while if it does not, the sum lies in $\mathfrak{a}_{t-1}$. Now for each $t \geqslant 0$, choose a minimal set B_t spanning $\mathfrak{a}_t$ over $\mathfrak{a}_t'$, i.e. representatives for a right k-basis of $\mathfrak{a}_t/\mathfrak{a}_t'$, and put $B = \bigcup B_t$. By induction on t it follows that every member of $\mathfrak{a}_t$ is right v-dependent on B, for all t. However, no element of B is right v-dependent on the rest, by the minimality condition in our choice of elements of B_t. Thus B is a weak v-basis for $\mathfrak{a}$.

Conversely, every weak v-basis of $\mathfrak{a}$ must have the property that the elements of degree t in it form a right k-basis of $\mathfrak{a}_t$ (mod $\mathfrak{a}_t'$); i.e. it can be obtained in the above way. Hence any two weak v-bases of a right ideal $\mathfrak{a}$ must have the same number of elements in each degree t, viz. $\dim_k (\mathfrak{a}_t/\mathfrak{a}_t')$. We shall call this the *number of v-generators of $\mathfrak{a}$ in degree t* and denote it by $r_t(\mathfrak{a})$; further we put $r(\mathfrak{a}) = \Sigma r_t(\mathfrak{a})$.

If $\mathfrak{a}$ and $\mathfrak{b}$ are two right ideals of R such that $\mathfrak{a} \subset \mathfrak{b}$, and s is the least degree for which $\mathfrak{a}_s \neq \mathfrak{b}_s$, then $r_t(\mathfrak{a}) = r_t(\mathfrak{b})$ for $t < s$, while for $t = s$, $r_t(\mathfrak{a}) < r_t(\mathfrak{b})$, provided that $r_t(\mathfrak{a})$ is finite.

So far R was any filtered k-ring, where $k = R_0$ is a field. We now impose the weak algorithm; then any weak v-basis is right v-independent, and hence any right ideal $\mathfrak{a}$ is free, with any weak v-basis as free generating set. Hence $\mathfrak{a}$ has unique rank. This proves

THEOREM 2.5. *Every filtered ring with weak algorithm is a right fir.* ∎

From the symmetry of the weak algorithm (to be proved in **2.3**) it will follow that a ring with weak algorithm is in fact a two-sided fir.

From Proposition **1.2.2** (or by an easy direct proof) we obtain the

COROLLARY. *Every ring satisfying the right division algorithm relative to a degree function is a principal right ideal domain.* ∎

By contrast we do not have a two-sided conclusion here, due to the asymmetry of the division algorithm.

We next turn to the case of a filtered ring with n-term weak algorithm. As expected, we find that this is an n-fir, but in addition we shall see that it always satisfies left and right ACC_n. The result will follow from a general property of weak v-bases in filtered rings:

LEMMA 2.6. *Let R be a filtered k-ring, where k is a field containing R_0. Then for any integer n, R satisfies ACC on right ideals $\mathfrak{a}$ such that $r(\mathfrak{a}) \leqslant n$.*

Proof. Let $\mathfrak{a}$ be a right ideal with $r(\mathfrak{a}) \leqslant n$, then $\mathfrak{a}$ has a weak v-basis $a_1, ..., a_m$ ($m \leqslant n$). We associate with $\mathfrak{a}$ the n-tuple $(v(a_1), v(a_2), ..., v(a_m), \infty, ..., \infty)$ (with $n - m$ ∞'s) as 'indicator'. Clearly the indicator is independent of the choice of basis. If $\mathfrak{a} \supset \mathfrak{b}$, the indicator of $\mathfrak{a}$ will be smaller, under the lexicographic ordering, than that of $\mathfrak{b}$. Since the set of these indicators is well-ordered, the ideals satisfy ACC. ∎

Although every right ideal $\mathfrak{a}$ in a filtered k-ring R has a weak v-basis, this basis need not be v-independent. But if $\mathfrak{a}$ happens to have a v-independent generating set, B say, then B must be a weak v-basis. For, given $a \in \mathfrak{a}$, let us write $a = \Sigma b_i c_i$ ($b_i \in B$, $c_i \in R$). By the v-independence of the b_i we have $v(a) = \max \{v(b_i) + v(c_i)\}$, and so $v(a - \Sigma b_i c_i) = -\infty$ is a relation of v-dependence of a on B. Hence all elements of $\mathfrak{a}$ are v-dependent on B, while clearly no element of B is v-dependent on the rest.

PROPOSITION 2.7. *Let R be a filtered ring with n-term weak algorithm, and in case $n = 1$, assume also that R_0 is a field†. Then R has left and right ACC_n.*

Proof. By Lemma 2.3 (and the proof of Theorem 2.4) we see that the n-generator right ideals of R will have v-independent sets of generators of cardinal at most n. These will be weak v-bases, hence by Lemma 2.6, such right ideals satisfy ACC. The assertion for left ideals follows by symmetry, if we take for granted the fact, soon to be proved (Proposition 3.1, Corollary) that the weak algorithm is left–right symmetric. ∎

Exercises 2.2

1. Let R be a filtered ring and $n > 1$. Show that R satisfies n-term weak algorithm if and only if R_0 is a field and in any right v-dependent family of at most n elements some member is right v-dependent on the rest.

2. Give an example of a ring with 1-term weak algorithm but not satisfying left or right ACC_1 (use Proposition 2.2).

3. Define filtered modules over a filtered ring and introduce the notion of weak algorithm for filtered modules. Show that every module satisfying the weak algorithm is free; what does the existence of a non-zero module with weak algorithm imply about R?

† For $n \geqslant 2$ this is automatic, but the proviso cannot be omitted (*cf.* Exercise 2).

4^0. Generalize Hasse's characterization (Exercise 1.5) of principal ideal domains to firs.

5^0. Investigate rings satisfying the weak algorithm relative to the trivial filtration.

6^0. Investigate the notion of weak algorithm relative to a function ϕ more general than a filtration.

2.3 The associated graded ring

By a *graded ring* H we understand a family of disjoint abelian groups $H_0, H_1, \ldots$ indexed by $\mathbf{N}$, with bilinear "multiplication" maps $H_i \times H_j \to H_{i+j}$ satisfying the obvious associativity relations on $H_i \times H_j \times H_k$, such that H_0 has a ring structure for which each H_i is an H_0-bimodule. More precisely, what we have defined is an *externally* graded ring. We write $H = \bigcup H_i$ and note that even the zero-elements of the H_i are taken to be distinct. Thus for each $a \in H$ there is a unique integer $d(a)$ which will be called the *degree* of a, such that $a \in H_{d(a)}$.

When we speak of a *relation*

$$\Sigma a_i b_i = 0 \tag{1}$$

holding in H, this will of course entail that the $d(a_i) + d(b_i)$ are equal for all i. A family of elements $a_i \in H$ satisfying such a relation, where not all the b_i vanish, is said to be *right linearly dependent*. The triviality of such relations is defined as for ordinary rings.

Let R be a filtered ring and for each $i \geqslant 0$, write $\mathrm{gr}_i R$ for the abelian group R_i / R_{i-1}; thus $\mathrm{gr}_0 R$ is simply R_0. This family has a natural graded ring structure which is defined as follows: given $\alpha \in \mathrm{gr}_i R$, $\beta \in \mathrm{gr}_j R$, take $a \in \alpha$, $b \in \beta$, then a is determined (mod R_{i-1}) and b (mod R_{j-1}), hence ab is an element of R_{i+j} which is determined (mod R_{i+j-1}). Hence the class in $\mathrm{gr}_{i+j} R$ obtained in this way depends only on α and β (not on a and b) and may be denoted by $\alpha\beta$. It is not hard to verify that with this multiplication the $\mathrm{gr}_i R$ become a graded ring, which we shall denote by $\mathrm{gr}\, R$. Given $a \in R^*$, the image of a in $\mathrm{gr}_{v(a)} R$ will be written $\bar{a}$ and will be called the *leading term* of a; $\bar{0}$ is not defined.

From the definitions in 2.2 it is clear that a family of elements of R is right v-dependent if and only if it contains 0 or the leading terms of its elements are right linearly dependent in $\mathrm{gr}\, R$. Further, an element of R is right v-dependent on a family if and only if it is 0 or its leading term is a right linear combination of the leading terms of the other members of the family.

This shows that the filtered ring R satisfies the (n-term) weak algorithm if and only if $\mathrm{gr}\, R$ satisfies the following condition:

Given any right linearly dependent sequence $a_1, ..., a_m$ (of at most n terms) with $d(a_1) \leqslant ... \leqslant d(a_m)$, some a_i is a right linear combination of $a_1, ..., a_{i-1}$.

In the next proposition we shall reformulate this condition in other equivalent ways, but in order to do so we need a matrix description of those operations on finite sequences of elements of H, which add to each term a right linear combination of the terms preceding it. Given a sequence $D = (d_1, ..., d_n)$ of non-negative integers $d_1 \leqslant ... \leqslant d_n$, let us define $\mathbf{Tr}_D(H)$ as the set of $n \times n$ matrices (x_{ij}) such that for $i < j$, x_{ij} is some element of $H_{d_j - d_i}$, for $i = j$ it is $1 \in H_0$ and for $i > j$ it is zero (considered as an operation from H_{d_i} to H_{d_j}). Such matrices clearly form a group which can act from the right on $H_{d_1} + ... + H_{d_n}$ in the desired manner.

Similarly, given $D' = (d_1', ..., d_n')$, where $d_1' \geqslant ... \geqslant d_n'$, the set of matrices which can act from the *left* on $H_{d_{1'}} + ... + H_{d_{n'}}$ by adding to each term a *left* linear combination of *subsequent* terms will form a group of upper triangular matrices, which we shall call $\mathbf{Tr}_{D'}{}^0(H)$. If $d = d_1 + d_1' = ... = d_n + d_n'$ (where the d_i are non-decreasing and the d_i' non-increasing) we see that $\mathbf{Tr}_D(H) = \mathbf{Tr}_{D'}{}^0(H)$; and if we further write the relation (1), where $d(a_i) = d_i$, $d(b_i) = d_i'$, in matrix form as $a \cdot b$, we see that for $\mu \in \mathbf{Tr}_D(H)$, $a\mu \cdot b = a \cdot \mu b$.

Let us call the relation (1) in H *well-arranged* if $d(a_1) \leqslant ... \leqslant d(a_n)$, or equivalently, $d(b_1) \geqslant ... \geqslant d(b_n)$. A well-arranged relation is said to be *Tr-trivializable* if there exists $\mu \in \mathbf{Tr}_D(H)$, $D = \big(d(a_1), ..., d(a_n)\big)$, such that

$$a\mu \cdot \mu^{-1} b = 0$$

is a trivial relation.

We can now state some equivalent ways of expressing the n-term weak algorithm in $\mathrm{gr}\, R$:

PROPOSITION 3.1. *Let H be a graded ring and n a positive integer. Then the following conditions are equivalent*:

(a) *Given any right linearly dependent sequence $a_1, ..., a_m \in H$ ($m \leqslant n$) such that $d(a_1) \leqslant ... \leqslant d(a_m)$, some a_i is a linear combination of $a_1, ..., a_{i-1}$;*

(b) *given $a_1, ..., a_m \in H$ ($m \leqslant n$) with $d(a_1) \leqslant ... \leqslant d(a_m)$, there exists an upper triangular matrix μ such that the non-zero terms of $(a_1, ..., a_m)\mu$ are right linearly independent;*

(c) *every well-arranged relation $\Sigma a_i b_i = 0$ of at most n terms holding in H is Tr-trivializable;*

(a*)–(c*) *The left–right analogues of* (a)–(c).

The proof is straightforward. ■

Definition. A graded ring H satisfying the equivalent conditions of Proposition 3.1 is said to possess the *n-term weak algorithm*. If H satisfies the *n*-term weak algorithm for all *n*, we say that it satisfies the *weak algorithm*.

The remarks preceding Proposition 3.1 show that the *n*-term weak algorithm (and hence the weak algorithm itself) holds in a filtered ring R if and only if it holds in the associated graded ring gr R. From the left–right symmetry of Proposition 3.1 we therefore obtain the

COROLLARY. *The n-term weak algorithm, for graded or filtered rings, is left–right symmetric.* ∎

While for an ordinary ring R the condition that R be a strong Tr_2-ring implies that R is a field, the conditions of Proposition 3.1 do not imply that all non-zero elements of H are invertible (in fact no elements other than those of H_0 can possibly be invertible). The reason lies in our restriction of the Tr-triviality assumption to well-arranged relations. In a graded ring with 2-term weak algorithm, the relation $a . 1 + (-1) . a = 0$ must be Tr-trivializable if $a \in H_0$ but not otherwise, so only H_0 need be a field.

Exercises 2.3

1. Develop a theory of graded modules over graded rings; in particular define the free module on a given set of free generators (with preassigned degrees) and prove the universal property of such modules.

2. Define a *graded semifir* as a graded ring in which every relation is trivializable. Show that (i) every filtered ring with a weak algorithm is a graded semifir (relative to the grading introduced by the filtration), and (ii) every graded semifir is a semifir.

3^0. Investigate rings graded by the positive real numbers.

2.4 Monomial bases in rings with a weak algorithm

In this section we shall prove an analogue of Theorem 1.2, describing the rings with a weak algorithm. If R is any filtered ring, we shall define the 'formal degree' of an expression $\Sigma_i a_{i1} \ldots a_{in_i}$ as

$$\max_i \{v(a_{i1}) + \ldots + v(a_{in_i})\}.$$

Clearly the actual degree of an element of R never exceeds the formal degree of any expression for it. We also note that the definition of v-independence of a family states that the degree of elements represented by certain expressions should equal the formal degrees of these expressions.

Given any filtered ring R for which R_0 is a field k, we shall construct a set X, the 'weak algebra basis', such that the monomials in X span R as right k-space and no element of X is right v-dependent on the rest.

For each $t > 0$ denote by R_t' the right k-subspace of R_t spanned by the products ab, where $a, b \in R_{t-1}$ and $v(a) + v(b) \leqslant t$. Now choose a minimal set X_t spanning R_t (mod R_t') over k, i.e. a set of representatives for a right k-basis of R_t/R_t', and put $X = \bigcup X_t$. This set X is called a *weak algebra basis* for R. To show that it has the properties stated above, suppose that an element $x \in X$ is right v-dependent on the rest:

$$x \equiv \Sigma x_j b_j \qquad (\mathrm{mod}\ R_{t-1}), \tag{1}$$

where $t = v(x)$. Any terms $x_j b_j$ with $v(x_j) \neq t$ lie in R_t', so we can write (1) as

$$x \equiv \Sigma x_j \beta_j \qquad (\mathrm{mod}\ R_t'),$$

where $\beta_j \in k$ and $v(x_j) = t$ whenever $\beta_j \neq 0$. But this contradicts the construction of X. Thus no element of X is right v-dependent on the rest. Now an easy induction on the degree shows that the monomials in X span R over k. More precisely, the monomials of formal degrees at most t span R_t. Thus we see that every filtered k-ring (with $R_0 = k$) has a weak algebra basis. As in the case of weak v-bases of right ideals, we see that the cardinal of a weak algebra basis (and more precisely, the number of elements of a given degree) is independent of the choice of basis.

We can now state conditions for the weak algorithm to hold in a filtered ring. Given any set X in a k-ring R, if R is generated by X over k, we can define a filtration on R by assigning to each $x \in X$ an arbitrary degree and to each element a of R the minimum of the formal degrees of expressions representing a. We shall denote by $x_I = x_1 x_2 \dots x_n$ the monomials in X.

THEOREM 4.1. *Let R be a filtered ring. Then R satisfies the weak algorithm if and only if R_0 is a field k and R has a right v-independent weak algebra basis X, whose elements have positive degrees.*

When this is so, the monomials in any weak algebra basis X form a right k-basis for R and the degree of any expression $\Sigma x_I \alpha_I$ is its formal degree.

Proof. Suppose that R satisfies the weak algorithm; then any weak algebra basis X of R is clearly right v-independent, and each $x \in X$ has a positive degree.

Conversely, let X be a weak algebra basis satisfying these conditions. Since X has no elements of degree 0, it follows that $R_0 = k$. Suppose that the monomials in X are linearly dependent, say $\Sigma x_I \alpha_I = 0$. By splitting off the left-hand factor in each x_I we can write this as

$$\Sigma x a_x + \alpha = 0 \qquad (x \in X, \quad a_x \in R, \quad \alpha \in k).$$

By the v-independence of X, each $a_x = 0$ and $\alpha = 0$. Now an induction on the formal degree shows that the given relation was trivial. Thus the monomials in X are right k-linearly independent, and hence form a k-basis for R.

We complete the proof by showing that R satisfies the left-hand analogue of the weak algorithm, which we know is equivalent to the weak algorithm itself. Let us consider how monomial *terms*, i.e. scalar multiples of monomials, multiply in R. The product $(x_1 \ldots x_i \alpha)(y_1 \ldots y_j \beta)$ can be written $(x_1 \ldots x_i)(\alpha y_1 \ldots y_j \beta)$. If we write the second factor as a right k-linear combination of monomials, little can be said about the terms that will occur, except that we know a bound for their degrees. However, in the product, all terms will clearly begin with $x_1 \ldots x_i$.

Let us fix a monomial $x_1 \ldots x_k$ of degree r, and let us define a right k-linear mapping $a \mapsto a^*$ of R into itself by sending every monomial of the form $x_1 \ldots x_k b$ to b and all other monomials to 0. Thus a^* is the 'right cofactor' of $x_1 \ldots x_k$ in the canonical expression for a. For any $a \in R$, $v(a^*) \leqslant v(a) - r$. Further, if $a, b \in R$,

$$(ab)^* = a^* b \qquad (\mathrm{mod}\ R_{v(b)-1}). \tag{2}$$

This is clear if a is a monomial term of degree at least r; in fact we have equality then. If a is a monomial term of degree less than r, the right-hand side of (2) is zero, and it holds as a congruence. Now the general case follows by linearity.

Assume now that $b_1, \ldots, b_n$ is a left v-dependent set, i.e.

$$v(\Sigma a_i b_i) < m = \max_i \{v(a_i) + v(b_i)\}.$$

Taking the b_i ordered so that $v(b_1) \geqslant \ldots \geqslant v(b_n)$, we must show that some b_i is left v-dependent on those that follow. By omitting terms if necessary we may assume that $v(a_i) + v(b_i) = m$ for all i; hence $v(a_1) \leqslant \ldots \leqslant v(a_n)$.

Let $x_1 \ldots x_k$ be a product of maximal degree $r = v(a_1)$ occurring in a_1 with a non-zero coefficient α and define the map * as above. Consider now $\Sigma a_i^* b_i$; the ith term differs from $(a_i b_i)^*$ by a term of degree less than $v(b_i) \leqslant v(b_1)$. Hence the sum will differ by a term of degree less than $v(b_1)$ from $(\Sigma a_i b_i)^*$ which has degree at most $v(\Sigma a_i b_i) - r < m - r = v(b_1)$. Therefore $v(\Sigma a_i^* b_i) < v(b_1)$ and this gives a relation of left v-dependence of b_1 on the remaining b_i, since $a_1^* = \alpha \in k^*$. ∎

An example of a filtered ring to which this theorem applies is the free associative algebra $F\langle X \rangle$ on any set X over a commutative field F, where we assign arbitrary positive integral degrees to the elements of X (e.g. we may take them all to be 1) and use the induced formal degree. This shows that free algebras over a field satisfy the weak algorithm. More precisely, we have

PROPOSITION 4.2. *Let R be an algebra over a commutative field F, with a filtration v such that $R_0 = F$. Then R is the free associative F-algebra on a right v-independent free generating set if and only if R satisfies the weak algorithm.*

For by hypothesis, F is contained in the centre of R and the form of the elements described in Theorem 4.1 shows that R is the free F-algebra on X. ∎

By Theorem 2.5 we obtain the

COROLLARY. *Any free associative algebra over a commutative field is a two-sided fir.* ∎

In connexion with Theorem 4.1 we note that, given a field k and a k-ring R with a subset X such that the monomials in X form a right k-basis for R, if we assign arbitrary degrees to members of X and give elements of R their formal degrees when expressed in terms of this basis, this will not necessarily define a filtered ring structure on R. The main reason is that for $\alpha \in k$, $x \in X$, αx when expressed as a right linear combination of monomials in X, may not have the same formal degree as x. In the above example of free associative algebras, this cannot happen. But in general it may not be possible to assign any suitable filtration. E.g. we may have $\alpha x = x^2 + y$ ($\alpha \in k$, $x, y \in X$). In the proof of Theorem 4.1, essential use was made of the fact that v is given as a filtration on R.

Here is another case where we can always assign a filtration in such a way as to satisfy the hypothesis of Theorem 4.1. Let k be any (skew) field and M a k-bimodule. Denote by M^r the tensor product over k of r copies of M; there is a natural mapping

$$M^i \otimes M^j \to M^{i+j},$$

in fact an isomorphism, and this can be used to define an associative multiplication on the direct sum

$$\mathbf{T}(M) = k \oplus M \oplus M^2 \oplus \dots . \tag{3}$$

The resulting ring is called the *tensor k-ring* on M. If X is a right k-basis of M, then it is clear from (3) that the monomials in X form a right k-basis of $\mathbf{T}(M)$. If we assign the degree 1 to each element of X, we obtain a filtration on $\mathbf{T}(M)$ by taking the formal degree of each expression. Clearly X is v-independent, and hence by Theorem 4.1, $\mathbf{T}(M)$ is a ring with a weak algorithm.

Exercises 2.4

1. State and prove analogues of Theorem 4.1 and Proposition 4.2 for graded rings with weak algorithm.

2. Give a direct proof that the centre of a non-Ore ring with weak algorithm is a field (*cf.* Chapter 6).

3. If R is a ring with weak algorithm, show that R_0 is complemented by a right ideal. Give examples to show that R_0 may not have a complement which is a two-sided ideal.

4. Let R be a k-ring with a subset X whose monomials form a right k-basis for R. Given a filtration on R such that $R_0 \subseteq k$, show that X is right v-independent if and only if every element of positive degree in R is v-dependent on X, and deduce that R satisfies the weak algorithm relative to this filtration.

5. Let $R = k\langle X \rangle$ be a free algebra; by regarding R as the universal associative envelope of the free Lie algebra on X, define a filtration on R for which the weak algorithm is not satisfied.

6. Let K be a field with a central subfield F, and consider the free product $R = K * F \langle X \rangle$ over F. Show that R is isomorphic to the tensor K-ring on $M = {}^X(K \otimes K)$, and deduce that R satisfies the weak algorithm.

7. Show that a skew polynomial ring $k[x; \alpha, \delta]$ is a 2-sided fir, even when α is not surjective.

2.5 The construction of rings with a weak algorithm

In Theorem 4.1 we obtained a description of filtered rings with a weak algorithm, in terms of a v-independent generating set, but this is not very explicit. Although it enables one to write down many examples, it does not provide a method for constructing *all* rings with a weak algorithm. We shall now describe such a method, but in order to do so we need another concept:

By a *truncated filtered ring of height h*, R_h, we shall mean a finite chain of abelian groups

$$0 = R_{-\infty} \subseteq R_0 \subseteq R_1 \subseteq \ldots \subseteq R_h,$$

with a function called *multiplication* defined in $\bigcup \{R_i \times R_j | i + j \leqslant h\}$ such that:

1. *For $i + j \leqslant h$, multiplication restricted to $R_i \times R_j$ is a bilinear function with values in R_{i+j},*

2. *For $i + j + k \leqslant h$ multiplication is associative on $R_i \times R_j \times R_k$,*

3. R_0 *contains a neutral element for multiplication, 1.*

Note that we have used the same symbol R_h, for the last term of the defining chain and the total structure.

By a morphism $f: R_h \to R_h{}'$ of truncated filtered rings of height h we mean a map f respecting addition, multiplication and unit element, such that $f(R_i) \subseteq R_i{}'$.

Let R_h be a truncated filtered ring, and let $a \in R_h$. We define the *degree*, $v(a)$ of a, as the least i such that $a \in R_i$. From the definition of multiplication in R_h we see that ab is defined precisely when $v(a) + v(b) \leqslant h$.

With every filtered ring and every integer h there is associated a truncated filtered ring of height h, obtained by 'forgetting the terms of degree greater than h'. This "truncation" functor has a left adjoint, the universal functor associating with any truncated filtered ring R_h the universal filtered ring $U(R_h)$ generated by R_h (*cf.* e.g. Cohn [65], Theorem III.4.2). More generally, there is a universal functor from h-truncated rings to g-truncated rings, where $g > h$.

Clearly we can define 'v-dependence' for a family of elements of a truncated filtered ring, and 'v-dependence' of an element on a family, exactly as for ordinary filtered rings. Of course the value of $\max_i \{v(a_i) + v(b_i)\}$ in the relations considered may not exceed h. Theorem 4.1 is then easily seen to go through in this context.

Given a truncated filtered ring R_h satisfying the weak algorithm, let us denote the field R_0 by k and let X be constructed as in Theorem 4.1. We denote by R_{h+1}^* the right k-space which has for right k-basis the monomials in X of formal degree at most $h + 1$ and for degree-function the formal degree induced by expressing elements in terms of this basis, using the degrees in R_h of the elements of X. We claim that R_{h+1}^* is the universal truncated filtered ring of height $h + 1$ for R_h.

Indeed, the space R_{h+1}^* will clearly have the desired universal property if it can be given a truncated filtered ring structure extending that of R_h. In defining this structure it is clear how we must define addition, right multiplication by elements of k and multiplication by elements of X, so it remains to define left multiplication of monomials in X by elements of k, i.e. to give a left k-space structure to R_{h+1}^*. Moreover, we are given this structure on R_h, so it suffices to define products αu, where $\alpha \in k$ and u is a monomial of degree $h + 1$ in X.

Such a monomial can be written as $u = st$, where s, t are monomials in X of lower degree. In R_h we can write αs as a linear combination $\alpha s = \Sigma s_i{}' \alpha_i{}'$, where $v(s_i{}') \leqslant v(s)$, and for each i let us write similarly $\alpha_i{}' t$ as $\alpha_i{}' t = \Sigma^j t_{ij}{}' \alpha_{ij}{}''$. We then put

$$\alpha st = \Sigma\Sigma s_i{}' t_{ij}{}' \alpha_{ij}{}''.$$

Routine calculations, using the ring structure of R_h, now show that this construction is independent of our choice of decomposition $u = st$, and leads to a truncated filtered ring structure of height $h + 1$ on our space $R_{h+1}^{\bullet}$. Here the main point requiring verification is the associativity for triple products of monomial terms. We leave these verifications to the reader.

Note that by Theorem 4.1 (in its extended form, for truncated filtered rings) $R_{h+1}^{\bullet}$ will now also satisfy the weak algorithm, and the h-truncation of R_{h+1} is just R_h. We now wish to find the *most general* $(h + 1)$-truncated filtered ring R_{h+1} with weak algorithm, having R_h for its h-truncation. If R_{h+1} is such a ring, it is clear from the method of construction of Theorem 4.1, that the set X constructed for R_h can be enlarged to a corresponding set for R_{h+1}. Since monomials in the elements of this set must be right k-linearly independent, R_{h+1} will have $R_{h+1}^{\bullet}$ embedded in it, i.e. the map given by the universal property will be injective.

Now let us take any k-bimodule R_{h+1} containing $R_{h+1}^{\bullet}$ as subbimodule, and extend v to it by setting it equal to $h + 1$ outside R_h. Then we have made R_{h+1} into a truncated filtered ring of height $h + 1$. For the only multiplications that need to be defined on the elements of degree $h + 1$ are their products with members of $R_0 = k$, and the conditions 1–3 that they must satisfy are just the conditions for a k-bimodule. Further, R_{h+1}, as a truncated filtered ring, will satisfy the weak algorithm, because on enlarging X by adjoining a minimal set of generators for the right k-space $R_{h+1} \pmod{R_{h+1}^{\bullet}}$ we get a generating set for R_{h+1} satisfying Theorem 4.1.

Since any filtered ring R with a weak algorithm can be written in a unique way as $\bigcup R_h$, where each R_h is a truncated filtered ring of height h satisfying the weak algorithm, and equal to the h-truncation of R_{h+1}, we find

THEOREM 5.1. *Any filtered ring with weak algorithm can be constructed by the following steps*:

(0) *Choose an arbitrary field k to be R_0,*

   …

$(h + 1)$ *Given R_h, form the universal extension $R_{h+1}^{\bullet}$ (explicitly described above), let R_{h+1} be any k-bimodule containing $R_{h+1}^{\bullet}$ as subbimodule, and consider R_{h+1} as $(h + 1)$-truncated filtered ring,*

   …

(∞) *Let R be the filtered ring $\bigcup R_h$.* ∎

Note that at step (1) we find $R_1^{\bullet} = R_0$ so that this step is simply: choose a k-bimodule containing k. Of course the structure of k-bimodules over a field k is itself a non-trivial topic. Thus k-bimodules are effectively $k^0 \otimes_{\mathbb{Z}} k$ modules and a ring of the form $k^0 \otimes k$ can have a highly non-trivial ideal structure.

As an illustration of Theorem 5.1, consider a skew polynomial ring $k[x; \alpha, \delta]$. Here R_{h+1} differs from R_{h+1}^{*} only for $h = 0$. More generally, if we apply the construction of Theorem 5.1 with $R_{h+1}^{*} = R_{h+1}$ for $h > 0$, the ring so obtained will be the universal k-ring on a certain k-bimodule R_1 containing k as subbimodule. In the special case where k is complemented (as subbimodule), say $R_1 = k \oplus M$, this is just the tensor k-ring on the k-bimodule M. Taking k to be commutative and M a k-vector space, with basis X, we get the usual tensor algebra on M, which may also be described as the free k-algebra on X, $k\langle X \rangle$.

The reader can easily provide his own examples of rings with a weak algorithm *not* generated entirely in degree 1.

Let R be a filtered k-ring, where k is a field, and form $\mathrm{gr}\, R = \{\mathrm{gr}_n R\}$, $\mathrm{gr}_n R = R_n/R_{n-1}$. If each $\mathrm{gr}_n R$ is finite-dimensional as right k-space, say $\dim_k (R_n/R_{n-1}) = \alpha_n$, we can form the power series

$$\gamma_R(t) = \Sigma \alpha_n t^n, \tag{1}$$

which is called the *gocha* of R (after Golod and Šafarevič). Let us calculate the gocha for a ring with weak algorithm, using Theorem 5.1. Clearly $\alpha_0 = 1$; at the nth stage in Theorem 5.1 we choose an extension R_n of R_n^{*}. Let us write $\lambda_n = \dim_k (R_n/R_n^{*})$. We assert that

$$\gamma_R(t) = (1 - \Sigma \lambda_n t^n)^{-1}. \tag{2}$$

To prove this formula we note that R_n/R_{n-1} is spanned (over k) by all monomials x_I of degree n. Each sequence $(n_1, ..., n_r)$ satisfying $n_1 + ... + n_r = n$ gives rise to $\lambda_{n_1} \lambda_{n_2} ... \lambda_{n_r}$ monomials of degree n, thus

$$\alpha_n = \Sigma \lambda_{n_1} \lambda_{n_2} ... \lambda_{n_r},$$

where the summation is over all ordered partitions of n. Hence

$$\gamma_R(t) = \Sigma \lambda_{n_1} t^{n_1} ... \lambda_{n_r} t^{n_r}, \quad \text{i.e. (2).}$$

Now let $\mathfrak{a}$ be a right ideal in R and define its gocha as

$$\gamma_R(t; \mathfrak{a}) = \Sigma \beta_n(\mathfrak{a}) t^n. \tag{3}$$

where $\beta_n(\mathfrak{a})$ is the number of v-generators of $\mathfrak{a}$ in degree n, as defined in **2.4**. This relative gocha has of course to be distinguished from the absolute gocha defined in (1). Thus when $\mathfrak{a} = R$, we have $\gamma_R(t; R) = 1$.

The quotient $R/\mathfrak{a}$ is a right R-module and in particular a right k-space. We choose a right k-basis for $R/\mathfrak{a}$ as follows: Put $u_0 = 1$; if $n > 0$ and a basis for $(R_{n-1} + \mathfrak{a})/\mathfrak{a}$ has been constructed, choose a set U_n of representatives for a right k-basis of $(R_n + \mathfrak{a})/(R_{n-1} + \mathfrak{a})$. The union of these sets,

$U = \bigcup U_n$ gives rise to a right k-basis for $R/\mathfrak{a}$. If $\dim_k (R_n + \mathfrak{a})/(R_{n-1} + \mathfrak{a}) = v_n$, we define the relative gocha of $R/\mathfrak{a}$ by

$$\gamma_R(t; R/\mathfrak{a}) = \Sigma v_n \, t^n. \tag{4}$$

We have the following relation between the gochas of a right ideal and its quotient module for a ring R with weak algorithm:

$$\gamma_R(t; R/\mathfrak{a}) = \gamma_R(t) \, (1 - \gamma_R(t; \mathfrak{a})). \tag{5}$$

To prove (5) we start from the direct sum representation

$$R = \Sigma u_i k + \Sigma e_\lambda R, \tag{6}$$

where $(u_i) = U$ is the k-basis constructed for $R/\mathfrak{a}$ and (e_λ) is a weak v-basis for $\mathfrak{a}$. Substituting for R on the right of (6), we find

$$R = \Sigma u_i k + \Sigma e_\lambda u_i k + \Sigma e_\lambda e_\mu R.$$

and by induction we see that the elements

$$u_i, \, e_\lambda u_i, \, e_\lambda e_\mu u_i, \, \ldots . \tag{7}$$

are right k-linearly independent. Moreover, their leading terms are still linearly independent. For if we had a relation involving some terms u_i, by taking this relation (mod $\mathfrak{a}$) we find that it contradicts the way U was constructed. If there are no terms in u_i, then by the v-independence of the e_λ the coefficient of each e_λ must vanish and we again obtain a relation involving terms in u_i alone, and so get a contradiction as before. This shows that the leading terms of the elements (7) are right k-linearly independent. Now let $a \in R$ be any element and express a in terms of the elements (7), say

$$a = \Sigma u_i \alpha_i + \Sigma e_\lambda u_i \alpha_{\lambda i} + \ldots + \Sigma e_{\lambda_1} \ldots e_{\lambda_n} \alpha_{\lambda_1 \ldots \lambda_n},$$

where all the α's lie in k except those in the last sum (which lie in R). Taking $n > v(a)$ we find that all coefficients of the final sum must be zero (by the v-independence of the e_λ), hence the elements (7) actually span R. Therefore they form a right k-basis of R; now the number of terms of degree n listed in (7) is the coefficient of t^n in

$$(\Sigma v_r t^r) \, (1 - \Sigma \beta_s(\mathfrak{a}) \, t^s)^{-1} = \gamma_R(t; R/\mathfrak{a}) \, (1 - \gamma_R(t; \mathfrak{a}))^{-1}.$$

Since (7) is a basis of R with k-independent leading terms, this must equal the gocha of R, and (5) follows.

We note in particular the special case of a free associative algebra. Let $R = F\langle x_1, \ldots, x_d \rangle$ where all the x_i are of degree 1, then $\gamma_R(t) = (1 - dt)^{-1}$. Suppose that $\mathfrak{a}$ is a right ideal of finite rank $r \neq 1$ and

$\dim_k (R/\mathfrak{a}) = n$, then on putting $t = 1$ in (5) we find $n = (1 - d)^{-1} (1 - r)$, i.e.

$$r - 1 = n(d - 1).$$

This formula, due to Lewin, is a precise analogue of Schreier's formula for groups.

Exercises 2.5

1. State and prove an analogue of Theorem 4.1 for truncated filtered rings.

2. For any filtered ring R write $T_h R$ for the h-truncation obtained from it, and for an h-truncated filtered ring R_h denote by UR_h the universal filtered ring. Given an h-truncated filtered ring R_h show that the canonical mapping $R_h \to T_h UR_h$ is surjective but not necessarily injective.

3. Let R, S be graded algebras over a commutative field F and $T = R \otimes S$ their tensor product. Prove that the gochas of R, S, T are related by the formula $\gamma_T = \gamma_R \gamma_S$.

4*. Let R be a graded algebra over a commutative field F. Show that γ_R has an inverse β_R (as a power series with integer coefficients) if and only if $\dim gr_0 R = 1$. Show that when $R = F[x]$ is a polynomial ring, $\beta_R = 1 - t$.

If R, S are graded algebras satisfying the above condition and U is their free product over F, show that U also satisfies the condition, and that $\beta_U = \beta_R + \beta_S - 1$.

5. Let F be a commutative field and R any F-algebra generated by d elements, and let $\mathfrak{a}$ be a right ideal of finite codimension n in R. Show that $\mathfrak{a}$ is finitely generated as a right ideal and can in fact be generated by $n(d - 1) + 1$ elements.

6. (Lewin [69]). Show that any two-sided ideal in a free algebra R has the same rank as right and as left R-module.

7*. Let $M = \mathbb{C} \otimes_R \mathbb{C}$ and consider the tensor $\mathbb{C}$-ring T on M. Show that T is the free product of an ordinary polynomial ring over $\mathbb{C}$ and a complex-skew polynomial ring over $\mathbb{C}$. How does this generalize to finite Galois extensions?

8. Let $R = k\langle x_1, ..., x_d \rangle$ be a free algebra and $\mathfrak{a}$ a non-zero principal right ideal. Then the Schreier–Lewin formula gives

$$\dim_k(R/\mathfrak{a}) = (1 - d)^{-1} (1 - 1) = 0.$$

Explain this apparent contradiction.

9. Examine the relation (5) between the gochas when $R = k[x; \alpha, \delta]$ is a skew polynomial ring. Do the same for the left gocha of this ring (i.e. the gochas of the opposite ring), when α is not surjective.

10. Define the gocha of a filtered module (over a filtered ring R) and when R satisfies a weak algorithm, derive an analogue of (5) for modules.

11⁰. (Ikeda [69]). In a free algebra of finite rank, is every ideal which is maximal as right ideal finitely generated as right ideal?

2.6 Generators and relations for GE$_2$(R)

Our next objective is to treat the analogue of the Euclidean algorithm that exists in rings with 2-term weak algorithm. One of the main consequences is that such a ring is always a strong GE_2-ring (*cf.* 1.3); moreover, there is a convenient normal form for the elements of **GE**$_2$(R), which leads to a presentation of this group. Since many of the formulae are valid in quite general rings, we shall digress in this section to discuss generators and relations in **GE**$_2$(R) for general rings.

For brevity let us write

$$P(x) = \begin{pmatrix} x & 1 \\ 1 & 0 \end{pmatrix}, \quad [\alpha, \beta] = \begin{pmatrix} \alpha & 0 \\ 0 & \beta \end{pmatrix}, \quad C(\alpha) = [\alpha, -\alpha^{-1}],$$

and observe the following relations between these matrices, valid over any ring R:

$$P(x + y) = P(x)\, P(0)\, P(y) \qquad x, y \in R, \tag{1}$$

$$P(\alpha)\, P(-\alpha^{-1})\, P(\alpha) = C(\alpha) \tag{2}$$

$$P(x)[\alpha, \beta] = [\beta, \alpha]\, P(\beta^{-1} x\alpha). \qquad \alpha, \beta \in U(R). \tag{3}$$

We observe that **GE**$_2$(R) is generated by the $P(x)$ and $[\alpha, \beta]$ since

$$\begin{pmatrix} 1 & x \\ 0 & 1 \end{pmatrix} = P(x)\, P(0) \quad \text{and} \quad \begin{pmatrix} 1 & 0 \\ x & 1 \end{pmatrix} = P(0)\, P(x).$$

From (1)–(3) we obtain the following consequences:

$$P(0)^2 = I, \quad [P(1)\, P(-1)]^3 = -I, \tag{4}$$

$$P(x)^{-1} = P(0)\, P(-x)\, P(0) = \begin{pmatrix} 0 & 1 \\ 1 & -x \end{pmatrix}, \tag{5}$$

$$P(x)\, P(y)^{-1} = P(x - y)\, P(0), \tag{6}$$

$$P(x)\, P(y)^{-1} P(z) = P(x - y + z), \tag{7}$$

$$P(x)\, P(\alpha)\, P(y) = P(x + \alpha^{-1})\, C(\alpha)\, P(y + \alpha^{-1}), \tag{8}$$

where $x, y, z \in R$, $\alpha \in U(R)$. Any element μ of **GE**$_2$(R) can be brought to the form

$$\mu = [\alpha, \beta]\, P(a_1) \dots P(a_r), \tag{9}$$

using (3) and (5). If $a_i = 0$ for some $i \neq 1, r$, this relation can be shortened by (1), while if $a_i \in U(R)$ for $i \neq 1, r$, it can be shortened by using (8) and then (3) to bring $C(\alpha)$ to the left. Thus in any ring R we can express any element μ of $\mathbf{GE}_2(R)$ in the form (9), where $\alpha, \beta \in U(R)$, $a_i \in R$ and such that for $1 < i < r$, a_i is not 0 or a unit. Moreover, if $r = 2$, we may assume, by (4), that a_1, a_2 are not both 0. This expression (9) for μ is called the *standard form* for μ. In the next section we shall see that in any ring with a 2-term weak algorithm the standard form is unique; more generally this holds for any ring with a degree function such that R_0 is a field (*cf.* Cohn [66']). Let R be any ring with unique standard form for GE_2, then in any relation the left-hand side can be brought to standard form, and by uniqueness, this must be I, i.e. any relation can be transformed to the trivial relation $I = I$ using only (1)–(3) and the obvious relation

$$[\alpha, \beta][\alpha', \beta'] = [\alpha\alpha', \beta\beta'].$$

This proves

PROPOSITION 6.1. *In any ring R with unique standard form for GE_2 the relations (1)–(3) and (10) form a complete set of defining relations for $\mathbf{GE}_2(R)$.* ∎

Let us return to the expression (9); in order to obtain explicit formulae for the product of the P's we define a sequence of polynomials p_n in non-commuting indeterminates $t_1, t_2, \ldots$ with integer coefficients. The p are defined by the recursion formulae:

$$p_{-1} = 0, \quad p_0 = 1,$$

$$p_n(t_1, \ldots, t_n) = p_{n-1}(t_1, \ldots, t_{n-1}) t_n + p_{n-2}(t_1, \ldots, t_{n-2}).$$

For $n \geqslant 0$, the suffix of p_n just indicates the number of arguments and so may be omitted when the arguments are given explicitly. We shall do so in what follows and only write the suffix when the arguments are omitted. We assert that

$$P(t_1) \ldots P(t_r) = \begin{pmatrix} p(t_1, \ldots, t_r) & p(t_1, \ldots, t_{r-1}) \\ p(t_2, \ldots, t_r) & p(t_2, \ldots, t_{r-1}) \end{pmatrix}.$$

This is clear for $r = 1$ and in the general case follows by induction, since on writing $p_i = p(t_1, \ldots, t_i)$, $p_i' = p(t_2, \ldots, t_{i+1})$ we have

$$\begin{pmatrix} p_{r-1} & p_{r-2} \\ p_{r-2}' & p_{r-3}' \end{pmatrix} \begin{pmatrix} t_r & 1 \\ 1 & 0 \end{pmatrix} = \begin{pmatrix} p_r & p_{r-1} \\ p_{r-1}' & p_{r-2}' \end{pmatrix}.$$

From the symmetry of (13) it is clear that the p's may also be defined by (11) and

$$p_n(t_1, \ldots, t_n) = t_1 p_{n-1}(t_2, \ldots, t_n) + p_{n-2}(t_3, \ldots, t_n).$$

Either definition shows that p_n may be described as the sum of $t_1 t_2 \ldots t_n$ and all terms obtained by omitting one or more pairs of adjacent factors

$t_i t_{i+1}$. This mode of forming p_n might be called the *leap frog construction*. Thus the first few polynomials are $p_1 = t_1$, $p_2 = t_1 t_2 + 1$, $p_3 = t_1 t_2 t_3 + t_1 + t_3$, $p_4 = t_1 t_2 t_3 t_4 + t_1 t_2 + t_3 t_4 + t_1 t_4 + 1$.

Equivalently, p_n may be described as the polynomial part of the formal product (when evaluated):

$$(t_1 + t_2^{-1})(t_2 + t_3^{-1}) \dots (t_{n-1} + t_n^{-1}) t_n. \tag{15}$$

When the t's are allowed to commute, the p's just reduce to the continuant polynomials, used in the study of continued fractions, and we shall use the term *continuant polynomial* also to describe the p's in the general case.

From (5) it follows that the inverse of $P(t_1) \dots P(t_n)$ is

$$P(0) P(-t_n) \dots P(-t_1) P(0) =$$

$$\begin{pmatrix} 0 & 1 \\ 1 & 0 \end{pmatrix} \begin{pmatrix} p(-t_n, \dots, -t_1) & p(-t_n, \dots, -t_2) \\ p(-t_{n-1}, \dots, -t_1) & p(-t_{n-1}, \dots, -t_2) \end{pmatrix} \begin{pmatrix} 1 & 0 \\ 0 & 1 \end{pmatrix}. \tag{16}$$

Clearly,

$$p(-t_1, \dots, -t_n) = (-1)^n p(t_1, \dots, t_n), \tag{17}$$

hence (16) reduces to

$$[P(t_1) \dots P(t_n)]^{-1} = (-1)^n \begin{pmatrix} p(t_{n-1}, \dots, t_2) & -p(t_{n-1}, \dots, t_1) \\ -p(t_n, \dots, t_2) & p(t_n, \dots, t_1) \end{pmatrix}. \tag{18}$$

Comparing this formula with (14) we obtain

LEMMA 6.2. *The continuant polynomials satisfy*

(i) $p(t_1, \dots, t_n) p(t_{n-1}, \dots, t_2) - p(t_1, \dots, t_{n-1}) p(t_n, \dots, t_2) = (-1)^n$,

(ii) $p(t_1, \dots, t_n) p(t_{n-1}, \dots, t_1) - p(t_1, \dots, t_{n-1}) p(t_n, \dots, t_1) = 0$. ∎

Of course this lemma can also be proved directly (by induction); it corresponds to the well known relations between successive convergents to continued fractions.

We shall use these formulae to analyse comaximal relations in GE$_2$-rings. We recall that in any ring R a relation

$$ab' = ba'$$

is comaximal if there exist $c, d, c', d' \in R$ such that $da' - cb' = ad' - bc' = 1$. Suppose now that R is weakly 2-finite. Then by Proposition 0.2.3, there exists $\mu \in \mathbf{GL}_2(R)$ such that

$$\mu = \begin{pmatrix} a & b \\ * & * \end{pmatrix} \qquad \mu^{-1} = \begin{pmatrix} * & -b' \\ * & a' \end{pmatrix}. \tag{19}$$

Similarly if R is 2-projective free, every equation of comaximality, $ad' - bc' = 1$, arises, by Proposition 0.2.5, from a pair of mutually inverse matrices

$$\mu = \begin{pmatrix} a & b \\ * & * \end{pmatrix} \qquad \mu^{-1} = \begin{pmatrix} d' & * \\ -c' & * \end{pmatrix}. \tag{20}$$

This leads to the following explicit formulae for comaximal relations in GE_2-rings:

PROPOSITION 6.3. *Let R be a ring, $x_1, \ldots, x_n, y, z \in R$ and $\alpha, \beta \in U(R)$.*

(i) *If R is a weakly 2-finite GE_2-ring, every comaximal relation in R has the form*

$$\alpha p(x_1, \ldots, x_n) \, p(x_{n-1}, \ldots, x_1) \, \beta = \alpha p(x_1, \ldots, x_{n-1}) \, p(x_n, \ldots, x_1) \beta. \tag{21}$$

(ii) *if R is a GE_2-ring which is 2-projective free, every equation of comaximality can be written*

$$\alpha p(x_1, \ldots, x_n) \, p(x_{n-1}, \ldots, x_2) \alpha^{-1} (-1)^n$$
$$- \alpha p(x_1, \ldots, x_{n-1}) \, p(x_n, \ldots, x_1) \alpha^{-1} (-1)^n = 1. \tag{22}$$

(iii) *if R is a strong GE_2-ring, every equation $rs = uv$, where r, s are not both 0 and u, v are not both 0, can be written*

$$yp(x_1, \ldots, x_n) \cdot p(x_{n-1}, \ldots, x_1) z = yp(x_1, \ldots, x_{n-1}) \cdot p(x_n, \ldots, x_1) z. \tag{23}$$

Proof. (i) In a weakly 2-finite ring every comaximal relation $ab' = ba'$ arises from a pair of mutually inverse matrices (19). Since R is a GE_2-ring, we can write μ in the form (9)

$$\begin{cases} \mu = [\alpha, \beta^{-1}] \, P(x_1) \ldots P(x_n), \\ \mu^{-1} = P(0) \, P(-x_n) \ldots P(-x_1) \, P(0) \, [\alpha^{-1}, \beta], \end{cases} \tag{24}$$

and now (21) follows on combining (19) and (24).

(ii) follows similarly from (24) and the form (20) of matrices arising from a relation of comaximality.

(iii) A strong GE_2-ring is in particular a 2-fir, so every relation $rs = uv$ is obtained from a comaximal relation $ab' = ba'$, in the form $ya \cdot b'z = yb \cdot a'z$. The result now follows by applying (i) and remembering that 2-firs are weakly 2-finite. ∎

The significance of this Proposition becomes clearer if we make the following definitions.

In any ring R, two elements a, a' are said to be *GL-related* if there exists $\mu \in \mathbf{GL}_2(R)$ such that μ has a in the $(1, 1)$-position and μ^{-1} has a' in the $(2, 2)$-position. In case such μ can be found in $\mathbf{GE}_2(R)$ we say that a, a' are *GE-related*. Thirdly, if there exist $x_1, ..., x_n \in R$ such that $a = p(x_1, ..., x_n)$, $a' = p(x_n, ..., x_1)$, then a, a' are said to be *E-related*.

From (18) it is clear that E-related elements are GE-related, and GE-related elements clearly are GL-related. In GE_2-rings we have the following converse:

PROPOSITION 6.4. *Let R be a GE_2-ring. Then for any a, $a' \in R$ the following three assertions are equivalent*:

   (i) *a is GL-related to a'*,

   (ii) *a is GE-related to a'*,

   (iii) *a is E-related to an associate of a'*.

The equivalence of (i) and (ii) is immediate from (9). When (ii) holds, then by (9), $a = \alpha p(x_1, ..., x_n)$, $a' = p(x_n, ..., x_1)\beta^{-1}$, whence (iii). ∎

In order to compare E-related elements we make use of the following formulae, which follow from the leap frog construction of continuants:

Let $t_1, ..., t_n$ be any elements of a ring and α a unit, then if n is odd,

$$p(t_1\alpha, \alpha^{-1}t_2, ..., \alpha^{-1}t_{n-1}, t_n\alpha) = p(t_1, ..., t_n)\alpha, \tag{25}$$

$$p(\alpha t_1, t_2\alpha^{-1}, ..., t_{n-1}\alpha^{-1}, \alpha t_n) = \alpha p(t_1, ..., t_n), \tag{26}$$

and if n is even,

$$p(t_1\alpha, \alpha^{-1}t_2, ..., t_{n-1}\alpha, \alpha^{-1}t_n) = p(t_1, ..., t_n), \tag{27}$$

$$p(\alpha t_1, t_2\alpha^{-1}, ..., \alpha t_{n-1}, t_n\alpha^{-1}) = \alpha p(t_1, ..., t_n)\alpha^{-1}. \tag{28}$$

Further, an easy calculation shows that for any n,

$$p(t_1, ..., t_n) = p_{n+1}(t_1, ..., t_{n-1}, t_n - 1, 1). \tag{29}$$

This formula allows us to change the parity of n in any representation of an element by a continuant, as in the proof of the next result:

PROPOSITION 6.5. *In any ring R, if a is E-related to a' and α is a unit, then (i) $a\alpha$ is E-related to $a'\alpha$, (ii) αa is E-related to $\alpha a'$ (iii) a is E-related to $\alpha^{-1}a'\alpha$.*

Proof. Let $a = p(x_1, ..., x_n)$, $a' = p(x_n, ..., x_1)$, then (i) follows by (25) if n is odd; if n is even we can replace it by $n + 1$, using (29) for a and the left-hand analogue of (29) for a', and then applying the previous argument. Similarly (ii) follows from (26). To prove (iii) we first ensure that n is even (using (29)) and then apply (27) and (28). ∎

We note that this Proposition can be used to give another proof of Proposition 6.4.

Finally we note the following consequence of (19):

PROPOSITION 6.6. *In any weakly 2-finite ring, two elements a, a' are GL-related if and only if they can be put in a comaximal relation*

$$ab' = ba'. ∎$$

Exercises 2.6

1. Prove the identity $p(t_1, ..., t_n) = p(t_1, ..., t_r) p(t_{r+1}, ..., t_n) + p(t_1, ..., t_{r-1}) p(t_{r+2}, ..., t_n)$, for any n and $1 \leqslant r < n$.

2. (Brenner [55]). Show that the matrices $\begin{pmatrix} 1 & 0 \\ \alpha & 1 \end{pmatrix}$ and $\begin{pmatrix} 1 & \alpha \\ 0 & 1 \end{pmatrix}$ generate a free group if α is a real number satisfying $\alpha \geqslant 2$.

3*. Find all pairs of algebraic integers α, β such that $(P(\alpha) P(\beta))^n = I$ for some n.

4*. Show that in $k\langle x, y, z, t \rangle$ the matrix $\begin{pmatrix} x & y \\ z & t \end{pmatrix}$ cannot be written as a product of elementary and diagonal matrices.

5*. Let $R = W_1(k)$ be the Weyl algebra on x and y over a field k of characteristic not two. Show that the matrix $\begin{pmatrix} yx & xy \\ y^2 & x^2 \end{pmatrix}$ is invertible but is not in $GE_2(R)$. (Hint: Use the filtration by degree in $W_1(k)$.)

6. Show that any local ring R is a GE-ring and that there is a relation of the form $[\alpha, \beta] P(a_1) P(a_2) P(a_3) P(a_4) = I$, where a_2, a_3 are neither zero nor units, unless R is a field.

7. Let R be a totally ordered ring. Given $a_1, ..., a_r \in R$ with $a_i > 0$ for $1 \leqslant i \leqslant r$, show that $p(a_1, ..., a_r) > 0$. Show that this still holds if $a_1 \geqslant 0$ and $a_i > 0$ for $2 \leqslant i \leqslant r$, provided that $r \geqslant 2$.

8*. Let R be a totally ordered ring such that $a > 0$ implies $a \geqslant 1$. Show that R has a unique standard form for GE_2.

9°. If two elements of a ring are E-related to a third, are they E-related to each other?

10*. If two elements of a ring are GL-related to a third, show that they are GL-related to each other (*cf.* 3.3).

11^0. In a free algebra, does there exist an infinite family of pairwise E-related elements?

12^0. (Bergman [67]). In a free algebra, is every element E-related to a square itself a square?

2.7 The 2-term weak algorithm

We shall now develop the usual Euclidean algorithm, using the 2-term weak algorithm. In particular, everything that is said will apply to classical Euclidean domains, whose algorithm is defined relative to a degree function.

Let R be any ring with $\lambda(R) > 2$ and take a filtration v for which the 2-term weak algorithm holds. Given any equation

$$ab' = ba' \neq 0 \tag{1}$$

in R, if we choose $q_1 \in R$ such that $v(a - bq_1)$ is minimal, we find by an easy induction (as in 2.1) that $v(a - bq_1) < v(b)$. Thus

$$a = bq_1 + r_1, \qquad v(r_1) < v(b). \tag{2}$$

Moreover, q_1, r_1 in (2) are unique, for if we also had

$$a = bq + r, \qquad v(r) < v(b),$$

then $b(q_1 - q) = r - r_1$, and if $q_1 \neq q$, then $v(b) \leqslant v(b(q_1 - q)) < v(b)$, a contradiction. Substituting from (2) into (1), we find $r_1 b' = (a - bq_1)b' = b(a' - q_1 b')$; if we put $r_1' = a' - q_1 b'$, this may be written

$$r_1 b' = b r_1'. \tag{3}$$

By (3) and (2), $v(b) + v(r_1') = v(r_1) + v(b') < v(b) + v(b')$, hence $v(r_1') < v(b')$, so there is complete left–right symmetry (as we know there must be, by Proposition 3.1, Corollary). It may happen that $r_1 = 0$, but by (3) this is so only if $r_1' = 0$. If this is not the case, we can apply the same reasoning to (3) and thus obtain the familiar chain of equations of the Euclidean algorithm. More precisely, we obtain two such chains, one for left and one for right division:

$$
\begin{array}{lll}
a = bq_1 + r_1 & a' = q_1 b' + r_1' & r_1 b' = b r_1', \\
b = r_1 q_2 + r_2 & b' = q_2 r_1' + r_2' & r_2 r_1' = r_1 r_2', \\
r_1 = r_2 q_3 + r_3 & r_1' = q_3 r_2' + r_3' & r_3 r_2' = r_2 r_3', \\
\cdots & \cdots & \cdots
\end{array}
\tag{4}
$$

Note that whereas the remainders r_i, r_i' on the two sides are in general distinct, the quotients q_i are the same. The degrees of the remainders decrease strictly,

$$v(b) > v(r_1) > \cdots \qquad v(b') > v(r_1') > \cdots \tag{5}$$

so the remainders must vanish eventually. Let n be the least integer such that $r_{n+1} = 0$. Since $r_{n+1} r_n' = r_n r_{n+1}'$, it follows that $r_{n+1}' = 0$; if we had $r_k' = 0$ for some $k \leqslant n$, then by symmetry $r_k = 0$, which contradicts the definition of n. Hence both chains in (4) terminate at the same step, i.e. r_{n+1}' is the first vanishing remainder of the right-hand division, and the last two rows of (4) read

$$r_{n-2} = r_{n-1} q_n + r_n \qquad r_{n-2}' = q_n r_{n-1}' + r_n' \qquad r_n r_{n-1}' = r_{n-1} r_n',$$
$$r_{n-1} = r_n q_{n+1} \qquad r_{n-1}' = q_{n+1} r_n' \qquad r_n = r_n' = 0. \tag{6}$$

From (4), (6) and the inequalities (5) we see that $v(q_i) > 0$ for $i = 2, 3, \ldots, n+1$, while $v(q_1) > 0$ if and only if $v(a) > v(b)$.

Let us again write $P(x) = \begin{pmatrix} x & 1 \\ 1 & 0 \end{pmatrix}$, for any $x \in R$; then we can express the equations (4) and (6) as follows:

$$(a\ b) = (r_n\ 0)\, P(q_{n+1})\, P(q_n) \ldots P(q_1), \tag{7}$$

$$\begin{pmatrix} a' \\ b' \end{pmatrix} = P(q_1)\, P(q_2) \ldots P(q_{n+1}) \begin{pmatrix} r_n' \\ 0 \end{pmatrix}. \tag{8}$$

These equations make it evident that r_n is a common left factor of a, b and since the P's are invertible, it is actually a highest common left factor (HCLF). Likewise r_n' is a HCRF of a' and b'. In particular, it follows that R is a strong GE_2-ring.

As in Proposition 6.3 (iii), we can use (7) and (8) to express the terms of an equation $ab' = ba'$ in terms of continuants; the above analysis of the division process showed that under the restriction (5) these expressions are unique. Thus we find

PROPOSITION 7.1. *In a ring R with 2-term weak algorithm, any relation $ab' = ba' \neq 0$, where $v(a) \geqslant v(b)$, can be written uniquely in the form*

$$yp(x_1, \ldots, x_n) \cdot p(x_{n-1}, \ldots, x_1) z = yp(x_1, \ldots, x_{n-1}) \cdot p(x_n, \ldots, x_1) z, \tag{9}$$

where $x_1, \ldots, x_n$, y, $z \in R^*$ and moreover, $x_2, \ldots, x_n$ are non-units; the p's are the continuant polynomials defined in 2.6. ∎

If we drop the condition $v(a) \geqslant v(b)$, we shall still get a unique form (9) but we must now allow x_n to assume the value 0 as well. Thus in every invertible matrix $\mu = \begin{pmatrix} a & b \\ c & d \end{pmatrix}$ we can reduce the first row by (7) and so write μ uniquely as $\begin{pmatrix} \alpha & 0 \\ u & \beta \end{pmatrix} P(x_1) \ldots P(x_n)$, where $\alpha, \beta \in U(R)$ because the right-hand side is invertible. Now $\begin{pmatrix} \alpha & 0 \\ u & \beta \end{pmatrix} = [\alpha, \beta] \, P(0) \, P(\beta^{-1} u)$, hence

$$\mu = [\alpha, \beta] \, P(0) \, P(\beta^{-1} u) \, P(x_1) \ldots P(x_n)$$

and this form is unique (with the proviso, that the first two P's are omitted if $u = 0$). Summing up, we have

PROPOSITION 7.2. *Any ring with 2-term weak algorithm is a strong* GE_2-*ring, and the standard form for* $\mathbf{GE}_2(R)$ *is unique.* ∎

Exercises 2.7

1. Let R be a ring with 2-term weak algorithm and X a generating set of R^* as semigroup. Write down a presentation for R^* in terms of X.

2. Let $R = k\langle x_1, x_2, \ldots \rangle$ where the free generating set is finite or countable. Show that $\mathbf{GL}_2(R) \cong \mathbf{GL}_2(k[x])$. (Hint: Show that a k-linear mapping preserves the defining relations.)

3. In a ring R with 2-term weak algorithm, show that a, b are right comaximal if there exist $u, v \in R$ and a unit λ such that $av - bu = \lambda$.

4. Let R be as in Exercise 3. Given two right comaximal elements a, b, show that the equation $ax - by = f$ has a solution (x, y) for any $f \in R$. More precisely, there exist $u', b', c', d' \in R$ such that $ab' = ba'$, $ad' - bc' = 1$, $v(c') < v(a') = v(a)$, $v(d') < v(b') = v(b)$, and the general solution (x, y) has the form $x = d'f + b'g$, $y = c'f + a'g$, where g is arbitrary in R.

5. In a ring with 2-term weak algorithm show that any two right comaximal elements may be written uniquely in the form $cp(a_1, \ldots, a_n)$, $cp(a_1, \ldots, a_{n-1})$, where c is a unit and $a_2, \ldots, a_n$ are not zero or a unit.

6*. (Bergman [71]). Let R be a ring with 2-term weak algorithm and S any semigroup of (ring-) endomorphisms of R. Show that the set of fixed points under the action of S is a strong GE_2-ring.

2.8 The inverse weak algorithm

The classical division algorithm, as defined for a polynomial ring $k[x]$ in 2.1, depended essentially on the degree function $d(a)$ defined in this ring. If instead we use the order function $o(a)$ (cf. 0.8) we have an analogous statement, with the opposite inequality:

Given $a, b \in k[x]$, such that $o(b) < o(a) < \infty$, there exist q and a_1 such that

$$a = bq + a_1 \qquad o(a_1) > o(a). \tag{1}$$

The process can be repeated, but since $\mathbf{N}$ has no maximal element, there is no reason why the process should terminate. However, we can pass to the completion of the ring $k[x]$, namely the formal power series ring $k[[x]]$. Here a repetition of the step (1) leads to a convergent process, and in fact one can make deductions about divisibility in $k[[x]]$ which are quite similar to (and often stronger than) the consequences of the classical division algorithm. E.g. the ring $k[[x]]$ displays such simple divisibility behaviour that its connexion with the algorithm (1) is usually forgotten. In the non-commutative case we do however obtain generally non-trivial results from the 'inverse algorithm'. For ease of comparison we shall state our results not in terms of the order function, but its negative.

By a *negative filtration* on a ring R we shall mean a function v such that

I.1. $-\infty < v(x) \leqslant 0$　for　$x \neq 0$,　　$v(0) = -\infty$,

I.2. $v(x - y) \leqslant \max\{v(x), v(y)\}$,

I.3. $v(xy) \leqslant v(x) + v(y)$.

By contrast the notion defined in **2.2** will be called a *positive filtration*. Writing again

$$R_n = \{x \in R \mid v(x) \leqslant n\},$$

we have the explicit filtration

$$R = R_0 \supseteq R_{-1} \supseteq R_{-2} \supseteq \dots \tag{2}$$

satisfying conditions (b), (c) as in **2.2**. The definition of right and left v-dependence of a family, or of an element on a family, also are identical to the definitions given in **2.2**. However, if an element a is dependent on a set B, it will be dependent on the members of B having degrees *at least* $v(a)$. In the case of a positive filtration we had the opposite inequality; the difference arises because there all non-zero elements have degree $\geqslant 0$, while here they have degree $\leqslant 0$. This accounts for the difference in the next definition:

Definition. A ring R is said to satisfy the *inverse weak algorithm* relative to a negative filtration v if, given any finite right v-dependent family $a_1, \dots, a_m$ with $v(a_1) \geqslant \dots \geqslant v(a_m)$, some a_i is right v-dependent on $a_1, \dots, a_{i-1}$. If R satisfies this condition for all families of at most n terms, we say that R satisfies the *n-term inverse weak algorithm*.

Given an inversely filtered ring R, let us define

$$\mathrm{gr}_i R = R_{-i}/R_{-i-1} \qquad (i \geqslant 0).$$

Then we obtain a graded ring $\mathrm{gr}\, R$, and we find that R satisfies the (n-term) inverse weak algorithm if and only if $\mathrm{gr}\, R$ satisfies the (n-term) weak algorithm. By the symmetry of the latter (Proposition 3.1, Corollary) we obtain

PROPOSITION 8.1. *A negatively filtered ring R has the $(n$-term$)$ inverse weak algorithm if and only if its opposite ring does.* ∎

As in **2.2**, we define the *inverse dependence number* $\mu_v(R)$ of a negatively filtered ring R as the least integer n for which the n-term inverse weak algorithm fails, or ∞ if no such n exists. The supremum of $\mu_v(R)$ over all negative filtrations v of R is called the *inverse dependence number* of R, written $\mu(R)$.

If R satisfies the 2-term inverse weak algorithm, then $\operatorname{gr}_0 R = R/R_{-1}$ is a field. This means that the following general exchange principle applies to such rings (in the case of the ordinary weak algorithm there is a corresponding principle, which we were able to use without stating it formally, because in that case our ring actually *contained* a field):

Exchange Principle. Let R be any negatively filtered ring such that R/R_{-1} is a field. Given a, $a' \in R$ and $A \subseteq R$, if $v(a) \geqslant v(a')$ and a is right v-dependent on $A \cup \{a'\}$, then either a is right v-dependent on A or a' is right v-dependent on $A \cup \{a\}$.

For by hypothesis, there exist $a_i \subset A$ and b_i, $b' \in R$ such that

$$v(a - \Sigma a_i b_i - a'b') < v(a), \tag{3}$$

where

$$v(a_i) + v(b_i) \leqslant v(a), \qquad v(a') + v(b') \leqslant v(a), \tag{4}$$

and (3) will not be affected if we omit any terms for which the corresponding inequality (4) is strict. If the term $a'b'$ survives, we have $v(a') + v(b') = v(a) \geqslant v(a')$, whence $v(b') = 0$; by hypothesis there exists $c \in R$ such that $v(b'c - 1) < 0$, and it follows that

$$v(a' + \Sigma a_i b_i c - ac) < v(a), \quad v(a_i) + v(b_i c) \leqslant v(a), \quad v(c) = 0. \text{∎}$$

Clearly the principle holds whenever $\mu_v(R) > 2$; by assuming it explicitly we shall find that some of our results can be extended to arbitrary negatively filtered rings.

In any negatively filtered ring R the chain (2) defines a topology on R in a natural way, and the completion of R in this topology (which always exists, *cf.* Bourbaki [61′]) will be denoted by $\hat{R}$. The ring $\hat{R}$ is again negatively filtered, and in fact the passage from R to $\hat{R}$ respects v-dependence:

THEOREM 8.2. *Let R be a negatively filtered ring, $\hat{R}$ its completion and let X be any subset of R. Then*

(i) *X is right v-dependent in R if and only if X is right v-dependent in $\hat{R}$,*

(ii) *Given $a \in R$, a is right v-dependent on X in R if and only if it is right v-dependent on X in $\hat{R}$,*

(iii) $\mu_v(R) = \mu_v(\hat{R})$; *in words, $\hat{R}$ satisfies the n-term inverse weak algorithm if and only if R does.*

Proof. (i) That a v-dependent set in R is also v-dependent in $\hat{R}$ is obvious. Let X be v-dependent in $\hat{R}$,

$$v(\Sigma x_i b_i) < \max_i \{v(x_i) + v(b_i)\} = n, \quad \text{say} \quad (x_i \in X, \ b_i \in \hat{R}). \tag{5}$$

By definition of $\hat{R}$ there exists $b_i' \in R$ such that $v(b_i - b_i') \leqslant n - 1$. Again we can drop any terms $x_i b_i$ from (5) for which the maximum is not reached (i.e. for which $v(x_i) + v(b_i) < n$). Then $v(b_i) \geqslant n$, and hence

$$v(x_i) + v(b_i') = v(x_i) + \max \{v(b_i), v(b_i - b_i')\} = n,$$

while

$$v(\Sigma x_i b_i') \leqslant \max \{v(\Sigma x_i b_i), v(\Sigma x_i(b_i' - b_i))\} < n.$$

This shows X to be v-dependent in R.

(ii) If a is v-dependent on X in R, it is also v-dependent on X in $\hat{R}$. Conversely, let a be v-dependent on X in $\hat{R}$, then there exist $x_1, \ldots, x_r \in X$, $c_1, \ldots, c_r \in \hat{R}$ such that

$$v(a - \Sigma x_i c_i) < v(a), \qquad v(x_i) + v(c_i) \leqslant v(a).$$

Again we can omit terms $x_i c_i$ for which $v(x_i) + v(c_i) < v(a)$, without affecting the dependence. Put $n = v(a) - 1$, then there exist $c_1', \ldots, c_r' \in R$ such that $v(c_i - c_i') \leqslant n$, hence for $i = 1, \ldots, r$,

$$v(x_i) + v(c_i') = v(x_i) + \max \{v(c_i), v(c_i' - c_i)\} = v(a),$$

and

$$v(a - \Sigma x_i c_i') \leqslant \max \{v(a - \Sigma x_i c_i), v(\Sigma x_i(c_i' - c_i))\}$$
$$\leqslant \max \{v(a) - 1, n\} < v(a).$$

This proves (ii); now (iii) is an immediate consequence. ∎

Before we can apply the inverse weak algorithm we still need two general reduction lemmas. To obtain the best results we have to take our rings complete:

LEMMA 8.3. *Let R be a complete negatively filtered ring and $a, a_1, \ldots, a_n$ any elements of R. Then there exist $b_1, \ldots, b_n \in R$ such that $v(b_i) \leqslant v(a) - v(a_i)$ and $a - \Sigma a_i b_i$ is either 0 or right v-independent of $a_1, \ldots, a_n$.*

Proof. Assume that we can find $b_i^{(k)}$ such that $v(b_i^{(k)}) \leqslant v(a) - v(a_i)$ and

$$v(a - \Sigma a_i b_i^{(k)}) \leqslant v(a) - k.$$

If $a - \Sigma a_i b_i^{(k)}$ is right v-independent of the a_i we are done; otherwise we can subtract a right linear combination of the a_i to get a still smaller degree. If this holds for all k, then $b_i^{(k)} \to b_i$ by completeness, and $a = \Sigma a_i b_i$. ∎

With the help of this lemma we can obtain an analogue of Lemma 2.3, but in a much stronger form:

LEMMA 8.4. *Let R be a complete negatively filtered ring such that R/R_{-1} is a field, and let A be a finite subset of R. Then there is an ordering of $A: a_1, \ldots, a_n$ and a matrix $\mu \in \mathrm{Tr}_n(R)$ such that $(a_1, \ldots, a_n)\mu = (a_1', \ldots, a_n')$ consists of a sequence of elements none of which is right v-dependent on the rest, followed by zeros.*

Proof. Let $a_1 = a_1'$ be any element of A of maximal degree. Applying Lemma 8.3, we can modify all other members of A by right multiples of a_1 so as to make them 0 or right v-independent of a_1'. This will not increase any of the degrees, so $v(a_1')$ will still be maximal in the resulting set. Let a_2' be of maximal degree among the resulting elements other than a_1'. By another application of Lemma 8.3 we can make all the elements other than a_1', a_2' zero or right v-independent of a_1', a_2'. Continuing this process, we get a sequence $a_1', a_2', \ldots, a_n'$ which will clearly be the image under a unitriangular matrix of a certain ordering of A. Since

$$v(a_1') \geqslant v(a_2') \geqslant \ldots \geqslant v(a_n')$$

by construction, all zeros will occur at the end. Now suppose some non-zero a_i' is right v-dependent on the remaining a_j'. By the exchange principle we conclude that some a_j' is right v-dependent on those preceding it. But this contradicts the construction. ∎

THEOREM 8.5. *Let R be a complete negatively filtered ring with n-term inverse weak algorithm. Then any sequence of at most n elements of R can be reduced by a permutation followed by an upper unitriangular matrix to a sequence of v-independent terms followed by zeros. In particular, R is a strong GE_n-ring and hence an n-fir.*

Proof. The case $n = 1$ is immediate; for $n > 1$ the result follows by Lemma 8.4, using the algorithm. ∎

"Weak v-bases" for ideals of negatively filtered rings can be defined as for positively filtered rings, and constructed similarly; the definition, the construction and Lemma 2.6 can all be stated in gr R, using the right ideal of leading terms of members of the right ideal $\mathfrak{a}$ under construction. However, if B is a weak v-basis of the right ideal $\mathfrak{a}$, it is no longer true that B generates $\mathfrak{a}$, but merely a dense right ideal in $\mathfrak{a}$. As in Proposition 2.7 we thus obtain

PROPOSITION 8.6. *Let R be a complete negatively filtered ring with n-term inverse weak algorithm, and in case $n = 1$ assume also that R/R_{-1} is a field. Then R has right (and left) ACC_n.* ∎

In the case of commutative negatively filtered rings or more generally, Ore rings, we again find that the inverse weak algorithm already follows from the 2-term inverse weak algorithm: Let us define a *right discrete valuation ring* (right DVR for short) as an integral domain R with a non-unit p such that every non-zero element of R has the form $p^r u$ ($r \geqslant 0$, $u \in U(R)$). When R is commutative this clearly reduces to the usual definition of DVR. Then we have the

COROLLARY. *Let R be a complete negatively filtered right Ore ring with 2-term inverse weak algorithm. Then R has inverse weak algorithm and is either a field or a right DVR.*

Proof. If $R_{-1} = 0$, R must be a field. Otherwise take $p \in R_{-1}$ such that $v(p)$ has its largest value; by hypothesis any two elements of R are right v-dependent, hence by Lemma 8.3, for any $a \in R$, either $a = pc$ or $p = ac$. In the second case, $v(c) = v(p) - v(a) \geqslant 0$, hence c is then a unit and we have in any case $a \in pR$. If we write

$$a = p^r u, \tag{6}$$

then $v(a) \leqslant rv(p)$, hence r is bounded, and if we choose it maximal in (6) u must be a unit. Thus R is then a right DVR, and this clearly satisfies the inverse weak algorithm. ∎

In 2.2 we saw that a positively filtered ring with weak algorithm is a fir (Theorem 2.5); this is not to be expected here. For a ring with inverse weak algorithm we find instead that it is a kind of "topological fir".

PROPOSITION 8.7. *Let R be a complete negatively filtered ring with inverse weak algorithm. Then any right ideal $\mathfrak{a}$ of R contains a right v-independent set B such that BR is dense in $\mathfrak{a}$.*

Proof. Let B be a weak v-basis for $\mathfrak{a}$. By the inverse weak algorithm it is right v-independent, and any $a \in \mathfrak{a}$ is right v-dependent on B. It follows that for any v, there exist elements $c_b^{(v)} \in R$ such that

$$v(a - \Sigma b c_b^{(v)}) < -v,$$

where b runs over B, and the sum is finite for any given v. As $v \to \infty$, $c_b^{(v)} \to c_b$ and we obtain

$$a = \Sigma b c_b,$$

where the sum on the right may be infinite, but is convergent in the sense that only finitely many terms of degree $> -v$ occur for any v. But this just means that BR is dense in $\mathfrak{a}$. ∎

The construction of the 'weak algebra basis' in the remarks preceding Theorem 4.1 was essentially carried out in gr R and so can be repeated here. But instead of finite sums we now must allow infinite convergent series of monomial terms, with coefficients chosen from a set of representatives of R/R_{-1}.

THEOREM 8.8. *Let R be a negatively filtered ring with inverse weak algorithm. Then $K = R/R_{-1}$ is a field and any weak v-basis for R_{-1} is v-independent.*

Moreover, if R is complete, $X = \{x_i\}$ is a weak v-basis for R_{-1} and $\overline{K}$ is a set of representatives for K in R (with 0 represented by itself, for simplicity), then every element of R can be uniquely represented as

$$\Sigma x_I \alpha_I \qquad (\alpha_I \in \overline{K}), \tag{7}$$

where $I = (i_1, \ldots, i_n)$ runs over all finite suffix-sets and $x_I = x_{i_1} \ldots x_{i_n}$ for short, and for any v the set of suffix-sets I such that $v(x_I) > -v$ and $\alpha_I \neq 0$ is finite. Conversely, all such expressions represent elements of R.

Proof. The first part follows from Proposition 8.7; the rest now follows as in the proof of Theorem 4.1. ∎

As a converse to this result we have

THEOREM 8.9. *Let R be a negatively filtered ring such that $R/R_{-1} = K$ is a field. Taking a set $\overline{K}$ of representatives of K in R (with 0 represented by itself), suppose that R has a subset X such that the monomials in X span a dense subspace of R (over K) and are right v-independent. Then R satisfies the inverse weak algorithm.*

The proof is similar to that of Theorem 4.1 and so may be left to the reader. ∎

The most important example of a ring with inverse weak algorithm is the power series ring in a number of non-commuting indeterminates over a field, and we indicate briefly how the inverse weak algorithm may be used to characterize such rings.

Let R be a negatively filtered K-ring, where K is a field. Then K may be considered as a subring of R, and since no element of R_{-1} is invertible, we have $R_{-1} \cap K = 0$. Thus R/R_{-1} may again be regarded as a K-ring; in the special case where this ring is K itself, R is said to be *connected*. This then means that

$$R = R_{-1} \oplus K.$$

Similarly a negatively filtered F-algebra (where F is any commutative field) is *connected* if $R = R_{-1} \oplus F$.

Let $R = F\langle X \rangle$ be the free associative algebra on a set X over a field F. For any $a \in R$, let m be the least integer such that a has a non-zero term of degree m, or ∞, if no such integer exists (i.e. if $a = 0$). Putting $v(a) = -m$, we obtain a negative filtration on R, as is easily checked. Moreover, since F is complemented by R_{-1}, R is connected. The completion $\hat{R}$ of R is called the *power series ring* in X over F and is denoted by $F\langle\!\langle X \rangle\!\rangle$; clearly it is again connected. This ring has the following characterization, analogous to the characterization of free algebras given in **2**.4.

PROPOSITION 8.10. *Let R be a complete negatively filtered connected F-algebra (where F is a commutative field). Then R is a power series ring if and only if it satisfies the inverse weak algorithm.*

Proof. If $R = F\langle\!\langle X \rangle\!\rangle$, then every element of R is unique of the form

$$\Sigma x_I \alpha_I \qquad (\alpha_I \in F), \tag{8}$$

where (8) is restricted as in (7), and relative to the filtration defined above, the monomials in X span a dense subspace of R and are right v-independent. Hence R satisfies the inverse weak algorithm, by Theorem 8.9.

Conversely, if R satisfies the inverse weak algorithm, let X be a weak v-basis for R_{-1}, then every element of R is unique of the form (8), with the same restriction as before. The subalgebra A generated by X over F is therefore free on X over F and clearly it is dense in R, whence R is isomorphic to the completion of A and the result follows. ∎

We have the following obvious consequences.

COROLLARY 1. *A negatively filtered complete algebra with an inverse weak algorithm is a power series ring if and only if it is connected.* ∎

COROLLARY 2. *If $F\langle\langle X\rangle\rangle$ is a power series ring over F, then any closed subalgebra of $F\langle\langle X\rangle\rangle$ satisfying the inverse weak algorithm is again a power series ring over F.* ∎

As a special case we obtain the following characterization of power series rings in a single variable.

PROPOSITION 8.11. *Let R be a complete negatively filtered connected F-algebra. Then R is a power series ring in a single variable over F if and only if $R \neq F$ and of any two non-zero elements of R, the one of smaller degree is v-dependent on the other.*

For clearly $F[[x]]$ satisfies the stated conditions. Conversely, when R satisfies all these conditions, it satisfies in particular the inverse weak algorithm and so, by Proposition 8.10, is of the form $F\langle\langle X\rangle\rangle$ for some v-independent set X. If X contains more than one element, let $x, y \in X$, $x \neq y$. By the v-independence of X, neither of x, y is v-dependent on the other. This contradicts the hypothesis and it follows that X has at most one element. Since $R \neq F$, X has exactly one element, as asserted. ∎

We conclude this section with an important relation between a ring with inverse weak algorithm and its completion. We shall need to assume our ring to be (internally) graded, thus if $G = (G_i)$ is an externally graded ring, then $R = \Sigma G_i$ is the corresponding internally graded ring and $\hat{R} = \Pi G_i$ its completion. Of course for a graded ring the weak algorithm and the inverse weak algorithm are equivalent, since both correspond to the weak algorithm for the externally graded ring.

THEOREM 8.12. *Let R be an internally graded ring with weak algorithm and $\hat{R}$ its completion, then R is totally inert in $\hat{R}$.*

Proof. Let A, B be families of rows and columns respectively, of length r over $\hat{R}$, such that $a \cdot b \in R$ for any $a \in A$, $b \in B$. We may assume that neither A nor B reduces to zero, and further that they are closed in the sense that each consists of the rows (respectively columns) which transport the columns (respectively rows) of the other into R.

Pick a non-zero $a \in A$, say $a = (a_1, ..., a_r)$, and let $v_1, v_2, ...$ be the (possibly infinite) sequence of degrees for which a component of one of the a_i is not a right linear combination of earlier ones. Let $a_1', a_2', ...$ be the corresponding components, then by the weak algorithm these components form a basis for the set of all the components of $a_1, ..., a_r$. Hence there is a matrix $U = (u_{\lambda j})$ of r columns such that

$$(a_1, ..., a_r) = (a_1', a_2', ...)U,$$

where $v(a_\lambda') \to -\infty$ as $\lambda \to \infty$ and each row of U contains at least one element with constant term 1 (and hence a unit in $\hat{R}$). For any $b \in B$, say $b = (b_1, \ldots, b_r)^T$, we have $a \cdot b = \Sigma a_i b_i \in R$, hence $\Sigma a_\lambda' u_{\lambda i} b_i \in R$, and so, for sufficiently large n,

$$\Sigma a_\lambda' (u_{\lambda i} b_i)_{n - v_\lambda} = 0,$$

where the suffix on the parenthesis denotes the homogeneous component. By the linear independence of the a_λ' over R we find that $\Sigma u_{\lambda i} b_i \in R$ for each λ. Hence $(u_{\lambda 1}, \ldots, u_{\lambda r}) \in A$ for each λ. Let us fix λ, say $\lambda = 1$, and apply column operations to A and the corresponding row operations to B so as to reduce $(u_{11}, \ldots, u_{1r})$ to $(1, 0, \ldots, 0)$; since some u_{1i} is a unit, this is possible. In the new set B we have $b_1 \in R$ for each $b = (b_1, \ldots, b_r)^T$. We now treat B in the same way A was treated before. Thus we can find a basis $b_1', b_2', \ldots$ of homogeneous elements for $b_1, \ldots, b_r$, consisting of components of these b_i, so that

$$\begin{pmatrix} b_1 \\ \vdots \\ b_r \end{pmatrix} = V \begin{pmatrix} b_1' \\ b_2' \\ \vdots \end{pmatrix},$$

where $V = (v_{i\mu})$ is a matrix of r rows, and each column contains a unit. We have $\Sigma a_i v_{i\mu} b_\mu' = \Sigma a_i b_i \in R$, hence (as before) $\Sigma a_i v_{i\mu} \in R$ and so $(v_{1\mu}, \ldots, v_{r\mu})^T \in B$. In particular, taking $a = (1, 0, \ldots, 0)$, we see that $v_{1\mu} \in R$. Now for some μ, $v_{1\mu}$ is a unit in $\hat{R}$; further, if we denote, for any $c \in R$, the highest degree occurring by $d(c)$ and the lowest degree by $o(c)$, we have

$$d(b_1) - o(b_1) \geqslant d(b_1) - d(b_\mu') \geqslant d(v_{1\mu}) \quad \text{for all} \quad \mu.$$

Hence we can find $v_{1\mu} \in R$ with constant term 1 and of degree not exceeding the minimum of $d(b_1) - o(b_1)$, as $b = (b_1, \ldots, b_r)^T$ ranges over B. Write $v = (v_{1\mu}, \ldots, v_{r\mu})^T$, then $v \in B$ and hence $vx = (v_{1\mu} x, \ldots, v_{r\mu} x)^T \in B$ for any $x \in R$. Thus if we put $B_1 = \{b_1 | (b_1, \ldots, b_r)^T \in B\}$ then $B_1 \supseteq v_{1\mu} R$; we claim that equality holds here.

If not, we can find $b = (b_1, \ldots, b_r)^T$ in B such that $b_1 \notin v_{1\mu} R$ and b_1 has least 'breadth' $d(b_1) - o(b_1)$ subject to this condition. Let b^* be the term of least degree in b_1, then $b - vb^* \in B$ and this column has a first component of higher order than b_1, but not of greater degree, hence it has smaller breadth, contradicting our choice of b. This shows that $B_1 = v_{1\mu} R$, and so $a \in A$, $b \in B$ is equivalent to $a_1 \in Rv_{1\mu}^{-1}$, $b_1 \in v_{1\mu} R$ and $\Sigma_2^r a_i b_i \in R$. Now the result follows by induction on r. ∎

The most important application of this result is to free algebras and their completions. In that case the theorem states that $F\langle X \rangle$ is totally inert in $F\langle\!\langle X \rangle\!\rangle$.

Exercises 2.8

1. State and prove analogues of Propositions 2.1 and 2.2 for the inverse dependence number.

2*. Let R be any ring and $n \geqslant 2$; show that the following two properties are equivalent where t is the trivial filtration:

(i) In a t-dependent family, any element a is t-dependent on those of degree $\leqslant t(a)$,

(ii) R is a local ring and an n-fir.

3. Find an extension of Lemma 8.4 to the case when A is infinite.

4. Verify the inverse weak algorithm for the following rings:

(i) $\mathbf{Z}$ with $v = -v_p$, where v_p is the p-adic valuation,

(ii) $k\langle X \rangle$ where k is a field and $-v$ is the degree of the least non-zero terms.

5. Show that in a complete negatively filtered ring R, where R/R_{-1} is a field, u is invertible if and only if $v(u) = 0$.

6. Show that every complete inversely filtered connected K-ring with inverse algorithm is a power series ring in one indeterminate x, with commutation rule

$$ax = xa^\alpha + x^2 a^{\delta_1} + \ldots$$

where α is an endomorphism of K and $(\delta_1, \delta_2, \ldots)$ is a higher α-derivation.

7. Let A be the group algebra over k of the free group on X and define an order function in terms of the total degree: $d(x_1^{\varepsilon_1} \ldots x_n^{\varepsilon_n}) = \Sigma \varepsilon_y$. Show that A has a completion $\hat{A}$ relative to this function, and that $k\langle\!\langle X \rangle\!\rangle$ can be embedded in $\hat{A}$.

8. Let R be a complete negatively filtered ring with $\mu_v(R) > 2$. Given a, b, a', $b' \in R^*$ satisfying $ab' = ba'$, $v(a) \leqslant v(b)$, show that there exists $c \in R$ such that $a = bc$, $a' = cb'$.

9. Let R be as in Exercise 8. If $a, b, a' \in R^*$ are such that $v(a) < 0$ and $ab = ba'$, find $u, v \in R$ and $r \geqslant 0$ such that $a = uv$, $a' = vu$, $b = d^r u = ua''$.

Deduce that in a power series ring, the eigenring of a non-zero element is an Artinian local ring.

10. If X is a finite set, the elements of $k\langle\!\langle X \rangle\!\rangle$ can be described as infinite power series $\Sigma x_I \alpha_I$, where x_I ranges over all monomials in X. Show that for infinite X this is no longer true, but that there is a ring whose elements are all the infinite power series in X, and which is obtained as the completion of $k\langle X \rangle$ by assigning suitable degrees to the elements of X.

11*. (Jooste [71]). Let R be a complete negatively filtered ring with $\mu_v(R) > n$. Show that the kernel of any derivation of R is an n-fir. (Hint: Use Exercise 2.)

12*. (Jooste [71]). Let R be a complete negatively filtered ring with inverse weak algorithm, and let d be any derivation in R, with kernel N. Show that N is a semifir and that R is flat as N-module.

13°. Investigate rings satisfying the inverse weak algorithm relative to the trivial filtration.

14°. Investigate negatively filtered F-algebras with inverse weak algorithm which are not connected.

15°. (Bergman) Let R be a free algebra and $\hat{R}$ its power series completion. If an element a of R is a square in $\hat{R}$, is it associated (in $\hat{R}$) to a square of an element of R?

2.9 The transfinite weak algorithm

In **2.1** we saw that the classical division algorithm can be defined for any ring R with a function from R^* to $\mathbf{N}$. All we used in fact was the well-ordering of $\mathbf{N}$ and this suggests that we consider a ring R with a function from R^* to the ordinals less than some ordinal τ. Using the notion of derived set introduced in **2.1**, let us again define sets S_α by putting

$$S_0 = R^*, S_{\alpha+1} = S_\alpha', \quad S_\lambda = \bigcap_{\alpha < \lambda} S_\alpha \text{ (at a limit ordinal } \lambda\text{).} \tag{1}$$

With $\theta(x)$ defined as in **2.1**, we see as before that

$$\theta(x) \geqslant \alpha \quad \text{for} \quad x \in S_\alpha,$$

while the existence of a (transfinite) division algorithm is now expressed by the fact that $S_\tau = \emptyset$ for some ordinal τ.

We note that for any $x \in R^*$ the ordinal $\min \{\alpha | x \notin S_\alpha\}$ is *not* a limit ordinal. For if λ is a limit ordinal and $x \notin S_\lambda$ but $x \in S_\alpha$ for all $\alpha < \lambda$, then $x \in \bigcap S_\alpha = S_\lambda$, a contradiction. Thus we can define a function ϕ by

$$\phi(x) + 1 = \min \{\alpha | x \notin S_\alpha\}.$$

As in **2.1** we see that $\phi(x) \leqslant \theta(x)$, if θ is any ordinal function for which the division algorithm holds, and $\phi(ab) \geqslant \phi(a)$. Thus we have

THEOREM 9.1. *Let R be a ring satisfying the division algorithm with respect to any function assuming ordinal values on R^*. Then there exists a function ϕ on R^* such that*

  (i) *$\phi(a)$ ranges over the ordinals $< \tau$, for some τ, as $a \in R^*$,*
  (ii) *$\phi(1) = 0$,*
  (iii) *$\phi(ab) \geqslant \phi(a)$,*

and R satisfies the division algorithm relative to ϕ. Moreover, ϕ is the 'least' function for which the division algorithm holds, in the sense that $\phi(a) \leqslant \psi(a)$ for any other function ψ satisfying the division algorithm. ∎

It is clear that the ordinal τ in this theorem is of cardinal not exceeding card (R).

We shall speak of the *transfinite division algorithm* when the value function is allowed to take ordinal values. As in **2.1**, we obtain the

COROLLARY. *A ring R satisfies the transfinite division algorithm (relative to some value function) if and only if, for some ordinal τ, $S_τ = \emptyset$, where $S_τ$ is the set defined by* (1). ■

No examples are known of commutative rings with transfinite division algorithm, for which the ordinary division algorithm does not hold. By contrast, in the non-commutative case there exist for every ordinal τ, rings for which the sequence (1) of derived sets terminates at stage τ (*cf.* Jategaonkar [69] and Chapter **8**). More generally, there is a transfinite *weak* algorithm; unlike the ordinary weak algorithm, this is not left–right symmetric and this makes it suited for studying rings which lack left–right symmetry in some respect. Instead of filtrations we need to consider more general functions whose precise form is suggested by Theorem 9.1.

Let *R* be a ring with a function *w* defined on it, satisfying the following conditions:

T.1. w maps R* to a well-ordered set, $w(0) = -\infty$,

T.2. $w(a - b) \leqslant \max \{w(a), w(b)\}$,

T.3. $w(ab) \geqslant w(a)$ *for* $b \neq 0$.

Clearly we may assume that *w* ranges over a set of ordinals: $0 \leqslant \alpha < \tau$. This imposes no restriction provided that τ is chosen greater than the cardinality of *R*; e.g. if *R* is the semigroup algebra of a semigroup *S* over a field *F*, and *w* is defined on *S* to satisfy *T.3*, we may set

$$w(\Sigma \, s\alpha_s) = \max \, \{w(s)|\alpha_s \neq 0\}. \tag{2}$$

In order to define the transfinite weak algorithm relative to such a function we shall need to modify the notions of dependence introduced in **2.1**. We emphasize that the definitions which follow will apply to functions *w* satisfying *T.1–3* (rather than filtrations) and in any case, will only be used in the present section.

Given a function *w* satisfying *T.1–3*, we shall say that a family (a_i) of elements of *R* is *right w-dependent* if there exist elements $b_i \in R$ almost all 0, such that

$$w(\Sigma \, a_i \, b_i) < \max_i \, \{w(a_i \, b_i)\},$$

or if some $a_i = 0$. Otherwise the family (a_i) is called right *w-independent*. An element $a \in R$ is said to be *right w-dependent* on a family (a_i) if $a = 0$ or if there exist $b_i \in R$, almost all 0, such that

$$w(a - \Sigma \, a_i \, b_i) < w(a), \quad w(a_i \, b_i) \leqslant w(a) \quad \text{for all} \quad i.$$

Otherwise *a* is *right w-independent* of (a_i). The notions of *left w-dependence* are defined analogously.

We shall only be concerned with *right* w-dependence and therefore we usually omit the distinguishing adjective. We note also that a is w-dependent on the empty set if and only if $w(au) < w(a)$ for some unit u.

Definition. A ring R with a function w satisfying T.1–3 is said to satisfy the *transfinite weak algorithm* if in any right w-dependent family some member is right w-dependent on the rest.

Suppose that $a_1, ..., a_n$ are right w-dependent and that a_i is right w-dependent on the rest, say

$$w(a_i - \Sigma a_j c_j) < w(a_i), \quad w(a_j c_j) \leqslant w(a_i) \quad (j \neq i);$$

take any j such that $c_j \neq 0$, then $w(a_i) \geqslant w(a_j c_j) \geqslant w(a_j)$, by T.3, hence a_i is already w-dependent on the a_j of value at most $w(a_i)$.

As before, we can define weak w-bases of a right ideal, and prove that they exist in any ring with a function satisfying T.1–3. Moreover, any two weak w-bases of a right ideal have the same cardinal (*cf.* **2.2**). Now assume the transfinite weak algorithm; then any weak w-basis B must be w-independent. For, given a dependence relation

$$w(\Sigma a_i b_i) < \max_i \{w(a_i b_i)\} \quad (a_i \in B, b_i \in R)$$

by omitting terms if necessary, we may assume that $w(a_i b_i) = \alpha$ for all i, say for $i = 1, ..., n$. Then the a's can be renumbered so that $w(a_1) \leqslant ... \leqslant w(a_n)$ and some a_i is w-dependent on $a_i, ..., a_{i-1}$, which contradicts the property of a w-basis. Thus every right ideal has a right w-independent generating set; a fortiori this is a free generating set, hence every right ideal is free of unique rank, and remembering Theorem **1.2.3**, we obtain

THEOREM 9.2. *Any ring with (right) transfinite weak algorithm is a right fir, and hence satisfies right* ACC_n, *for all n.* ∎

On the other hand, a ring with right transfinite weak algorithm need not be a left fir. This is shown by the following construction, which in fact provides us with an example of a right but not left fir.

Let F be a commutative field and R the F-algebra on $y, x_1, x_2, ...$ with the defining relations

$$x_i = yx_{i+1} \quad (i = 1, 2, ...). \tag{3}$$

Let us define x_{-i} for $i \geqslant 0$ by the equation $x_{-i} = y^{i+1} x_1$. Then R has an F-basis consisting of all products $v = u(x)y^s$, where $u(x)$ runs over all products in the x_i ($i \in \mathbf{Z}$). If $u(x)$ has degree r (in the usual sense) we assign the 'degree' (r, s) to v; clearly these values are well-ordered in the lexicographic ordering and satisfy T.3. We can therefore extend this degree function

w to the whole of R by (2) so as to satisfy $T.1$–3. Let us verify that R has a transfinite weak algorithm relative to w.

Given products $a_1 b_1, \ldots, a_n b_n$ of maximal degree $\lambda = (r, s)$, suppose that

$$w(\Sigma\, a_v b_v) < \lambda. \tag{4}$$

Clearly we can omit any terms of degree less than λ. First assume that for some v, a_v is of degree $(r, *)$; then b_v must have degree $(0, *)$, say $b_v = \Sigma\, \beta_{vt}\, y^t$. Thus

$$w(\Sigma'\, a_v \beta_{vt}\, y^t + \Sigma''\, a_v b_v) < \lambda,$$

where Σ' is the sum over the terms with $w(a_v) = (r, *)$ and Σ'' is the sum over the remaining terms. In Σ'' the terms of highest degree in each b_v have degree (r', s), where $r' > 0$. So these terms may be written as $b_v' y^s$, and keeping only these terms in Σ'' we find

$$w(\Sigma'\, a_v \beta_{vt}\, y^t + \Sigma''\, a_v b_v' y^s) < \lambda. \tag{5}$$

Let a_1 have maximal degree among the a's, say $w(a_1) = (r, p)$ and put $q = s - p$; then $\beta_{1t} = 0$ for $t > q$, while $\beta_{1q} \neq 0$, and equating coefficients of y^q in (5) we obtain (after dividing by β_{1q}) a w-dependence of a_1 on the remaining a's.

There remains the case where each a_v is of degree less than $(r, 0)$. Then the highest term in each b_v is of the form $b_v' y^s$, where b_v' is a homogeneous expression in the x's, and hence

$$w(\Sigma\, a_v b_v') < (r, 0). \tag{6}$$

For any expression f in the x's (without constant term) denote by $[f]_t$ the expression obtained by decreasing the suffix of each left-hand factor x_i in f by t, i.e. if $f = \Sigma\, x_i f_i$, then $[f]_t = \Sigma\, x_{i-t} f_i$. If the term of maximum degree in a_v is $a_v' y^{q_v}$, then

$$a_v' y^{q_v} b_v' = a_v' [b_v']_{q_v}. \tag{7}$$

Each term on the right of (7) has degree r in the x's and by (6), the sum of these terms (over all v) has degree less than r. By the weak algorithm in the free algebra on the x's it follows that some a_v' is right d-dependent on the rest, d being the x-degree. If we take n to be minimal it follows that each a_v' of maximal degree is right d-dependent on the rest. Let a_1 be an element of maximal w-degree among the a_v, then (i) a_1' has maximal d-degree, r' say, among the a_v' and (ii) for all v with $d(a_v') = r'$ we have $q_v \leqslant q_1$; we shall denote by Σ' the sum over these v and by Σ'' the sum over the remaining terms. By hypothesis, we have

$$a_1' = \Sigma\, a_v'\, c_v,$$

where $d(c_v) > 0$ for any v from Σ''. Hence

$$a_1 = a_1'y^q = \Sigma'\, a_v\, y^{q_1 - q_v}\, c_v + \Sigma''\, a_v[c_v]_{qv}\, y^{q_1} + \text{lower terms}, \qquad (8)$$

where we have used the fact that in Σ', c_v commutes with y, because $c_v \in F$ when $d(a_v') = r'$. Now (8) shows a_1 to be right w-dependent on the remaining a's; thus the transfinite weak algorithm holds in R.

By Theorem 9.2, R is a right fir; however, as the chain

$$Rx_1 \subset Rx_2 \subset \ldots$$

shows, R does not satisfy left ACC_1 and so cannot be a left fir.

Exercises 2.9

1. Show that in a ring with transfinite weak algorithm any element of least value is a unit, and any element whose value is the successor of the least value is an atom.

2. Let R be a ring with transfinite weak algorithm, and let 0 be the least value. If $p \in R$ is an element of value 1, and $\mathfrak{a}$ is any proper right ideal containing p, show that $\mathfrak{a}$ has a weak w-basis including p.

3. Prove an analogue of Lemma 2.6 for rings with a function satisfying $T.1$–3.

4*. Let R be the k-algebra on generators $x_1, x_2, \ldots, y_1, y_2, \ldots, z$ with defining relations $x_i = zx_{i+1}, y_i = zy_{i+1}$ $(i = 1, 2, \ldots)$. Show that R has transfinite weak algorithm and hence is a right fir. Show that $\bigcap z^n R = x_1 R + y_1 R$, and deduce that the lattice of principal right ideals containing $x_1 R$ is not complete.
Show also that z is a right large non-unit in R.

5⁰. (Samuel [68]). Let R be a ring with transfinite algorithm. Does R have an integer valued algorithm? What is the answer if the residue class field mod pR, for each atom p, is known to be finite?

Notes and comments on Chapter 2

The familiar Euclidean algorithm occurs in Euclid, Book VII, Propositions 1 and 2, as a method for finding the HCF of two integers. The extension to polynomial rings was not undertaken until the 16th century, when Simon Stevin, in his Arithmetic (1585), Book II, Problem LIII, uses it to find the HCF of two polynomials. He remarks that this application is probably new, since Pedro Nuñez, writing only a few years earlier (Libro de Algebra, 1567) attempts to treat the same problem, but does not get beyond a few generalities.

There is a very extensive literature dealing with the Euclidean algorithm in algebraic number fields; most of this does not concern us, but Motzkin [49] who determines the imaginary quadratic extensions admitting a Euclidean algorithm with respect to any function, introduces the notion of a derived set and proves Theorem 1.1. Theorem 1.2 is due to Jacobson [34] and was found again by Cohn [61.]

The notion of weak algorithm was introduced by Cohn [61, 63'] as a simplified and abstract version of what was observed to hold in the free product of fields (*cf.* Cohn [60]). It was rediscovered by Bergman [67]. Our presentation is based on all these sources; in particular the original definitions have been modified as suggested by Bergman so as to be closer to the usual notion of dependence. The n-term weak algorithm was introduced by Cohn [66, 69''], where Proposition 2.7 was proved. The proof given in the text, using weak v-bases, is due to Bergman [67]. The device of 'equating right cofactors', used in the proof of Theorem 4.1, is a useful tool in studying free algebras. For its use in power series rings (in the form of 'transductions') see Nivat [68] and Fliess [70].

The weak algorithm for graded rings (and the material in **2.3**) is taken from Bergman [67]. The results of **2.4**, characterizing free algebras by the weak algorithm, are due to Cohn [61,] though the presentation largely follows Bergman [67]. The problem of constructing all rings with weak algorithm (**2.5**) was raised by Cohn [63'] and solved by Bergman [67]. The analogue of Schreier's formula in **2.5** was obtained by Lewin [69] as a corollary of the theorem that submodules of free modules over a free ring are free; Lewin proves this by a Schreier-type argument. The proof given here is taken from Cohn [69]; its extension to rings with weak algorithm is new.

The Euclidean algorithm in a (non-commutative) integral domain with a division algorithm was treated by Wedderburn [32]; the presentation in **2.7** (valid for any ring with 2-term weak algorithm) follows Cohn [63'], and the description of $GE_2(R)$ in **2.6** is taken from Cohn [66'].

The inverse weak algorithm (**2.8**) was introduced by Cohn [62], where 8.1–2, 8.7–11 are proved; 8.3–6 and the 'exchange principle' are due to Bergman [67,] who also proved Theorem 8.12 in the case of 1 inertia; the case of total inertia is new (*cf.* Cohn [71''']) for applications). See also Bowtell [67] for an application of the n-term inverse weak algorithm.

The transfinite algorithm is discussed by Motzkin [49] in the commutative case; the generalization in **2.9** (taken from Cohn [69']) was suggested by a method in Skornyakov [65,] which is used there to construct the right but not left fir given in the text.

3. Factorization

After reviewing the familiar notion of a commutative unique factorization domain in 3.1, we study its generalization obtained by looking at the lattice of factors; the resulting notion of non-commutative UFD is mainly of interest in the case of 2-firs. In the remaining sections we look at different aspects of factorization, including the notion of rigid UFD, which is the non-commutative analogue of a discrete valuation ring.

3.1 The commutative case

As is well known, a commutative integral domain is called a *unique factorization domain* (UFD for short) if it is atomic and any expression of an element as a product of atoms is unique except for the order of the factors and the presence of unit factors. This definition makes it clear that unique factorization is a property of the multiplicative semigroup of the ring, so let us restate the definition in terms of semigroups.

Any commutative semigroup S has a preordering by divisibility:

$$a|b \quad \text{if and only if} \quad b = ac \quad \text{for some} \quad c \in S. \tag{1}$$

If $a|b$ and $b|a$ in a cancellation semigroup, then

$$a = bu \quad \text{for some unit } u \text{ of } S, \tag{2}$$

i.e. a and b are associated. Clearly (1) defines a partial ordering of S if and only if 1 is the only unit of S. A cancellation semigroup with this property is said to be *conical*. With every commutative cancellation semigroup S, having a group U of units, we associate a conical semigroup S_U whose elements are the classes of associated elements of S. Since the relation (2) between a and b clearly defines a congruence on S, the set of these classes forms a semigroup in a natural way.

A cancellation semigroup S will be called a *UF-semigroup* if the associated conical quotient semigroup S_U is free commutative. With this definition it is clear that a commutative ring is a UFD if and only if its non-zero elements form a *UF*-semigroup under multiplication. In studying unique factorization in commutative rings we can therefore limit ourselves to *UF*-semigroups.

112

To state the conditions for unique factorization in semigroups succinctly, let us define a *prime* of a commutative semigroup S as an element p of S which is a non-unit and such that

$$p|ab \quad \text{implies} \quad p|a \text{ or } p|b.$$

Clearly any associate of a prime is again prime. Further, a prime in a cancellation semigroup is necessarily an atom. For if $p = ab$, then $p|a$ or $p|b$, say the former, so $a = pq$, $p = ab = pqb$ and by cancellation, $qb = 1$, i.e. b is a unit. The converse is false: an atom need not be prime, e.g. consider the semigroup generated by a and b with the defining relation $a^2 = b^2$. Here a is an atom, but not prime. In fact the converse, together with a finiteness condition, ensures that we have a UF-semigroup.

THEOREM 1.1 *Let S be a commutative cancellation semigroup. Then S is a UF-semigroup if and only if S satisfies* ACC$_1$, *together with one of the following three conditions, equivalent under* ACC$_1$

(i) *every atom in S is prime,*
(ii) *any two elements in S have an* HCF,
(iii) *any two elements in S have an* LCM.

For later applications it is useful to prove this result in a slightly more general form. A semigroup S is said to be *invariant*, if

$$aS = Sa \quad \text{for all} \quad a \in S.$$

Clearly this includes commutative semigroups and it shares with them the property that the preorderings defined (on any semigroup) by left and right divisibility coincide. For if $c = ab$, then also $c = ba_1$ for some $a_1 \in S$ and $c = b_1 a$ for some $b_1 \in S$. Therefore we can define primes as in commutative semigroups and we can again associate a conical semigroup S_U with S, whose elements are the classes of associated elements. An invariant cancellation semigroup S is called a UF-*semigroup* if its associated conical quotient semigroup S_U is free commutative. This clearly generalizes the previous definition and the following result includes Theorem 1.1 as a special case:

THEOREM 1.2. *Let S be an invariant cancellation semigroup. Then S is a UF-semigroup if and only if S satisfies* ACC$_1$, *together with one of the following three conditions, equivalent under* ACC$_1$:

(i) *every atom of S is prime,*
(ii) *any two elements of S have an* HCF,
(iii) *any two elements of S have an* LCM.

Proof. None of the conditions is affected if we pass to the associated conical quotient semigroup, so we may assume that 1 is the only unit in S. Clearly a free commutative semigroup is conical and satisfies ACC_1 as well as (i)–(iii). Conversely, assume that S is a conical cancellation semigroup satisfying ACC_1 and (i). If a, b are distinct atoms, let $ab = ba_1$ say, then $a|ba_1$ but $a \nmid b$. Since a is prime by (i), it follows that $a|a_1$, say $a_1 = au$. Thus $ab = bau$; by symmetry $ba = abv = bauv$, hence $uv = 1$, i.e. $u = v = 1$ and $ab = ba$. This shows that the subsemigroup generated by the atoms is commutative. In fact it is the whole of S, for if not, let aS be a maximal ideal such that a is not a product of atoms, then $a = bc$, where $aS \subset bS$, cS. Hence b, c are products of atoms and so a is too, which contradicts our assumptions.

The uniqueness proof follows a well known pattern. If

$$c = p_1^{\alpha_1} \dots p_n^{\alpha_n} = p_1^{\beta_1} \dots p_n^{\beta_n} \tag{3}$$

where the p_i are distinct atoms and $\alpha_1 < \beta_1$ say, then $p_1|c$ but $p_1 \nmid p_2^{\alpha_2} \dots p_n^{\alpha_n}$, hence $p_1|p_1^{\alpha_1}$, therefore $\alpha_1 > 0$. Now cancel p_1 in (3) and use induction on $\Sigma \alpha_i$ to complete the proof that S is a free commutative semigroup.

It remains to show that (i)–(iii) are equivalent, and from what has been proved, it is enough to show that (ii) $\Rightarrow$ (iii) $\Rightarrow$ (i). Assume (ii): given a, $b \in S$, there is a common multiple, viz. ab. Let m be a common multiple of a, b for which mS is maximal; if m' is any other common multiple of a, b we claim that $m|m'$, for otherwise the HCF, d say, of m and m', is again a common multiple of a and b and has the property $mS \subset dS$. Thus m is in fact a *least* common multiple.

Finally assume (iii) and let p be an atom such that $p|ab$. Denote the LCM of p and a by m, then $m = ap_1 = pa_1$ say, and since ap is a common multiple, $ap = md = ap_1 d$ say. Thus $p = p_1 d$, but p is an atom, so either (α) $d = 1$ and $p = p_1$ or (β) $p_1 = 1$ and $d = p$ (since 1 is the only unit). Case (α): $m = ap$ is an LCM, hence $ab = me = ape$, therefore $b = pe$, $p|b$. Case (β): $m = a$ is an LCM, hence $p|a$. Thus p is a prime, i.e. (i) holds, as asserted. ∎

In any invariant cancellation semigroup S, if a, b have an LCM, m say, then

$$aS \cap bS = mS. \tag{4}$$

Conversely, if (4) holds, then m is an LCM of a and b. Thus we have the

COROLLARY. *An invariant cancellation semigroup is a UF-semigroup if and only if it satisfies ACC_1 and the intersection of any two principal ideals is principal.* ∎

This result cannot be dualized, replacing $aS \cap bS$ by $aS \cup bS$, or even by the ideal generated by aS and bS.

It is clear how Theorem 1.2 may be used to study unique factorization in rings whose non-zero elements form an invariant cancellation semigroup. More generally, let R be any ring; an element $c \in R$ is said to be *invariant* in R if it is a non-zerodivisor and $cR = Rc$. The set $I = \mathbf{I}(R)$ of invariant elements of R is closed under multiplication and contains all units of R, in fact it is always an invariant cancellation semigroup. This follows from the fact that if any two elements of the equation

$$ab = c$$

are invariant in R, then so is the third.

For an invariant element the notions of left and right divisibility by a given element a coincide; we can therefore define a *prime* in R as an invariant non-unit p satisfying

$$p|ab \quad \text{implies} \quad p|a \quad \text{or} \quad p|b.$$

This is to be distinguished from the notion of a prime in I, which will be called an *I-prime* for contrast. Thus every prime lies in I and hence is an I-prime, but not conversely. Similarly an atom in I will be called an *I-atom*, so that an invariant atom in R is an I-atom, but not conversely (in general). The ring R is said to have *unique factorization of invariant elements* if $\mathbf{I}(R)$ is a UF-semigroup. By applying Theorem 1.2 we thus obtain

THEOREM 1.3. *Let R be any ring and $I = \mathbf{I}(R)$ the set of its invariant elements. Then R is a ring with unique factorization of invariant elements if and only if R satisfies* ACC *on invariant principal ideals or* $\mathbf{I}(R)$ *is atomic, and one of the following three conditions holds:*

(i) *every I-atom is an I-prime,*
(ii) *any two invariant elements have an* HCF *in I,*
(iii) *any two invariant elements have an* LCM *in I.* ∎

We note that e.g. (ii) certainly holds if any two invariant elements have an HCF in R and this HCF is invariant. If we merely know that $a, b \in I$ have an HCF d in R, we cannot assert that d is invariant, though there is an important case in which this holds, namely when R is a 2-fir (*cf.* 6.2 below).

Frequently one needs to know the effect of localizing on unique factorization. Again we begin by setting out the problem in terms of semigroups. Let S be an invariant cancellation semigroup, T a subsemigroup of S and define a relation on S by putting

$$a \equiv b \,(\text{mod } T) \quad \text{if and only if} \quad ar = bs \quad \text{for some} \quad r, s \in T. \tag{5}$$

It is easily checked that this is an equivalence on S which is compatible with multiplication, i.e. it is a congruence. The semigroup of congruence classes is again called the *quotient of S by T* and is written S_T. We note that S_T is conical whenever T contains all the units of S. The following criterion for S to be a UF-semigroup generalizes a theorem of Nagata:

THEOREM 1.4. *Let S be an invariant cancellation semigroup and T a subsemigroup of S generated by primes. Moreover, assume that S satisfies* ACC_1, *or that S is atomic, and that the quotient S_T is a UF-semigroup. Then S is itself a UF-semigroup.*

Proof. In order to apply Theorem 1.2 we need only verify that every atom of S is prime. Denote the canonical mapping $S \to S_T$ by $x \mapsto x'$ and let p be any atom of S. Clearly p' is either an atom of S_T and hence a prime, or a unit; we treat these cases separately.

(α) p' is a prime. If $p|ab$, say $pc = ab$, then $p'c' = a'b'$, hence $p'|a'$ or $p'|b'$, say the former, $p'e' = a'$. This means that

$$per = as \qquad (r, s \in T).$$

No prime factor of s can be an associate of p, otherwise p' would be a unit. Hence the prime factors of s divide e or r, and on cancelling them one by one we obtain the equation $pe_1 r_1 = a$, i.e. $p|a$.

(β) p' is a unit in S_T. Then $pr = us$ where $r, s \in T$ and u is a unit. Again we can cancel the prime factors of s one by one, but this time we find that one of them is associated to p, for u is a unit, while p is not. Hence p is prime. ■

Applied to rings, Theorem 1.4 yields

THEOREM 1.5. *Let R be any ring, $I = \mathbf{I}(R)$ the set of invariant elements and T a subsemigroup of I generated by I-primes. If R satisfies ACC on principal invariant ideals (or is atomic) and the image of I in the localization R_T is a UF-semigroup, then I is itself a UF-semigroup.* ■

Taking R to be commutative, we obtain the following slight generalization of Nagata's theorem:

COROLLARY. *Let R be a commutative integral domain with ACC on principal ideals. If T is a subsemigroup of R^* generated by primes, and the localization R_T is a UFD, then R is a UFD.* ■

Exercises 3.1

1. Show that every Noetherian integral domain is atomic.

2. Let k be a commutative field and $R = k[x, y, z, t]$, $\mathfrak{a} = (xy - zt) R$. Show that the ring $R/\mathfrak{a}$ (the coordinate ring of a quadric) is an atomic integral domain, but not a UFD.

3. If an invariant cancellation semigroup satisfies ACC_1 and the join of any two principal ideals is principal, show that it is a *UF*-semigroup. Show that the converse is false, by considering the multiplicative semigroup of a suitable UFD.

4. Let R be a commutative integral domain. If $a, b \in R$ have an LCM, show that they have an HCF. Show that the converse is false, by considering the subring of $\mathbf{Z}[x]$ consisting of polynomials with even coefficient of x.

5. Show that in a commutative integral domain, the LCM of any pair of elements exists if and only if the HCF of any pair of elements exists.

6*. (Amitsur [48]) Show that a commutative UFD with DCC on the ideals containing any fixed non-zero ideal is a principal ideal domain.

7*. In the ring of integral quaternions, show that

$$i(1 + ai + bj + ck) - (1 + ai + bj + ck)(qi + rj + sk),$$

where $q = 1 - (b^2 + c^2)u$, $r = (ab - c)u$, $s = (ac + b)u$ and $u - 2(1 + a^2 + b^2 + c^2)^{-1}$. Hence find all invariant elements.

8⁰. Let S be an invariant cancellation semigroup and S_U the associated conical semigroup. Is S_U necessarily commutative? (S_U is commutative provided it is atomic, cf. the proof of Theorem 1.2.)

3.2 The category of strictly cyclic modules

In order to study the factorizations of an element c in a non-commutative ring R it is more convenient to take the module R/cR, which turns out to reflect all the properties of factorizations of c itself. To avoid ambiguity we shall take c to be a non-zerodivisor.

Given any ring R, an R-module is said to be *strictly cyclic* or a *𝒞-module*, if it has a presentation of the form

$$M \cong R/cR \qquad \text{where } c \text{ is a non-zerodivisor.} \qquad (1)$$

A cyclic presentation of a 𝒞-module need not have the form (1), but if R is an integral domain such that

$$R \oplus P \cong R^2 \quad \text{implies} \quad P \cong R, \qquad (2)$$

then every cyclic presentation of a 𝒞-module has the form (1). For if M, given by (1), also has the presentation $M \cong R/\mathfrak{a}$, for some right ideal $\mathfrak{a}$ of R, then by Schanuel's lemma $R \oplus \mathfrak{a} \cong R^2$, hence by (2), $\mathfrak{a} \cong R$, i.e. $\mathfrak{a}$ is principal, generated by a non-zero element. Thus every cyclic presentation of M has the form (1). We note in particular, that this is true for 2-firs.

The category of all strictly cyclic right R-modules and all homomorphisms between them is denoted by $\mathscr{C}_R$ and the corresponding category of left R-modules is written $_R\mathscr{C}$. In general these categories are not additive, let alone abelian, but there is a remarkable duality between them, valid over any ring.

THEOREM 2.1. *For any ring R, the categories $\mathscr{C}_R$ and $_R\mathscr{C}$ are dual, i.e. there exist contravariant functors $D_1: \mathscr{C}_R \to {}_R\mathscr{C}$, $D_2: {}_R\mathscr{C} \to \mathscr{C}_R$ such that D_1D_2 and D_2D_1 are each naturally equivalent to the identity functor.*

Proof. Since natural equivalence is all that is asserted, we need only define $D_1(M)$ up to isomorphism. Given $M \in \mathscr{C}_R$, say $M \cong R/aR$, we put $D_1(M) = R/Ra$; of course we shall need to show that this depends only on M and not on the presentation used. Given any R-homomorphism

$$f: R/aR \to R/bR, \tag{3}$$

we recall that f is determined by an element $c \in R$ such that for some $c' \in R$,

$$ca = bc'. \tag{4}$$

In fact, any pair (c, c') satisfying (4) defines a homomorphism (3) by the rule

$$x \ (\text{mod } aR) \mapsto cx \ (\text{mod } bR).$$

Now c is determined by f up to an element of bR; thus if $c_1 = c + bu$, then by (4), $c_1a = ca + bua = b(c' + ua)$. This shows that c' is determined by f up to an element of Ra, and by (4) it defines a unique R-homomorphism

$$D_1(f): R/Rb \to R/Ra.$$

The correspondence $f \to D_1(f)$ is easily seen to be a contravariant functor. In particular, if f is an isomorphism, so is $D_1(f)$, so that $D_1(M)$ depends only on M and not on a. Now $D_2: {}_R\mathscr{C} \to \mathscr{C}_R$ is defined by symmetry and a routine verification shows that $D_1D_2 \simeq 1$, $D_2D_1 \simeq 1$. ∎

The duality described here will appear again in Chapter 5, in a more general form. For the moment we note the following consequences.

COROLLARY 1. *Let R be any ring and a a non-zerodivisor. Then aR and Ra have isomorphic eigenrings:*

$$\mathbf{E}(aR) \cong \mathbf{E}(Ra). \quad \blacksquare$$

Of course the fact that we have an isomorphism rather than an anti-isomorphism arises from the notational convention of writing homomorphisms of left and right R-modules on opposite sides.

COROLLARY 2. *Let a, a' be non-zerodivisors in a ring R. Then $R/aR \cong R/a'R$ if and only if $R/Ra \cong R/Ra'$.* ∎

In view of this result we define for any non-zerodivisors a, a': a is *similar* to a', in symbols $a \sim a'$, if and only if $R/aR \cong R/a'R$. Now Corollary 2 shows that this relation is left–right symmetric. In particular, in an integral domain this definition applies to all non-zero elements.

Theorem 2.1 may be interpreted as follows. Given a factorization of an element c into non-zerodivisors:

$$c = a_1 \ldots a_r \qquad (5)$$

we can associate with (5) the series of right ideals of R,

$$R \supseteq a_1 R \supseteq a_1 a_2 R \supseteq \ldots \supseteq a_1 \ldots a_r R = cR,$$

and the corresponding quotient modules in $\mathscr{C}_R$:

$$R/a_1 R,\ a_1 R\ a_1 a_2 R \cong R/a_2 R,\ \ldots,\ R/a_r R;$$

we also have the series of left ideals of R,

$$R \supseteq Ra_r \supseteq Ra_{r-1} a_r \supseteq \ldots \supseteq Ra_1 \ldots a_r = Rc,$$

and the quotient modules in $_R\mathscr{C}$:

$$R/Ra_r,\ R/Ra_{r-1},\ \ldots,\ R/Ra_1.$$

Clearly it does not matter which we choose, and Theorem 2.1 is a general way of expressing this symmetry; it may be described as the *factorial duality*.

Two factorizations of c are said to be *isomorphic* if they have the same length and their quotients are pairwise isomorphic (not necessarily in the given order). More precisely, if

$$c = b_1 \ldots b_s \qquad (6)$$

is a second factorization of c into non-zerodivisors, then (5) and (6) are isomorphic if and only if $r = s$ and there is a permutation $i \mapsto i'$ of $1, \ldots, r$ such that $b_i \sim a_{i'}$. It is clear that this notion of isomorphism between factorizations is left–right symmetric.

Now a non-commutative *unique factorization domain* (UFD for short) may be defined as an integral domain R which is atomic, and such that any two atomic factorizations of any given non-zero element are isomorphic. To justify this definition we must show that it reduces to the usual definition in the case of commutative rings. This will follow if we can show that in a commutative ring, similar elements are associated. But when R is commutative, aR is determined by the module R/aR as the annihilator of that module. Hence if $R/aR \cong R/a'R$, then $aR = a'R$, and so a, a' are associated.

For non-commutative UFD's, the only case of importance so far is that of 2-firs.

THEOREM 2.2. *An atomic 2-fir is a UFD.*

Proof. In a 2-fir R, the set $L(cR, R)$ of principal right ideals containing a given non-zero element c is closed under addition and intersection. The former holds by definition, the latter by Theorem 1.5.1, Corollary. Thus $L(cR, R)$ is a sublattice of the lattice of all right ideals and therefore modular. By atomicity, $L(cR, R)$ has finite length, and now the Jordan–Hölder theorem shows that any two factorizations of c are isomorphic. ∎

COROLLARY. *Any left and right fir is a UFD.*

For if $c \neq 0$, R/cR satisfies ACC on strictly cyclic submodules, as does R/Rc, hence by duality, R/cR satisfies DCC, so that every chain in R/cR is finite, i.e. R is atomic. ∎

In general, neither a submodule nor a quotient module of a $\mathscr{C}$-module need be in $\mathscr{C}_R$, even over a 2-fir. But if

$$0 \to M' \to M \to M'' \to 0$$

is a short exact sequence of cyclic modules over a 2-fir R and any two of M, M', M'' lie in $\mathscr{C}_R$, then so does the third. This follows from the more precise

PROPOSITION 2.3. *The category $\mathscr{C}_R$ of strictly cyclic modules over a 2-fir R is a category with kernels, images and cokernels, i.e. given $f: M \to N$ in $\mathscr{C}_R$, then ker f, im f and coker f lie in $\mathscr{C}_R$.*

Proof. Given $f: M \to N$ in $\mathscr{C}_R$, write im $f = M''$ and let $N = R/aR$, then $M'' = (aR + bR)/aR \cong bR/(aR \cap bR)$. If $aR \cap bR = 0$, then M'' is free of rank 1, contradicting the fact that R is weakly 2-finite (which implies that no proper homomorphic image of R can be $\cong R$). Hence $aR \cap bR \neq 0$ and $aR + bR = dR \supseteq aR$, whence coker $f = N/M'' = R/dR$ and im $f = M'' \cong dR/aR$. By duality, ker f is also in $\mathscr{C}_R$. ∎

This result has numerous applications. In the first place we can derive analogues of Schur's and Fitting's lemmas, using the notion of simple or indecomposable object in the category $\mathscr{C}_R$. A module $M \in \mathscr{C}_R$ is said to be $\mathscr{C}$-*simple* if it is a simple object in $\mathscr{C}_R$, i.e. if $M \neq 0$ and no submodule of M (apart from 0 and M) lies in $\mathscr{C}_R$. For example, if $R = k\langle x, y \rangle$ is a free algebra, then R/yR is $\mathscr{C}$-simple, though of course far from simple. If we write down the factorization corresponding to a chain of submodules in $\mathscr{C}_R$ we see that over a 2-fir, a $\mathscr{C}$-module R/cR is $\mathscr{C}$-simple if and only if c is an atom.

PROPOSITION 2.4. *In a 2-fir, the eigenring of an atom is a field.*

For a is an atom if and only if R/aR is $\mathscr{C}$-simple. By Proposition 2.3, any non-zero endomorphism must have kernel 0 and image R/aR, and hence be an automorphism. ∎

In some cases a more precise statement is possible. Thus let R be a 2-fir which is an algebra over a field k. Then the eigenring of any element of R is again an algebra over k. Let a be an atom, so that $\mathbf{E}(aR)$, its eigenring, is a field over k. If λ is any element of $\mathbf{E}(aR)$ which is transcendental over k, then the elements $(\lambda - \beta)^{-1}$ for $\beta \in k$ all belong to $\mathbf{E}(aR)$ and are linearly independent over k. This shows that the dimension of $\mathbf{E}(aR)$ over k is at least card (k). If the dimension of R over k is less than card (k) this cannot happen, and so we have proved

PROPOSITION 2.5. *If R is a 2-fir which is an algebra over a field k, of dimension less than card (k), then the eigenring of any atom of R is algebraic over k.* ∎

As an immediate consequence we have the

COROLLARY. *If R is a 2-fir which is an algebra of countable dimension over an uncountable field k which is algebraically closed, then the eigenring of any atom of R is k itself.* ∎

We now turn to Fitting's lemma (*cf.* Jacobson [43] Chapter 1). This states that if M is an indecomposable R-module satisfying both chain conditions, then $\text{End}_R(M)$ is a *completely primary ring*, i.e. a local ring in which the unique maximal ideal is nilpotent. A module $M \in \mathscr{C}_R$ is said to be $\mathscr{C}$-*indecomposable* if it is indecomposable as object in $\mathscr{C}_R$, i.e. if it is non-zero and cannot be written as the direct sum of two non-zero $\mathscr{C}$-modules. It is easily seen that R/aR is $\mathscr{C}$-indecomposable if and only if a is *indecomposable* in the sense that any two proper factorizations of a:

$$a = bc' = cb'$$

are not both left and right coprime. With this definition we have an analogue of Fitting's lemma in the following form:

PROPOSITION 2.6. *In an atomic 2-fir the eigenring of an indecomposable element is a completely primary ring.*

The proof is very similar to that of Fitting's lemma (using Proposition 2.3 above) and may therefore be omitted. ■

Exercises 3.2

1. Show that in any integral domain the following are equivalent:
   (i) right ACC_1,
   (ii) ACC on right factors of a fixed non-zero element,
   (iii) DCC on left factors of a fixed non-zero element.

2. In a 2-fir with left ACC_1 any family of elements, not all 0, has a highest common left factor.

3. Let R be an integral domain with right and left ACC_1. Show that every non-zero element is either a unit or a product of atoms. Show that conversely, any 2-fir satisfying the conclusion has right and left ACC_1.

4. Show that any UFD has right and left ACC_1.

5. Let R be the subring of $\mathbf{Q}[x, y]$ consisting of all polynomials in which the coefficient of $x^r y^s$ is integral whenever $rs = 0$. Show that R is atomic, but does not satisfy ACC_1.

6*. Let $R = \mathbf{W}_1(k)$ be the Weyl algebra on x, y over k. Verify that $xR \cap yR$ is non-principal; deduce that the strictly cyclic module R/xR has a cyclic presentation which needs at least two defining relations.

7. If R is any ring such that $R \oplus M \cong R^2$ implies $M \cong R$, show that every cyclic presentation of a $\mathscr{C}$-module has the form R/cR, where c is a left non-zerodivisor.

8. Let R be a 2-fir and $M = R/cR$ a $\mathscr{C}$-module. Show that every finitely generated submodule of M is in $\mathscr{C}_R$ if and only if c is right large. Further, the qualifier 'finitely generated' can be omitted if R is atomic.

9. Show that any element similar to a right large element in a 2-fir is again right large.

10. Over a 2-fir R, show that R/cR is simple as R-module if and only if c is a right large atom.

11. Let R be a 2-fir, and $a, b \in R$. Show that there is an automorphism of R as right R-module which maps aR to bR if and only if a is associated to b.

12°. For which rings is any direct summand of a strictly cyclic module again strictly cyclic?

13. Let R be a 2-fir. Show that $\mathscr{C}_R$ has pullbacks (pushouts) if and only if R is a right (left) Ore ring. Interpret also the following properties of R in terms of $\mathscr{C}_R$: R has right (left) ACC_1.

14. Let A be the group algebra of the free group on x, y, z over a field k, and let R be the subalgebra generated by $z^{-n}x$, $z^{-n}y$, z ($n = 1, 2, \ldots$). Show that R is a right fir, but that the family $\{z^{-n}x, z^{-n}y\}$ has no highest common left factor.

15. Let R be an atomic 2-fir in which every principal right ideal is an intersection of maximal principal right ideals. Show that every $\mathscr{C}$-module is a direct sum of $\mathscr{C}$-simple modules. Show also that there are infinitely many pairwise non-associated elements similar to a given atom.

16. Show that an atomic right Bezout domain is a principal right ideal domain. Deduce that a ring is a principal ideal domain if and only if it is an atomic (left and right) Bezout domain.

17. In the Weyl algebra $W_1(k)$ on x and y, verify that $1 + xy$ is an atom, and is similar to xy. Deduce that $xyx + x$ has two atomic factorizations of different lengths.

18. Show that $Z\langle x, y \rangle$ is not a UFD, by considering factorizations of $xyx + 2x$. Show that the eigenring of any atom is either a field or a free algebra over a finite field.

3.3 Similarity

For a closer study of factorization we shall need criteria for similarity of elements in a ring. Although the main application is to 2-firs, it is best to develop this notion in a somewhat wider context.

Two non-zerodivisors a, a' of a ring R were called similar in 3.2 when $R/aR \cong R/a'R$. This may be regarded as a special case of the similarity of right ideals. Two right ideals $\mathfrak{a}$, $\mathfrak{a}'$ of a ring R are said to be *similar* if

$$R/\mathfrak{a} \cong R/\mathfrak{a}', \tag{1}$$

as right R-modules. It is clear from the definition that similar right ideals have isomorphic eigenrings. A criterion for similarity is given by

PROPOSITION 3.1. *Let $\mathfrak{a}$, $\mathfrak{a}'$ be two right ideals in a ring R. Then $\mathfrak{a}$ is similar to $\mathfrak{a}'$ if and only if there exists $b \in R$ such that*

(i) $\mathfrak{a} + bR = R$,

(ii) $\mathfrak{a}' = \{x \in R | bx \in \mathfrak{a}\}$.

In an abbreviated notation, (ii) may be expressed as $\mathfrak{a}' = b^{-1}\mathfrak{a}$.

Proof. Suppose that $\mathfrak{a}$ is similar to $\mathfrak{a}'$ and let $1 \pmod{\mathfrak{a}'}$ correspond to b $\pmod{\mathfrak{a}}$ in the isomorphism (1). Then b generates $R \pmod{\mathfrak{a}}$, hence (i) holds. Further $x \mapsto bx$ and so $bx \in \mathfrak{a}$ if and only if $x \in \mathfrak{a}'$, i.e. (ii). Conversely, if (i) and (ii) hold,

$$R/\mathfrak{a} \cong (\mathfrak{a} + bR)/\mathfrak{a} \cong bR/(\mathfrak{a} \cap bR) \cong R/\mathfrak{a}'. \quad \blacksquare$$

Note that whereas the relation of similarity is clearly symmetric, the criterion of Proposition 3.1 is not, so there are two ways of applying the criterion, once as it stands and once with $\mathfrak{a}$ and $\mathfrak{a}'$ interchanged. We also see from (ii) that $\mathfrak{a}'$ is determined in terms of $\mathfrak{a}$ and b. For example, if $\mathfrak{a}$ is a maximal right ideal, then (i) holds provided that $b \notin \mathfrak{a}$, and it follows that the right ideal $\mathfrak{a}'$ determined by (ii) is also maximal, because of (1). We state this as

COROLLARY 1. *If $\mathfrak{a}$ is a maximal (proper) right ideal of a ring R and $b \in R$, $b \notin \mathfrak{a}$, then the set $b^{-1}\mathfrak{a} = \{x \in R | bx \in \mathfrak{a}\}$ is a right ideal similar to $\mathfrak{a}$ and hence also maximal.* ∎

If b is a left non-zerodivisor, (ii) takes on the form

$$\mathfrak{a} \cap bR = b\mathfrak{a}'. \tag{2}$$

In particular, when R is an integral domain and $\mathfrak{a}$ is a proper right ideal, then $b \neq 0$ by (i) and so (ii) can be put in the form (2) in this case. Setting $\mathfrak{a} = 0$, we thus obtain

COROLLARY 2. *In an integral domain the only right ideal similar to 0 is 0.* ∎

We now consider the case of principal ideals. In this case the criterion of Proposition 3.1 can be transformed as follows:

THEOREM 3.2. *Let R be any ring and a, $a' \in R$. Then $aR \sim a'R$ if and only if there exists $b \in R$ such that the following three equivalent conditions hold:*

(α) *$aR + bR = R$, $aR \cap bR = ba'R$ and $1 + c'b \in a'R$ for some $c' \in R$,*

(β) *$aR + bR = R$ and $a'R = \{x \in R | bx \in aR\}$,*

(γ) *there exist matrices*

$$\mu = \begin{pmatrix} a & b \\ * & * \end{pmatrix} \quad \text{and} \quad v = \begin{pmatrix} * & * \\ * & a' \end{pmatrix} \quad \text{satisfying} \quad \mu v = \begin{pmatrix} 1 & 0 \\ * & * \end{pmatrix}, \quad v\mu = \begin{pmatrix} * & * \\ 0 & 1 \end{pmatrix},$$

where stars stand for unspecified elements.

Proof. By Proposition 3.1, (β) just expresses the similarity of aR and $a'R$, so we need only prove the equivalence of (α), (β), (γ).

(α) $\Rightarrow$ (β). By (α), $ba'R \subseteq aR$, whence

$$a'R \subseteq \{x \in R | bx \in aR\}. \tag{3}$$

Conversely, let $bx \in aR$, then by (α), $bx \in ba'R$, say $bx = ba'y$, i.e.

$$b(x - a'y) = 0. \tag{4}$$

By (α), $1 + c'b = a'd$ for some $d \in R$, i.e. $a'd - c'b = 1$, and so

$$x - a'y = (a'd - c'b)(x - a'y) = a'd(x - a'y),$$

by (4). Hence $x = a'(y + d(x - a'y)) \in a'R$, and equality in (3) follows.

(β) $\Rightarrow$ (γ). We must find elements c, d, b', c', $d' \in R$ such that

$$ad' - bc' = 1 \tag{5}$$

$$ba' = ab', \tag{6}$$

$$c'a = a'c, \tag{7}$$

$$a'd - c'b = 1, \tag{8}$$

for then $\mu = \begin{pmatrix} a & b \\ c & d \end{pmatrix}$ and $v = \begin{pmatrix} d' & -b' \\ -c' & a' \end{pmatrix}$ satisfy the given conditions.

Since $aR + bR = R$, there exist c', $d' \in R$ such that (5) holds. Further, $ba' \in aR$, so there exists $b' \in R$ such that (6) holds. Now multiply (5) by a: $ad'a - bc'a = a$, i.e. $bc'a \in aR$, whence $c'a \in a'R$ and we obtain (7). Similarly if we multiply (5) by b, we get $ad'b - bc'b = b$, i.e. $b(1 + c'b) \subset aR$, whence $1 + c'b \in a'R$ and this yields (8).

(γ) $\Rightarrow$ (α). Assume (5)–(8), then $aR + bR = R$ by (5), and by (8), $1 + c'b \in a'R$. Further, by (6), $ba'R \subseteq aR \cap bR$; conversely, if $m \in aR \cap bR$, say $m = ax = by$, then by (8) and (7), $y = a'dy - c'by = a'dy - c'ax = a'dy - a'cx = a'z$ say. Hence $m = by = ba'z$, and this shows that $aR \cap bR = ba'R$, i.e. (α). ∎

With a mild restriction on a and a' we can show that the matrices μ and v are mutually inverse.

COROLLARY 1. *If a, a' are left non-zerodivisors, then $aR \sim a'R$ if and only if there exists $\mu = \begin{pmatrix} a & * \\ * & * \end{pmatrix}$ with inverse $\mu^{-1} = \begin{pmatrix} * & * \\ * & a' \end{pmatrix}$.*

Proof. Assume that $aR \sim a'R$ and let μ, v be as in the proof of Theorem 3.2. Put $e_1 = (1 \ 0)$, $e_2 = (0 \ 1)$, then $e_1(\mu v - I) = e_2(v\mu - I) = 0$ and we must show that $v = \mu^{-1}$. Let us write $\mu v - I = \begin{pmatrix} 0 & 0 \\ u & v \end{pmatrix}$, and note that $e_2 v(\mu v - I) = e_2(v\mu - I)v = 0$. But $e_2 v = (-c', a')$, hence $(a'u, a'v) = 0$, so $u = v = 0$ and $\mu v = I$. Similarly $v\mu = I$. ∎

By comparing conditions α and γ of Theorem 3.2 in the case of an integral domain we obtain the following result which is sometimes useful.

COROLLARY 2. *Two elements of an integral domain R may be taken as the first row of an invertible matrix over R if and only if they are right comaximal and have a* LCRM. ∎

Condition (γ) of Theorem 3.2 states that a, a' are GL-related (*cf.* **2**.6), thus we have

COROLLARY 3. *In any ring R, let a, $a' \in R$ be left non-zerodivisors. Then aR is similar to $a'R$ if and only if a and a' are GL-related.* ∎

In particular, for non-zerodivisors this shows that similarity between elements, as defined in 3.2, coincides with the notion of being GL-related. Since we are mainly concerned with integral domains, the notions of being similar and being GL-related will be coextensive in much of what follows.

The following obvious property of GL-related elements is often useful.

PROPOSITION 3.3. *Let a, a' be GL-related elements of a ring R. Then the images of a and a' under any homomorphism are again GL-related.* ∎

For example, it is easily verified that any element GL-related to a unit is itself a unit. Hence we obtain the

COROLLARY. *Let a, a' be GL-related elements of a ring R. If under some homomorphism, a maps to a unit, then so does a'.* ∎

In 2-firs it is possible to simplify the notion of similarity still further. We recall (Proposition **2**.6.6) that in any weakly 2-finite ring, two elements are GL-related if and only if they can be put in a comaximal relation

$$ab' = ba'. \tag{9}$$

Such a relation is clearly *coprime*, in the sense that a, b have no common left factor and a', b' no common right factor (apart from units). In a 2-fir, conversely, every coprime relation is comaximal, for excluding trivial cases, we may take $ab' = ba' \neq 0$, then $aR + bR$ is principal, and hence equal to R, by coprimality. Similarly, $Ra' + Rb' = R$. If we recall (from Theorem **1**.1.1, Corollary 2) that a 2-fir is always weakly 2-finite, we therefore obtain

PROPOSITION 3.4. *In a 2-fir two elements a, a' are similar if and only if they can be put in a coprime relation $ab' = ba'$.* ∎

Using the notion of similarity we can describe right denominator sets in 2-firs as follows:

PROPOSITION 3.5. *Let R be a 2-fir and S a subsemigroup of R^* such that* (i) *each element of S is right large,* (ii) *$ab \in S$ implies $b \in S$ and* (iii) *if $a \in S$, any element similar to a is in S. Then S is a right denominator set.*

Proof. By definition (0.5) we need only show that for any $a \in R$, $u \in S$, $aS \cap uR \neq \emptyset$. When $a = 0$, this is clear, so let $a \neq 0$, then $aR \cap uR \neq 0$, by (i), hence $aR + uR = dR$, say $a = da_1$, $u = du_1$. Here $u_1 \in S$ by (ii) and there is a comaximal relation $a_1 u' = u_1 a'$ with u' similar to u_1, and hence $u' \in S$ by (iii). Thus $au' = ua'$ is the required common multiple. ∎

Exercises 3.3

1. (a) Let a, b be two right ideals in a ring R such that $a + b = R$. Verify that the sequence
$$0 \rightarrow a \cap b \xrightarrow{\lambda} a \oplus b \xrightarrow{\mu} R \rightarrow 0$$
is split exact, where $\lambda(x) = (x, x)$, $\mu(x, y) = x - y$.

(b) Let a, a' be similar right ideals of R and assume that the criterion of Proposition 3.1 holds with a non-zerodivisor b. Show that
$$a \oplus R \cong a' \oplus R.$$

(c) Give an example of a ring containing similar right ideals for which this isomorphism does not hold.

(d) Give examples of similar right ideals that are not isomorphic.

2. In a 2-fir show that any right ideal similar to a principal right ideal is again principal. More generally, show that this holds in any integral domain in which any two comaximal principal right ideals have a principal intersection.

3. (Bowtell [67]) If $\alpha = \begin{pmatrix} a & b \\ c & d \end{pmatrix}$ and $\beta = \begin{pmatrix} d' & -b' \\ -c' & a' \end{pmatrix}$ are two mutually inverse matrices, show that $aR \cap bR = ad'bR + bc'aR$. Deduce that if R is an integral domain in which the sum of any two principal right ideals with non-zero intersection is principal, then R is a 2-fir.

4. Let R be a 2-fir. Show that the following conditions on a pair of elements a, b of R are equivalent:

(i) a, b are right comaximal,
(ii) a, b are left-hand factors in a coprime relation,
(iii) a, b form the first row of an invertible 2×2 matrix.

5. In a 2-fir, let $a = bc$ and $a \sim a'$. Show that there exist $b' \sim b$, $c' \sim c$ such that $a' = b' c'$. Does this hold in more general rings?

6. In any ring, if $\alpha = \begin{pmatrix} a & * \\ * & * \end{pmatrix}$ has the inverse $\begin{pmatrix} * & * \\ * & a' \end{pmatrix}$, find a matrix β such that $\begin{pmatrix} a & 0 \\ 0 & 1 \end{pmatrix} = \beta \begin{pmatrix} 1 & 0 \\ 0 & a' \end{pmatrix} \alpha$.

7. Show that two non-zerodivisors a, a' in a ring R are similar if and only if $\begin{pmatrix} a & 0 \\ 0 & 1 \end{pmatrix}$ and $\begin{pmatrix} a' & 0 \\ 0 & 1 \end{pmatrix}$ are associated in R_2.

8. Let R be a ring and R_* the ring of infinite matrices over R which differ from a scalar matrix only in a finite square. Show that any two non-zerodivisors in R_* that are similar are associated.

9*. (Fitting [36]) Let R be any ring and $a, b \in R$. Show that aR and bR are similar if and only if the matrices $\alpha = \begin{pmatrix} a & 0 & 0 & 0 \\ 0 & 1 & 0 & 0 \end{pmatrix}$ and $\beta = \begin{pmatrix} b & 0 & 0 & 0 \\ 0 & 1 & 0 & 0 \end{pmatrix}$ are associated, i.e. there exist a 2×2 matrix μ and a 4×4 matrix ν which are invertible and satisfy $\mu \alpha \nu = \beta$.

10. Show that any element GL-related to a zero-divisor is itself a zero-divisor, and that any element GL-related to a unit is a unit. Explicitly, if μ, ν are mutually inverse matrices as in Theorem 3.2, and a is a unit, then $a'^{-1} = d - ca^{-1}b$.

11. In a weakly 1-finite ring, show that any element GL-related to 0 is 0. Is there a converse?

12. Let k be a field and $a, b \in k$. Show that $t - a$ and $t - b$ are similar in $k[t]$ if and only if a, b are conjugate in k (i.e. $a = u^{-1}bu$ for some $u \in k^*$).

13. Let k be a commutative field of characteristic 2 and α the mapping which sends each element to its square. Show that in the skew polynomial ring $k[x; \alpha]$ there are just two similarity classes of elements linear in x.

14. Show that in a skew polynomial ring, similar elements have the same degree.

15. In a free algebra, if two homogeneous elements are similar, show that they are associated.

16*. (Bergman) In the complex-skew polynomial ring, show that $x^4 + 1$ can be written as a product of two atomic factors in infinitely many different ways. By considering the factors, deduce the existence of a similarity class of elements which contains infinitely many elements that are pairwise non-associated.

17*. In a 2-fir, if $au - bv = c$ and a, c are right comaximal, show that $ua' - b'w = c'$, where $a' \sim a$, $b' \sim b$ and $c' \sim c$. (Hint: let $ac' = ca'$ be a comaximal relation; multiply the given equation by a' on the right and use the fact that $aR \cap bR$ is principal.)

Deduce that $xa - by = c$ has a solution (x, y) provided that there exist comaximal relations $a_1c = c_1a, a_1b = b_2a_2$.

18. In a 2-fir R, show that for any $a, b \in R$, $1 + ab \sim 1 + ba$.

3.4 Rigid factorizations

Let us return to the definition of a UFD and consider more closely in what respects it differs from the definition in the commutative case. As we saw, given two atomic factorizations,

$$c = a_1 \dots a_r \tag{1}$$

and

$$c = b_1 \dots b_r \tag{2}$$

of an element c (necessarily of the same length), there is a permutation $i \mapsto i'$ such that

$$R/a_i R \cong R/b_{i'} R. \tag{3}$$

However, we note that

(i) a_i and $b_{i'}$ are not necessarily associated, and

(ii) in general we do not obtain c on writing the b's in the order $b_1, b_2, \dots b_{r'}$.

To this extent unique factorization in non-commutative domains, as here defined, is a more complicated phenomenon than in the commutative case. For this reason another more restrictive definition is sometimes useful.

An element c of an integral domain R is said to be *rigid*, if

$$c = ab' = ba' \quad \text{implies} \quad aR \subset bR \quad \text{or} \quad bR \subseteq aR.$$

In other words, c is rigid if the lattice $\mathbf{L}(cR, R)$ is a chain. If every non-zero element of R is rigid, R is said to be a *rigid* integral domain. In that case R^* is a rigid semigroup in the sense of **0.5**, and by the result proved there we have

PROPOSITION 4.1. *Let R be a rigid integral domain. Then R^* can be embedded in a group.* ∎

We observe that a rigid integral domain R is necessarily a 2-fir, by Theorem **1.5.2**. Now a *rigid* UFD is defined as an atomic rigid integral domain. For example, in the commutative case, a rigid integral domain is a valuation ring and a rigid UFD is a discrete valuation ring. However, a non-commutative rigid domain can be much more general than a non-commutative valuation ring.

Our main source of rigid UFD's is the following result.

THEOREM 4.2. *Any complete negatively filtered ring with 2-term inverse weak algorithm is a rigid UFD.*

Proof. Let R be a ring satisfying the hypothesis. By Theorem **2.8.5**, R is a 2-fir; if $ab' = ba' \neq 0$ and $v(a) \leqslant v(b)$ say, then a, b are right v-dependent and so, by the same Theorem, $aR \subseteq bR$; hence R is rigid. By Proposition 2.8.6 it has right and left ACC_1 and so is atomic. Thus R is a rigid UFD. ∎

An obvious example of a rigid UFD (other than a valuation ring) is a free power series ring $k\langle\!\langle X \rangle\!\rangle$.

We go on to describe rigid domains in more detail; for this we need the following lemma which is also useful elsewhere. In any ring R, a pair of elements u, v (in that order) is said to be *comaximally transposable* if there exist $u', v' \in R$ such that $uv = v'u'$ is a comaximal relation.

LEMMA 4.3. *Let R be a 2-fir. Then the elements $u, v \in R$ are comaximally transposable if and only if there exist $x, y \in R$ such that*

$$xu - vy = 1. \qquad (4)$$

Proof. Suppose we have a comaximal relation

$$uv = v'u'. \qquad (5)$$

Then $uR + v'R = R$ and $uR \cap v'R = uvR$, hence by Theorem 3.2, there are mutually inverse matrices $\mu = \begin{pmatrix} v' & u \\ * & * \end{pmatrix}$ and $\mu^{-1} = \begin{pmatrix} * & * \\ * & v \end{pmatrix}$, and equating the $(2, 2)$-entries in the matrix equation $\mu^{-1}\mu = I$ we obtain an equation (4). Conversely, given (4), we can by Proposition 0.2.5 find a matrix $\mu = \begin{pmatrix} * & v \\ * & * \end{pmatrix}$ with inverse $\mu^{-1} = \begin{pmatrix} u & * \\ * & * \end{pmatrix}$, and on equating the $(1, 2)$-entries of $\mu^{-1}\mu = I$, we obtain (5). Moreover, by looking at the $(1, 1)$- and $(2, 2)$-entries we see that this is a comaximal relation. ∎

As a first consequence we determine the radical of a 2-fir.

THEOREM 4.4. *Let R be a 2-fir. Then $\mathbf{J}(R)$, the Jacobson radical of R, consists of those elements of R that cannot be comaximally transposed with any non-unit of R.*

Proof. Suppose that c cannot be comaximally transposed with any non-unit. For any $x \in R$ we have the comaximal relation

$$c(xc + 1) = (cx + 1)c,$$

hence $cx + 1$ is a unit for all $x \in R$, and this shows that $c \in \mathbf{J}(R)$. Conversely, assume that $c \in \mathbf{J}(R)$ and suppose that b, c can be comaximally transposed. Then by Lemma 4.3, $cx + yb = 1$, and by the choice of c, $yb = 1 - cx$ is a unit, hence b is a unit. ∎

Secondly we shall obtain sufficient conditions for an element of a 2-fir to be rigid:

PROPOSITION 4.5. *Let R be a 2-fir and let $c \in R$ be such that any two neighbouring non-unit factors occurring in a factorization of c generate a proper ideal in R; then c is rigid. More precisely, it is enough to check a maximal family of factorizations of c such that of any two, one is a refinement of the other. In particular, it is enough to check a single atomic factorization of c, if one exists.*

Proof. If c is not rigid, then $c = ab' = ba'$, where neither of a, b is a left factor of the other, say $a = da_1$, $b = db_1$, where a_1, b_1 are both non-units. Let $a_1 b_1' = b_1 a_1'$ be an LCRM, then $a' = a_1' e$, $b' = b_1' e$ and $a_1 b_1' = b_1 a_1'$ is a comaximal transposition occurring in $c = da_1 b_1' e$. A corresponding transposition will occur in some member of our maximal family. By Lemma 4.3, $xa_1 - b_1' y = 1$, so the ideal generated by a_1, b_1' is improper, a contradiction. ∎

Now rigid domains are described by

THEOREM 4.6. *An integral domain is rigid if and only if it is a 2-fir and a local ring.*

Proof. We have seen that any rigid domain is a 2-fir, and clearly no two non-units can be comaximally transposable, hence by Theorem 4.4 the non-units of R form an ideal, i.e. R is a local ring. Conversely, in a 2-fir which is a local ring, *any* two non-units generate a proper ideal, hence by Proposition 4.5 every element is rigid. ∎

Adding the condition of atomicity, we obtain the

COROLLARY. *A ring is a rigid UFD if and only if it is an atomic 2-fir and a local ring.* ∎

The description of commutative rigid UFD's, namely as discrete valuation rings, can be extended to right Ore rings.

It is easily verified that any right DVR is a rigid UFD and conversely a rigid UFD is a right DVR if and only if any two atoms are right associated.

THEOREM 4.7. *Let R be a rigid UFD. Then R is a right discrete valuation ring if and only if it contains a right large non-unit.*

Proof. Assume that the rigid UFD R contains a right large non-unit c. Given any atom $p \in R$, we have $pR \cap cR \neq 0$, because c is right large, hence $cR \subseteq pR$ or $pR \subseteq cR$. Since c is a non-unit, the second alternative would mean that $cR = pR$, hence in any case $cR \subseteq pR$, i.e. $c = pc_1$. Now write

$$c = p^k u, \tag{6}$$

where $k \geq 1$ is chosen maximal (clearly k is bounded by the length of c). Then u must be a unit, for by (6) p is right large, and so $pR \cap uR \neq 0$; if u were not a unit, we would have $u = pu_1$, which inserted in (6) would contradict the maximality of k. If q is any atom of R, then since p is right large, $pR \cap qR \neq 0$ and hence as before, $q = pq_1$. Since q is an atom, q_1 must be a unit and so p and q are right associated. Therefore R is a right DVR; conversely, in a right DVR, the unique atom is a right large non-unit. ∎

From the normal form of the elements it is clear that in a right DVR every right ideal is two-sided. By symmetry we obtain

COROLLARY 1. *If R is a rigid UFD with a left and right large non-unit, then R has a unique atom p (up to unit factors) and every left or right ideal is two-sided, of the form $p^n R = Rp^n$.* ∎

Exercises 3.4

1. Show that a direct limit of rigid UFD's is again rigid, but not necessarily a UFD.

2*. Let $R = k[x; \alpha]$, where k is a commutative field of characteristic p and α is the pth power mapping. Find all elements of low degree in x that are rigid.

3. Let R be a rigid domain with right ACC_1 and a left and right large non-unit. Show that R is a left and right DVR.

4. (Koševoi [66]) An ideal $\mathfrak{p}$ in a ring R is said to be *strongly prime* if $R/\mathfrak{p}$ is an integral domain. Let R be an atomic 2-fir and $\mathfrak{p}$ a strong prime in R. Show that $\mathfrak{p}$ contains with any atom a, all atoms similar to a. Let $c \in R$ have the atomic factorizations $c = a_1 \ldots a_n = b_1 \ldots b_n$ and let $a_{i_1}, \ldots, a_{i_r}, b_{j_1}, \ldots, b_{j_s}$ be those atoms that lie in $\mathfrak{p}$. Show that $r = s$ and $a_{i_v} \sim b_{j_v}$.

5. Let k be a field with an endomorphism α which is not surjective, and put $R = k[[x; \alpha]]$. Show that R is a local ring with maximal ideal xR, and is a right DVR, but that R has a left ideal isomorphic to R^N. (It can be shown that R^N is not free, hence this provides another example of a right but not left fir. *Cf.* Chase [62], Cohn [66′′′]).

6*. Let R be a right hereditary local ring but not a DVR. Show that its centre is a field (*cf.* 6.4).

7^0. Find an example of a fir which is also a local ring, but not principal.

8^0. In a rigid UFD, is the intersection of the powers of the maximal ideal necessarily zero? (*cf.* **5.7**, **8.6**.)

9^0. Can every UFD be embedded in a group?

3.5 Factorization in 2-firs: a closer look

We shall now examine the relation between different factorizations of an element c in a 2-fir R. This is essentially a study of the lattice $L(cR, R)$, but we shall sometimes want to express the results directly in terms of factorizations. For brevity, in speaking of the "left factors" of an element we shall tacitly understand the equivalence classes under right multiplication by units. In this way the *left* factors of an element c correspond to the principal *right* ideals containing cR:

$$c = ab \quad \text{if and only if} \quad cR \subseteq aR.$$

Similarly, a chain of principal right ideals from cR to R corresponds to a factorization of c which is determined up to unit factors, thus we shall not distinguish between the factorizations

$$c = a_1 a_2 \dots a_r \quad \text{and} \quad c = b_1 b_2 \dots b_r,$$

if $b_i = u_{i-1}^{-1} a_i u_i$ (u_i unit, $u_0 = u_r = 1$).

With these conventions a rigid UFD may be described as an integral domain in which each element has essentially only one atomic factorization. In a general UFD the atomic factorizations are of course by no means unique; but neither can the factors be interchanged at will. Let us compare the different factorizations of an element in a 2-fir. A factorization

$$c = a_1 \dots a_r \tag{1}$$

is said to be a *refinement* of another,

$$c = b_1 \dots b_s, \tag{2}$$

if (2) can be obtained from (1) by bracketing some of the a's together, in other words, if (1) arises from (2) by factorizing the b's further. By the Schreier refinement theorem for modular lattices (*cf.* Appendix 1) we obtain

THEOREM 5.1. *In a 2-fir, any two factorizations of a non-zero element have isomorphic refinements.* ∎

Looking at the proof of the lattice–theoretic result quoted here, we find that we can go from the refinement of one chain to that of the other by a series of steps, which each change the chain at a single point, from

$$\dots \geqslant x_\vee y \geqslant x \geqslant x_\wedge y \geqslant \dots \quad \text{to} \quad \dots \geqslant x_\vee y \geqslant y \geqslant x_\wedge y \geqslant \dots$$

This corresponds to a change in the factorizations of the form

$$a_1 \ldots a_i \, a_{i+1} \ldots a_r \to a_1 \ldots a_i' \,_{+1} \, a_i' \ldots a_r,$$

where $a_i \, a_{i+1} = a_i' \,_{+1} \, a_i'$ is a comaximal relation. Calling such a modification of factorizations a *comaximal transposition* of terms, we get the following more precise form of Theorem 5.1.

THEOREM 5.2. *In a 2-fir, any two factorizations of a given element have refinements which can be obtained from one another by a series of comaximal transpositions of terms.* ■

As a simple illustration, suppose that $pq = rs$ are two factorizations of an element in a 2-fir R. Write $pR + rR = cR$, $p = ca$, $r = cb$, so that $aR + bR = R$, and similarly $Rq + Rs = Rd$, $q = b'd$, $s = a'd$. Then our two factorizations become $(ca)(b'd) = (cb)(a'd)$, where $ab' = ba'$ is a comaximal relation.

In an integral domain it is not necessarily true that a factorization $c = ab$ of an element induces a corresponding factorization of a similar element c'—the submodule of $R/c'R$ corresponding to aR/cR in R/cR will again be cyclic, but need not have a cyclic inverse image in R. However, in a 2-fir, the principal right ideals between cR and R are characterized by the fact that they give rise to strictly cyclic submodules of R/cR; now an application of the parallelogram law for modular lattices gives

PROPOSITION 5.3. *In a 2-fir R, let c and c' be similar elements, then the lattices* $\mathbf{L}(cR, R)$ *and* $\mathbf{L}(c'R, R)$ *are isomorphic, and the right ideals corresponding to each other under this isomorphism are similar.*

The last statement follows because corresponding right ideals are endpoints of perspective intervals. ■

The isomorphism in Proposition 5.3 can be described explicitly as follows: If $ab' = ba'$ is a comaximal relation for a and a', then to each left factor a_1 of a corresponds the left factor a_1' of a' given by $a_1 R + bR = ba_1' R$ and to each left factor a_1' of a' corresponds the left factor a_1 of a given by

$a_1 R = aR + ba_1' R$. These maps are inverse to one another, and induce an isomorphism between the lattices of left factors of a and a'.

However, we note that the actual lattice-isomorphism we get may depend on our choice of comaximal relation—or equivalently, on our choice of the isomorphism between R/aR and $R/a'R$. For example, take the ring $R = C[x; \bar{\ }]$ of complex-skew polynomials; this is a principal ideal domain and hence a 2-fir. The automorphism of the lattice of factors of $x^2 - 1$ induced by the comaximal relation

$$x(x^2 - 1) = (x^2 - 1)x$$

interchanges the factorizations $(x + i)(x + i)$ and $(x - i)(x - i)$ and leaves the factorization $(x + 1)(x - 1)$ fixed. The automorphism induced by the relation

$$1(x^2 - 1) = (x^2 - 1)1$$

is of course the identity, while the automorphism induced by

$$i(x^2 - 1) = (x^2 - 1)i$$

interchanges the factorizations $(x + 1)(x - 1)$ and $(x - 1)(x + 1)$.

In any 2-fir, the factorizations of a given element are closely related to those of its factors; this is best understood by looking first at the situation in lattices. In any lattice L a *link* or *minimal interval* is an interval $[a, b]$ in L consisting of just two elements, namely its end-points, and no others, thus $a < b$ and no $x \in L$ satisfies $a < x < b$.

In any modular lattice L of finite length, there are only finitely many projectivity classes of links, and the homomorphic images of L that are subdirectly irreducible are obtained by collapsing all but one equivalence class of links; these are in fact the simple homomorphic images of L. Here we count two homomorphic images as the same if and only if they are induced by the same homomorphism. If the distinct simple images are $L_1, ..., L_r$ we have a representation of L as a subdirect product of $L_1, ..., L_r$. In the case of distributive lattices of finite length, all links are projective, so the only simple homomorphic image is the 2-element lattice $[0, 1]$, also written **2**. We state these results as

THEOREM 5.4. (i) *Any modular lattice L of finite length can be expressed as a subdirect product of a finite number of simple modular lattices, viz. the simple homomorphic images of L.*

(ii) *A distributive lattice of finite length can be expressed as a subdirect power of* **2**. ∎

To apply this result to 2-firs one would need to know that the result of adjoining the inverse of an element to a 2-fir is again a 2-fir. But this is not true in general, as is shown by examples of 2-firs that are not embeddable in fields (Cohn [69″]).

There is another representation of modular lattices, to some extent dual to that of Theorem 5.4, which is of use in studying factorizations.

THEOREM 5.5. *Let L be a modular lattice. Given $a, b \in L$, there is a lattice embedding $[a, a \vee b] \times [b, a \vee b] \to L$ given by*

$$(x, y) \to x \wedge y. \tag{3}$$

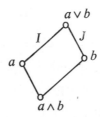

Proof. Clearly the map (3) is meet-preserving. Let us write $I = [a, a \vee b]$, $J = [b, a \vee b]$, $I' = [a \wedge b, b]$, $J' = [a \wedge b, a]$, so that $I \cong I'$, $J \cong J'$. The map $I \times J \to I' \times J' \to L$ given by $(x, y) \mapsto (x \wedge b, y \wedge a) \mapsto (x \wedge b) \vee (y \wedge a)$ preserves joins, for this clearly holds for the second part, and the first part is an isomorphism. Applying modularity we find, since $x \geqslant a \geqslant y \wedge a$, that

$$(x \wedge b) \vee (y \wedge a) = x \wedge (b \vee (y \wedge a)) = x \wedge y.$$

Hence the two maps of $I \times J$ into L are the same and so (3) is a lattice-homomorphism. It is injective because we can recover x and y from $x \wedge y$: $(x \wedge y) \vee a = x \wedge (y \vee a) = x \wedge (a \vee b) = x$; similarly $b \vee (x \wedge y) = y$. ∎

In general the embedding of Theorem 5.5 is not an isomorphism, even when $L = [a \wedge b, a \vee b]$, as the example of the 5-element modular but non-distributive lattice shows (*cf.* Appendix 1 Fig. 3). But if we exclude this case, by taking L distributive, we get the

COROLLARY. *Let L be a distributive lattice. Then for any $a, b \in L$,*

$$[a, a \vee b] \times [b, a \vee b] \cong [a \wedge b, a \vee b],$$

the isomorphism being given by (3).

For the map $z \mapsto (z \vee a, z \vee b)$, where $a \wedge b \leqslant z \leqslant a \vee b$, is easily seen to be an inverse to (3) in a distributive lattice. ∎

The translation of Theorem 5.5 into factorizations reads as follows:

PROPOSITION 5.6. *In a 2-fir R, any comaximal relation*

$$c = ab' = ba' \tag{4}$$

gives rise to an embedding of $\mathbf{L}(aR, R) \times \mathbf{L}(bR, R)$ *in* $\mathbf{L}(cR, R)$. *If* $\mathbf{L}(cR, R)$ *is distributive, this is an isomorphism.* ∎

This result gives us a powerful tool relating the factorizations of a and b to those of c. For example suppose we have a comaximal relation (4) in which a has a factorization xyz and b a factorization uv. This gives us chains of lengths 3 and 2 in the lattice of left factors of a, b respectively, and applying Proposition 4.6 we find that the lattice of left factors of c will have a sublattice of the form shown in Fig. 1. Here intervals are marked with the factor of c to which they correspond.

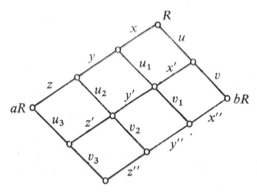

FIGURE 1.

Factorizations of c are given by paths from the top to the bottom of this diagram; every rectangle corresponds to a comaximal relation; not only the minimal rectangles, giving relations such as $yu_2 = u_1 y'$, but also larger ones, such as $(y'z')v_3 = v_1(y''z'')$. Thus in these various factorizations of c any factor from a and factor from b are comaximally transposed.

Further if, say, x and y are comaximally transposable: $xy = \bar{y}\bar{x}$, giving the subdiagram of left factors of a shown in Fig. 2, then from Proposition 4.6 we get a corresponding expanded diagram of factors of c, including comaximal rectangles $x'y' = \bar{y}'\bar{x}'$ etc.

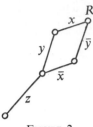

FIGURE 2.

However, when $L(cR, R)$ is not distributive, the embedding of Proposition 5.6 may not be an isomorphism. In terms of factorizations this means that some factors of a and b may be comaximally transposable in more ways than the one induced by the comaximal relation (4). E.g. in Fig. 1 any of the rectangles may be replaced by the diagram of Fig. 3. An example of such behaviour occurs in the complex-skew polynomial ring. Here $x^2 - 1 = (x + 1)(x - 1) = (x - 1)(x + 1)$ is a comaximal relation in which each factor is an atom, yet its full diagram of factorizations is of the form of Fig. 4: for any u on the complex unit circle, $(x + u)(x - \bar{u})$ is a factorization of $x^2 - 1$.

FIGURE 3.

FIGURE 4.

Of course, when $L(cR, R)$ is distributive, it can be expressed as a subdirect power of **2**. This case will be studied in more detail in **4.4**.

Exercises 3.5

1. Show that the group algebra (over a field) of the additive group of rationals (written multiplicatively) is a non-atomic Bezout domain (Hint: write the elements as $\Sigma c_\alpha x^\alpha$, $\alpha \in \mathbf{Q}$, and express the ring as a direct limit of polynomial rings.)

2. Show that the ring of power series $\Sigma c_\alpha x^\alpha$ ($\alpha \in \mathbf{Q}$, $\alpha \geqslant 0$) with well-ordered support is an atomless Bezout domain.

3^0.Give an example of an atomless semifir which is not an Ore ring.

4. In a 2-fir, if an element c can be written as a product of 2 atoms in at least 3 essentially different ways, show that all atomic factors of c are similar. Generalize this result to n factors.

5. Over a 2-fir R consider an equation

$$\begin{pmatrix} p & 0 \\ 0 & q \end{pmatrix} = \alpha\beta,$$

where p, q are dissimilar atoms in R and α, β non-invertible matrices. Show that this factorization is equivalent to either $\begin{pmatrix} p & 0 \\ 0 & 1 \end{pmatrix}\begin{pmatrix} 1 & 0 \\ 0 & q \end{pmatrix}$ or $\begin{pmatrix} 1 & 0 \\ 0 & q \end{pmatrix}\begin{pmatrix} p & 0 \\ 0 & 1 \end{pmatrix}$. What are the possibilities when q is similar to p?

6°. Develop a theory of UFD's that are not necessarily 2-firs.

7°. In a 2-fir, if $a \sim au$, is u necessarily a unit? (For atomic 2-firs the answer is 'yes', by unique factorization.)

8°. Investigate 2-firs in which any two elements without a common left or right factor can be comaximally transposed (note that this includes all commutative Bezout domains).

3.6 The primary decomposition

Besides the multiplicative decomposition of elements there is the primary decomposition of ideals which plays a role in commutative Noetherian rings. Much of this can be formulated in terms of lattices and by applying it to $\mathscr{C}_R$ we obtain various types of decomposition of elements in 2-firs.

In any ring R, an element $c \neq 0$ is said to be *decomposable* if it has two proper factorizations

$$c = ab' = ba', \tag{1}$$

which are left and right coprime; otherwise c is *indecomposable*. If c has two proper factorizations (1) which are right (left) coprime, c is said to be *right (left) decomposable*, otherwise *right (left) indecomposable*. Clearly any decomposable element is both right and left decomposable; hence any element which is either right or left indecomposable is indecomposable. E.g. in an atomic 2-fir c is right indecomposable if and only if every atomic factorization of c ends in the same right factor (*cf.* Proposition 6.6 below). In any 2-fir the definitions may be rephrased as follows:

(a) $c \in R^*$ is right decomposable if and only if there exist a, $b \in R$ such that

$$cR = aR \cap bR, \quad \text{where} \quad cR \neq aR, bR. \tag{2}$$

(b) $c \in R^*$ is decomposable if and only if there exist a, $b \in R$ such that

$$cR = aR \cap bR, \quad \text{and} \quad aR + bR = R, \quad \text{where} \quad cR \neq aR, bR. \tag{3}$$

To show that the definitions depend only on the similarity class of c, we express them as conditions on R/cR:

PROPOSITION 6.1. *Let R be a 2-fir, then an element c of R^* is right decomposable if and only if R/cR is an irredundant subdirect sum of two $\mathscr{C}$-modules, and c is decomposable if and only if R/cR is a direct sum of two non-zero $\mathscr{C}$-modules. In particular if c is (left, right) decomposable, so is any element similar to c.*

Proof. Let c be right decomposable, say (2) holds, then there is a monomorphism

$$R/cR \to R/aR \oplus R/bR. \tag{4}$$

The result of composing this with the projection onto either of the summands is surjective, but neither is injective, hence R/cR is an irredundant subdirect sum of R/aR and R/bR. Conversely, given any subdirect sum representation of R/cR, the kernels of the projections are again $\mathscr{C}$-modules, say R/aR, R/bR, which will satisfy (2).

Next let c be decomposable as in (3), then

$$R/cR = (aR + bR)/cR \cong aR/cR \oplus bR/cR,$$

where the last sum is direct because $aR \cap bR = cR$, the summands are in $\mathscr{C}_R$ and are not zero. The converse is clear. ∎

Let R be a 2-fir and $c \in R^*$, and suppose that we have an irredundant representation

$$cR = a_1 R \cap \ldots \cap a_n R, \tag{5}$$

where each a_i is right indecomposable. This means that the natural mapping

$$R/cR \to \Sigma \oplus R/a_i R \tag{6}$$

obtained by composing the mappings $R/cR \to R/a_i R$ $(i = 1, \ldots, n)$ is injective. Thus we have subdirect sum representation of R/cR. Such a representation certainly exists if R is atomic. The irredundancy of (5) means that no term on the right of (6) can be omitted, while the right indecomposability of the a_i shows that each module $R/a_i R$ is subdirectly irreducible, by Proposition 6.1. If we now apply the Kuroš–Ore theorem for modular lattices (*cf.* Appendix 1) we obtain

THEOREM 6.2. *Let R be an atomic 2-fir. Then for each $c \in R^*$, cR has an irredundant representation*

$$cR = a_1 R \cap \ldots \cap a_n R, \tag{7}$$

where each a_i is right indecomposable, and if a second such decomposition of cR is given,

$$cR = b_1 R \cap \ldots \cap b_m R, \tag{8}$$

then $m = n$ and the $b_j R$ may be exchanged against the $a_i R$, i.e. after suitably renumbering the b's we have for $i = 1, \ldots, n$,

$$cR = a_1 R \cap \ldots \cap a_i R \cap b_{i+1} R \cap \ldots \cap b_n R. \quad \blacksquare$$

Similarly, cR has a representation (5) where each a_i is indecomposable and such that

$$a_i R \neq R, \quad a_i R + \left(\bigcap_{j \neq i} a_j R \right) = R \qquad (i = 1, \ldots, n). \tag{9}$$

Let us call such a representation (5) a *complete direct decomposition* of cR. When (9) holds, the mapping (6) is an isomorphism, so that a complete direct decomposition of cR corresponds to a direct sum representation

$$R/cR \cong R/a_1 R \oplus \ldots \oplus R/a_n R. \tag{10}$$

The first condition in (9) shows that each term on the right is non-zero, while the fact that each a_i is indecomposable means (by Proposition 6.1) that each module $R/a_i R$ is indecomposable in the category $\mathscr{C}_R$. We shall express this by saying that each $R/a_i R$ is *$\mathscr{C}$-indecomposable*. It is clear that in a 2-fir R, $a \in R$ is indecomposable if and only if R/aR is $\mathscr{C}$-indecomposable. If we now apply the Krull–Schmidt theorem for modular lattices (*cf.* Appendix 1) to the lattice of principal right ideals containing cR, we obtain

THEOREM 6.3. *Let R be an atomic 2-fir. Then for each $c \in R^*$, cR has a complete direct decomposition (7), and if a second such decomposition (8) is given, then $m = n$ and the b's may be exchanged against the a's. Moreover, the a's and b's are similar in pairs, in fact, after suitably renumbering the b's, we have a set of coprime relations*

$$ua_i = b_i u_i, \tag{11}$$

where u corresponds to a unit in the eigenring $\mathbf{E}(cR)$.

Only the last part needs proof. Let $u \in \mathbf{I}(cR)$ correspond to the automorphism of R/cR transforming (7) into (8), then there exists a coprime relation $uc = cu'$ say, and if $c = b_i b_i' = a_i a_i'$, then we have $ua_i a_i' = b_i b_i' u'$, and if the b's are renumbered so that $R/a_i R \to R/b_i R$ in the isomorphism, then we have the coprime relations (11) for some $u_i \in R$. $\quad \blacksquare$

The Krull–Schmidt theorem for modules can either be proved by lattice theory, as above, or by Fitting's lemma (*cf.* 3.2). This lemma is often useful quite apart from the Krull–Schmidt theorem.

Since every element of a completely primary ring is either a unit or nilpotent, we have the following consequence of Proposition 2.6:

PROPOSITION 6.4. *Let R be an atomic 2-fir and c an indecomposable element of R. Given $d \in \mathbf{I}(cR)$, the idealizer of cR, either d is right comaximal with c or there exists $e \in R$ and $n \geqslant 0$ such that*

$$ce = d^n. \blacksquare$$

In particular, assuming d to be an atom, we find the

COROLLARY. *Let R and c be as in Proposition 6.4. If $\mathbf{I}(cR)$ contains an atom p then either p is right comaximal with c or there exists $e \in R$ and $n \geqslant 0$ such that $ce = p^n$. In particular, all atomic factors of c are similar in this case. $\blacksquare$*

Of course in general the atomic factors of an indecomposable element need not all be similar, *cf. xy in $k\langle x, y\rangle$.*

In conclusion we note the following criterion for one-sided indecomposability:

PROPOSITION 6.5. *Let R be a 2-fir and $c \in R$ any element of finite length. Then the following are equivalent:*

  (i) *c is right indecomposable,*
  (ii) *the lattice $\mathbf{L}(cR, R)$ has a unique minimal element covering cR,*
  (iii) *the lattice $\mathbf{L}(Rc, R)$ has a unique maximal element covered by R,*
  (iv) *the right atomic factor of c is unique up to left associates.*

The proof may be left to the reader. $\blacksquare$

Using (iii) we obtain the

COROLLARY. *In an atomic 2-fir, if c is right indecomposable, so is any right factor of c. $\blacksquare$*

Another type of decomposition, sometimes of interest, exists for certain elements, namely those whose associated $\mathscr{C}$-modules are completely reducible (or semisimple). An element c of a 2-fir R is said to be *fully reducible* if

$$cR = \bigcap p_i R, \tag{12}$$

where the p_i are atoms. By passing to the corresponding $\mathscr{C}$-module we see that c is fully reducible if and only if R/cR is a subdirect product of $\mathscr{C}$-simple modules. This shows that being fully reducible depends only on the similarity class of c. If further, R is atomic, it is enough to take finitely many atoms on the right of (12). Taking the number of terms to be minimal, we obtain R/cR as a subdirect sum of the $R/p_i R$ and by minimality the sum is actually direct:

$$R/cR \cong R/p_1 R \oplus \ldots \oplus R/p_n R. \tag{13}$$

By duality, being fully reducible in an atomic 2-fir is a left–right symmetric property. Further, any factor of a fully reducible element is again fully reducible, as the representation (13) shows.

In order to relate full reducibility to factorizations we need a definition. A factorization

$$c = a_1 \dots a_r \qquad (14)$$

of an element $c \in R^*$ is said to be a *cleavage* if every atomic factorization of c can be obtained by refining (14). A non-unit with no cleavage into more than one non-unit is said to be *uncleft*. For example, take $R = \mathbf{Z}$, or more generally any commutative principal ideal domain. If p is an atom (i.e. a prime number in case $R = \mathbf{Z}$), the element $p^2(p + 1)$ is uncleft, but it has a cleft factor p^2. Let us call an element c *totally uncleft* if all its factors are uncleft. Thus in a commutative principal ideal domain the totally uncleft elements are just the 'squarefree' elements. The connexion with fully reducible elements is given by

PROPOSITION 6.6. *Let R be a 2-fir. Given $c \in R$ of finite length, the following three conditions on c are equivalent*:

(i) *c is totally uncleft*,

(ii) *c is fully reducible*,

(iii) *any two neighbouring atomic factors in an atomic factorization of c can be comaximally transposed*.

Proof. (i) $\Rightarrow$ (ii). Let c be totally uncleft, then not every complete factorization of c ends in the same atom, say $c = c_1 p_1 = c_2 p_2$, where $Rp_1 \neq Rp_2$. Then $cR = c_1 R \cap c_2 R$; now c_1, c_2 as factors of c are again totally uncleft and by induction on the length of c, each c_i is fully reducible, say $c_i = \bigcap_j p_{ij} R$. Hence $cR = \bigcap_{ij} p_{ij} R$, which shows c to be fully reducible.

(ii) $\Rightarrow$ (iii). Assume c fully reducible, take a complete factorization $c = a_1 \dots a_r$ and fix i in the range $2 \leqslant i \leqslant r$. Then $a_{i-1} a_i$ is fully reducible, as factor of c, hence there is a comaximal relation $a_{i-1} a_i = a_i' a'_{i-1}$.

(iii) $\Rightarrow$ (i). This follows because a cleavage of any factor of c, on being refined to a complete factorization, leads to a pair of neighbouring atomic factors which cannot be comaximally transposed. ∎

In a rigid UFD, clearly any complete factorization is a cleavage. Generally, let us say that a factorization of a certain type, e.g. into maximal uncleft factors, is *rigidly unique*, if it is the only one of this type (up to unit factors).

THEOREM 6.7. *Let R be a 2-fir. Then every $c \in R$ of finite length has a factorization*

$$c = a_1 \ldots a_r \tag{15}$$

into maximal uncleft factors, and this is rigidly unique.

Proof. Let a_1 be an uncleft left factor of c of maximal length. If a_1' is another maximal uncleft factor of c, say $c = a_1 b_1 = a_1' b'$, let

$$a_1 d = a_1' d' \tag{16}$$

be a LCRM of a_1 and a_1'. If d is not a unit, then $a_1 d$ is cleft, say $a_1 d = uv$, and every complete factorization is obtained by refining uv, whence either $a_1 = ue$, $v = ed$, or $u = a_1 e$, $d = ev$. In the former case, $a_1 = ue$ is a cleavage, contradicting the fact that a_1 is uncleft, hence this case is ruled out and $u = a_1 e$, $d = ev$. Similarly, $u = a_1' e'$, $d' = e'v$, but by (16), d, d' are right coprime, so v is a unit, again a contradiction. This shows d to be a unit, similarly d' is a unit and $a_1 R = a_1' R$. Thus a_1 in (15) is determined up to a right unit factor, and an induction on the length of c shows the uniqueness of (15). ∎

In general (15) need not be a cleavage, but this is the case when the factors on the right of (15) are atoms. By Proposition 6.6 this happens precisely when no two factors in (15) are comaximally transposable. In terms of lattices this means that $L(cR, R)$ is a chain. Thus we have the

COROLLARY. *An element c of finite length in a 2-fir has a cleavage into atomic factors if and only if $L(cR, R)$ is a chain, or equivalently, if c is rigid.* ∎

Exercises 3.6

1. In the complex-skew polynomial ring, obtain all possible irredundant representations of $(x^2 - 1)R$.

2. Show that in any commutative principal ideal domain that is not a local ring there are elements which have no rigidly unique factorization into totally uncleft factors.

3. In an atomic 2-fir, show that any factor of a fully reducible element is fully reducible.

4. An element c of a ring R is said to be *primary* if $cR = \bigcap p_i R$, where the p_i are pairwise similar atoms. Show that in an atomic 2-fir every fully reducible element can be written uniquely as a LCRM of primary elements.

5. Let R be a 2-fir and let $a = a_1 \ldots a_k$ be a decomposition into maximal uncleft factors. If b is a left factor of a, with decomposition $b = b_1 \ldots b_h$ into maximal uncleft factors, show that $b_1 \ldots b_i$ is a left factor of $a_1 \ldots a_i$ ($i = 1, \ldots, h$).

Notes and comments on Chapter 3

The first non-commutative UFD to have been studied is the ring $K[D; 1, ']$ of linear differential operators. It is discussed at some length by Schlesinger [97], who proves that it is an integral domain. Landau [02] shows that all complete factorizations of a given operator have the same length, and corresponding irreducible factors have the same order (= degree in D). Loewy [03] shows that corresponding factors are 'equivalent' operators, in a sense introduced by Poincaré, and this turns out to correspond to the notion of similarity. A large number of papers on the subject appeared at this time. The first abstract account of this ring was given by Ore [32], who also introduced the notion of 'eigenring'. A further generalization, to principal ideal domains, is undertaken by Asano [38]; this and much other work is summarized in chapter 3 of Jacobson [43]. The general notion of non-commutative UFD is defined in Cohn [63], and it is shown that free algebras have this property. Rigid UFD's are also defined there *a propos* of the example of free power series rings.

Nagata [57] proved Theorem 1.4 for commutative Noetherian rings; the other results of 3.1 are mostly well known, while much of 3.2 follows Cohn [63, 69, 70]. An interesting generalization of UFD, to include the case of free algebras over the integers, has recently been proposed by Brungs [69] (for another approach, see Cohn [70]). For a detailed study of similarity of ideals, see Fitting [36]; section 3.3 contains a simplified account of what is relevant to us in the present context. Proposition 3.3, Corollary generalizes a result of Schwarz [49]. For 3.4, see Cohn [62] and Bowtell [67]. Proposition 3.4.5 generalizes a result on free algebras by Koševoi [66]. Section 3.5 is due to Bergman [67], and 3.6 generalizes earlier results of Ore [33], *cf.* Feller [60], Johnson [65] and Cohn [69, 70].

4. Rings with a distributive factor lattice.

This chapter examines more closely those 2-firs in which the lattice of factors of any non-zero element is distributive. It is shown in **4**.3 that this holds for free algebras and the consequences are traced out in **4**.2 and **4**.4–5, while **4**.6 describes the form eigenrings take in this case.

4.1 Endomorphisms of distributive modules

In this section $\mathscr{A}$ will be an abelian subcategory of $\mathscr{M}_R$, the category of all right R-modules over a given ring R. This means that $\mathscr{A}$ is a subcategory of $\mathscr{M}_R$ with kernels, cokernels and finite direct sums. All notions such as submodule, quotient, simple etc. will be understood as referring to $\mathscr{A}$.

Since $\mathscr{A}$ is an abelian category, the set $\text{Lat}_{\mathscr{A}}(M)$ of submodules of a given $\mathscr{A}$-module M is a lattice, necessarily modular. If $\text{Lat}_{\mathscr{A}}(M)$ is distributive, the module M is said to be *distributive*. The following are some examples of distributive modules:

(i) $\mathscr{A} = \mathscr{M}_R$ where R is a commutative Bezout domain. Any cyclic R-module is distributive (Jensen [63]).

(ii) $\mathscr{A} = \mathscr{M}_R$, where R is semisimple Artinian. An R-module M of finite length is distributive if and only if each simple module occurs at most once in M. If each simple module occurs exactly once, then M is faithful. For this reason R is called *distributively representable* by Behrens [65].

(iii) In **4**.3 we shall see that any strictly cyclic module M over a free associative algebra R is distributive. Here it is essential to consider M as object of $\mathscr{C}_R$ rather than $\mathscr{M}_R$.

In the rest of this section we shall examine the structure of a distributive module, in preparation for what follows. If M is a distributive module, then clearly any submodule and any quotient of M are again distributive. The following property of distributive modules is basic:

PROPOSITION 1.1. *Let* $M = M_1 \oplus M_2$ *be a distributive module expressed as a direct sum, then* $\text{Hom}(M_1, M_2) = 0$.

Proof. Take $\alpha \in \text{Hom}\,(M_1, M_2)$ and let Γ be the graph of α, i.e. the submodule $\Gamma = \{(x, \alpha(x))|x \in M_1\}$ of M, then $\Gamma \cap M_2 = 0$, but by distributivity, $\Gamma = (\Gamma \cap M_1) + (\Gamma \cap M_2)$, hence $\Gamma = \Gamma \cap M_1 \subseteq M_1$, i.e. $\alpha = 0$. ∎

As a consequence we have

THEOREM 1.2. *Let* $M = M_1 \oplus \ldots \oplus M_r$ *be any direct decomposition of a distributive module, then*

$$\text{End}\,(M) \cong \Pi\,\text{End}\,(M_i).$$

For any endomorphism α of M is represented by a matrix (α_{ij}) with $\alpha_{ij} \in \text{Hom}\,(M_i, M_j)$ and Proposition 1.1, applied to $M_i \oplus M_j$, shows that $\alpha_{ij} = 0$ for $i \neq j$. ∎

If M satisfies both chain conditions, we can write M as a direct sum of a finite number of indecomposable modules. Now when N is indecomposable, then by Fitting's lemma, $E = \text{End}\,(N)$ is completely primary. Hence we obtain

THEOREM 1.3. *Let* M *be a distributive module with both chain conditions, then* End (M) *is a direct product of a finite number of completely primary rings.* ∎

For a closer analysis of distributive modules we shall need a number of technical results, which are contained in Propositions 1.4–6.

PROPOSITION 1.4. *Given any module* M, *let* U, V *be submodules of* M, *neither contained in the other, and suppose that* U, V *each have a unique maximal submodule* U', V' *respectively, say. Then*

$$(U + V)/(U' + V') \cong (U/U') \oplus (V/V'). \tag{1}$$

Proof. We have $(U + V)/(U + V') = ((U + V') + V)/(U + V') \cong V/V \cap (U + V') = V/(V \cap U) + V' = V/V'$, where the last step follows because $U \cap V$ is a proper submodule of V and hence contained in V'. Similarly, $(U + V)/(U' + V) \cong U/U'$, hence we can represent $U + V$ as a subdirect sum of U/U' and V/V'. The kernel is

$$(U + V') \cap (U' + V) = V' + (U \cap (U' + V))$$
$$= V' + ((U \cap V) + U') = U' + V',$$

and this shows that the left-hand side of (1) is isomorphic to a submodule of the right-hand side. Now the module on the right has length 2, so to establish the isomorphism we need only show that the module on the left also has length 2. We have seen that $(U + V)/(U + V')$ is simple and $(U + V')/(U' + V') \cong U/(U \cap (U' + V')) \cong U/((U \cap V) + U') = U/U'$, hence the left-hand side of (1) has length 2 and the isomorphism follows. ∎

COROLLARY. *Let M be a module with a unique maximal submodule M'. Then for any two homomorphisms $\alpha, \beta: M \to N$ into a distributive module N, either $\alpha M \subseteq \beta M$ or $\beta M \subseteq \alpha M$.*

Proof. We may assume $\alpha, \beta \neq 0$; then $\ker \alpha$ is a proper submodule of M, hence $\ker \alpha \subseteq M'$, $\alpha M \cong M/\ker \alpha$ and it follows that αM has a unique maximal submodule, namely $\alpha M'$. Similarly βM has a unique maximal submodule $\beta M'$, and if αM, βM are incomparable, then by Proposition 1.4,

$$(\alpha M + \beta M)/(\alpha M' + \beta M') \cong (\alpha M/\alpha M') \oplus (\beta M/\beta M') \cong (M/M')^2.$$

The left-hand side is a factor of N and hence distributive, but the right-hand side cannot be so, by Proposition 1.1. This contradiction shows that either $\alpha M \subseteq \beta M$ or $\beta M \subseteq \alpha M$. ∎

PROPOSITION 1.5. *Let M be a distributive module with DCC and let α be any homomorphism from a submodule N of M into M. Then $\alpha N \subseteq N$.*

Proof. If the conclusion is false then there exists a submodule N of M and a homomorphism $\alpha: N \to M$ such that $\alpha N \not\subseteq N$. Let us choose the pair (N, α) in such a way that N is minimal with this property. We set $D = N \cap \alpha N$ and consider

$$(N + \alpha N)/D \cong (N/D) \oplus (\alpha N/D).$$

As a factor of M this is again distributive and by Proposition 1.1,

$$\text{Hom } (N/D, \alpha N/D) = 0. \tag{2}$$

If $D \subset N$, then by the choice of N, $\alpha D \subseteq D$, hence α induces a homomorphism $N/D \to \alpha N/D$, which must be 0, by (2), and so $\alpha N = D \subset N$, which contradicts the definition of N. The alternative is that $D = N$; this means that $N \subseteq \alpha N$ and here the inclusion is strict, by the choice of N. Consider the module $\alpha^{-1} N = \{x \in N | \alpha x \in N\}$; since $\alpha N \supset N$, it follows that $\alpha^{-1} N \subset N$, therefore by hypothesis, $\alpha \cdot \alpha^{-1} N \subseteq \alpha^{-1} N$, but $\alpha \cdot \alpha^{-1} N = N$, so $N \subseteq \alpha^{-1} N$, which is again a contradiction. Hence $\alpha N \subseteq N$ for every $\alpha: N \to M$. ∎

Here the DCC certainly cannot be omitted. Thus take the rational numbers $\mathbf{Q}$, considered as $\mathbf{Z}$-module. This is distributive, but it has $\mathbf{Z}$ as submodule which is not mapped into itself by the endomorphism $x \mapsto x/2$.

Another way of expressing Proposition 1.5 is as follows: if M is a distributive module with DCC, and P is a submodule of M which is isomorphic to a factor N/N' of M, then $P \subseteq N$. Suppose now that $\alpha, \beta: N \to M$ are homomorphisms such that $\ker \alpha \subseteq \ker \beta$. Then $\beta N \cong N/\ker \beta$ is a quotient of $N/\ker \alpha \cong \alpha N$, hence $\beta N \subseteq \alpha N$, and we obtain the

COROLLARY. *Let M be a distributive module with* DCC *and let α, $\beta: N \to M$ be homomorphisms such that* $\ker \alpha \subseteq \ker \beta$; *then $\beta N \subseteq \alpha N$.* ∎

Sometimes we shall need the duals of these results; we state them as follows:

PROPOSITION 1.5*. *Let M be a distributive module with* ACC *and let $\alpha: M \to Q$ be any homomorphism from M to a quotient module $Q = M/N$. Then $\ker \alpha \subseteq N$, so that α is induced by an endomorphism of Q. Moreover, if $\alpha, \beta: M \to Q$ are homomorphisms and $\beta M \subseteq \alpha M$, then $\ker \alpha \subseteq \ker \beta$.* ∎

Let M again be a distributive module with both chain conditions and take a composition series for M:

$$M = M_0 \supset M_1 \supset \ldots \supset M_n = 0. \tag{3}$$

By the *factors* of this series we understand the modules

$$A_i = M_{i-1}/M_i \qquad (i = 1, \ldots, n).$$

By the Jordan–Hölder theorem (for abelian categories) these factors depend only on M and not on the choice of the series; moreover, since each A_i is simple, its endomorphism ring is a field (by Schur's lemma). If $\alpha \in \text{End}(M)$, then by Proposition 1.5 (or 1.5*) $\alpha M_i \subseteq M_i$ and so α induces an endomorphism α_i of A_i. It is easily seen that the mapping $\alpha \mapsto \alpha_i$ is a ring-homomorphism. In this way we obtain a homomorphism

$$\text{End}(M) \to \Pi \, \text{End}(A_i). \tag{4}$$

The endomorphism $\alpha_i: A_i \to A_i$ induced by α is 0 precisely when $\alpha M_{i-1} \subseteq M_i$. Hence the kernel of the mapping (4) consists of all endomorphism α such that

$$\alpha M_{i-1} \subseteq M_i \qquad (i = 1, \ldots, n). \tag{5}$$

Any endomorphism α satisfying (5) is nilpotent, in fact $\alpha^n = 0$, by an easy induction. Conversely, if α is a nilpotent endomorphism, then each α_i is nilpotent and hence 0. Thus α satisfies (5), so that the kernel of (4) consists precisely of the nilpotent endomorphisms of M. We summarise the result as

THEOREM 1.6. *Let M be a distributive module with both chain conditions and let $A_1, \ldots, A_n$ be the simple modules (with their proper multiplicities) occurring in a composition series for M. Then there is a homomorphism*

$$\text{End}\,(M) \to \prod_1^n \text{End}\,(A_i),$$

whose kernel N is the Jacobson radical of $\text{End}\,(M)$. Moreover, N consists of all nilpotent endomorphisms of M and satisfies $N^n = 0$. ∎

We conclude with a converse of Proposition 1.5; note that no chain condition is needed here.

PROPOSITION 1.7. (i) *Let M be a module such that any homomorphism α from a submodule N of M into M satisfies $\alpha N \subseteq N$; then any two isomorphic submodules of M are equal,*

(ii) *Suppose that a module M and all its quotients have the property that any two isomorphic submodules are equal, then M is distributive.*

Proof. (i) Let $N_1 \cong N_2$ then the isomorphism $\alpha : N_1 \to N_2$ is a homomorphism of N_1 into M, hence $N_2 = \alpha N_1 \subseteq N_1$ and similarly $N_1 \subseteq N_2$.

(ii) If M is not distributive it contains a sublattice of the form shown. Here all segments are projective, and so correspond to isomorphic but distinct submodules of a quotient of M. ∎

Exercises 4.1

1. (Behrens [65]) Find the distributive modules over a semisimple Artinian ring.

2. (Jensen [63]) Let M be a distributive R-module. If $a, b \in M$, show that either $aR \subseteq bR$ or $bxR \subseteq aR$ for some $x \in R$ such that $1 - x$ is a non-unit. Deduce that if R is a local ring, $\text{Lat}(M)$ is totally ordered.

3. Let M be a distributive module. Show that if a perspectivity interchanges two simple factors P, Q, then P, Q cannot be isomorphic.

4*. Let M be a distributive module of finite composition length. By the Jordan–Hölder–Schreier theorem, any two composition series are projective, in the sense that we can pass from one to the other by a series of perspectivities. Show that any such projectivity preserves the order of the simple factors of a given isomorphism type.

5^0. Consider the following properties of a module M:

(a) Any homomorphism α from a submodule N of M into M satisfies $\alpha N \subseteq N$,

(b) M and all its quotients have the property that any two isomorphic submodules are equal,

(c) M is distributive.

In Proposition 1.7 it was shown that (a) $\Rightarrow$ (b) $\Rightarrow$ (c), and if M satisfies DCC, (c) $\Rightarrow$ (a). An example of a module with ACC, satisfying (c) but not (a) was given. Show that this example does not satisfy (b). Find examples satisfying (b) but not (a). Find also examples separating the assertions dual to (a)–(c) (cf. Proposition 1.5*).

6^0. Let M be a distributive module with DCC. Is End (M) (i) left invariant, or (ii) uniserial?

4.2 Distributive factor lattices

From Theorem 1.5.1 we see that a 2-fir may also be defined as an integral domain R such that for any $c \in R^*$ the set $L(cR, R)$ is a sublattice of the lattice of all right ideals. In the commutative case this condition simply states that the principal ideals form a sublattice of the lattice of all ideals. In that case we can go over to the field of fractions and consider the principal fractional ideals; by what has been said they form a modular lattice with respect to the ordering by inclusion. Clearly they also form a group under multiplication, and the group operations respect the ordering. Thus we have a lattice-ordered group; such a group is always distributive, as a lattice (Birkhoff [67], p. 292). This suggests that we single out 2-firs with the corresponding property, and we make the following

Definition. An integral domain R is said to have a *distributive factor lattice*, if for each $c \in R^*$, the set $L(cR, R)$ is a distributive sublattice of the lattice of all right ideals in R.

From the definition it is clear that a ring with distributive factor lattice is a 2-fir. Moreover, since $L(cR, R)$ is anti-isomorphic to $L(Rc, R)$ (Theorem 3.2.1), the definition is left–right symmetric. We shall reformulate this condition below in a number of ways, in terms of strictly cyclic modules.

We begin with a couple of technical lemmas.

LEMMA 2.1. *Let M, N be strictly cyclic right R-modules over a 2-fir R. Then the direct sum $M \oplus N$ is strictly cyclic if and only if there is a left comaximal pair of elements a', $b' \in R$ such that*

$$M \cong R/a' R, \qquad N \cong R/b' R, \qquad (1)$$

Proof. Assume (1) for some pair a', b' of left comaximal elements. Then a', b' are non-zero and hence left commensurable, say

$$m = ab' = ba' \neq 0, \tag{2}$$

where m is a LCLM of a' and b'. It follows that (2) is left coprime; since R is a 2-fir, a, b are right comaximal and $R/mR = (aR + bR)/mR = aR/mR + bR/mR = R/b'\,R + R/a'\,R$, where the last sum is direct, because $aR \cap bR = mR$. Conversely, assume that $M \oplus N \cong R/mR$, and suppose that in the iso-morphism, M corresponds to bR/mR and N to aR/mR, where $aR \cap bR = mR$. Then (2) holds for some right coprime pair a', b'. It follows again that a', b' are left comaximal, and

$$M = bR/mR = R/a'\,R, \qquad N = aR/mR = R/b'\,R.$$

Thus (1) holds for a left comaximal pair a', b'. ∎

LEMMA 2.2. *Let R be a 2-fir and $a \in R^*$. Then a is right comaximal with an element similar to a if and only if a is right comaximal with ba, for some $b \in R$.*

Proof. The similarity of a and a' can be expressed by the existence of two mutually inverse matrices

$$\mu = \begin{pmatrix} a & b \\ c & d \end{pmatrix} \qquad \mu^{-1} = \begin{pmatrix} d' & -b' \\ -c' & a' \end{pmatrix}.$$

Let us replace these matrices by $\tau\mu$, $\mu^{-1}\tau^{-1}$, where

$$\tau = \begin{pmatrix} 1 & 0 \\ t & 1 \end{pmatrix},$$

then the equation of comaximality obtained by equating $(1, 1)$-elements in $\tau\mu \,.\, \mu^{-1}\tau^{-1}$ is

$$a(d' + b'\,t) - b(c' + a'\,t) = 1. \tag{3}$$

By hypothesis a and a' are right comaximal, say

$$au - a'v = 1, \tag{4}$$

hence $a'vc' + c' = auc'$, and taking $t = vc'$ in (3) we find

$$a(d' + b'\,vc') - bauc' = 1,$$

which shows a and ba to be right comaximal. Conversely, if a and ba are right comaximal, say $ad' - bac' = 1$, then by Lemma 3.4.3, taking this relation in the form $a \,.\, d' - b \,.\, ac' = 1$, we have a comaximal relation $ac' \,.\, a = a_1 v$, where $a_1 \sim a$. Hence a is right comaximal with a_1 as asserted. ∎

More generally, two elements similar to a are right comaximal if and only if there is an equation $xay + ua'v = 1$, with $a' \sim a$. For this equation is equivalent to $a_1 y_1 x_1 + a_1' v_1 u_1 = 1$, with $a_1' \sim a_1 \sim a$ (by Exercise 3.3.18).

We now list some conditions for distributivity of $L(cR, R)$:

PROPOSITION 2.3. *Let R be a 2-fir and $c \in R^*$. Then the following conditions are equivalent:*

(a) *R/cR is not distributive (as a strictly cyclic module),*

(b) *$c = amb$ where $R/mR \cong M^2 \neq 0$ for some strictly cyclic module M,*

(c) *$c = amb$ where $m = a_1 a_2 = a_3 a_4$ is a comaximal relation in which $a_1, ..., a_4$ are all similar non-units,*

(a*)–(c*). *The left–right analogues of (a)–(c).*

Proof. (a) holds precisely when $L(cR, R)$ contains a 5-element sublattice of length 2, and this entails (c). Clearly (c) $\Rightarrow$ (b), and when (b) holds, R/mR cannot be distributive, by Proposition 1.1, hence neither is R/cR, i.e. (a) is satisfied. Now the left–right symmetry holds because (c) is symmetric. ∎

Next we have the following conditions for global distributivity:

THEOREM. 2.4. *Let R be a 2-fir, then the condition*

(a) *a, a' similar, $bac \in a'R$ ($c \neq 0$) implies $ba \in a'R$,*

implies

(b) *a, a' similar and $aR \cap a'R \neq 0$ implies $aR = a'R$,*

and this implies the following conditions, which are equivalent among themselves and to their left–right analogues:

(c) *for each $c \in R^*$, the lattice $L(cR, R)$ is distributive,*

(d) *there is no comaximal relation $a_1 a_2 = a_3 a_4$, where $a_1, ..., a_4$ are all similar non-units,*

(e) *if a is a non-unit, $(R/aR)^2$ is not strictly cyclic,*

(f) *if a, a' are similar non-units, there is no equation $xay + ua'v = 1$.*

Moreover when R satisfies right ACC_1, all these conditions (a)–(f) are equivalent.

Proof. Consider the assertion (a). We observe that $R/a'R$ is isomorphic to R/aR which is a quotient of R/acR; since $bac \in a'R$, left multiplication by b defines a homomorphism $R/acR \to R/a'R$. Thus (a) asserts that any homomorphism $R/acR \to R/aR$ is induced by an endomorphism of R/aR.

Similarly (b) asserts that isomorphic quotient modules of a strictly cyclic module have the same kernel. Hence (a) $\Rightarrow$ (b) $\Rightarrow$ (c), by the dual of Proposition 1.7, and (c) $\Rightarrow$ (a) in the presence of ACC for strictly cyclic modules, i.e. right ACC_1 in R, by Proposition 1.5*.

Now (c) $\Leftrightarrow$ (d) $\Leftrightarrow$ (e) by Proposition 2.3, and (d) $\Leftrightarrow$ (f) by Lemma 2.2 and the remark following it. ∎

We observe that every commutative 2-fir, and in particular, every commutative principal ideal domain, satisfies (a) and hence also (b)–(f).

By Theorem 1.3 we find the

COROLLARY. *Let R be an atomic 2-fir with distributive factor lattice. Then the eigenring of any non-zero element is a direct product of a finite number of completely primary rings.* ∎

Let us return to the condition (b) of Theorem 2.4; a 2-fir with the property that any two similar elements that are right commensurable are right associated will be called *right uniform*. Left uniform rings are defined similarly and rings satisfying both conditions are called *uniform*. By Theorem 2.4, any right uniform 2-fir has a distributive factor lattice, and the converse holds in the presence of the right ACC_1. Whether this condition can be omitted is not known. Some properties of uniform 2-firs are listed in the exercises; here we shall consider 2-firs that are not uniform and show that whenever an atom has two similar atomic right factors that are not left associated, then it has (generally) infinitely many; more precisely, we give a lower bound to the number of classes.

Let R be a 2-fir, $c \in R^*$ and let $c = ab$. Then $M = R/cR$ has the submodule $N = aR/abR \cong R/bR$. If $c = a'b'$ is another factorization in which $b' \sim b$, then $N' = a'R/a'b'R \cong R/b'R$ is a submodule of M which is isomorphic to N, and $N' = N$ if and only if $aR = a'R$, or equivalently $Rb = Rb'$, i.e. b, b' are left associated. Suppose now that $N \neq N'$, say $N' \nsubseteq N$; since N, N' are isomorphic cyclic modules, they have generators u, u' respectively, which correspond under this isomorphism. Given any $\alpha \in \text{End}(N)$, $\alpha u + u'$ generates a submodule N_α of M which is a homomorphic image of N, for the mapping $f_\alpha : ux \mapsto (\alpha u + u')x$ clearly defines a homomorphism. We claim that $f_\alpha \neq 0$; for if $f_\alpha = 0$, then $\alpha u = -u'$, hence $N' = \alpha N \subseteq N$, a contradiction.

Assume now that $N \cap N' = 0$; this means that $aR \cap a'R = cR$ or equivalently, $Rb + Rb' = R$, i.e. b and b' are left comaximal. Then the submodules N_α defined by the different endomorphisms of N are all distinct; for if $N_\alpha = N_\beta$, then $\alpha u + u' = (\beta u + u')x$, i.e. $u'(1 - x) \in N \cap N' = 0$, so $u' = u'x$, hence $u = ux$ and $\alpha u = \beta ux = \beta u$, therefore $\alpha = \beta$. Thus there are as many different submodules N_α as there are elements in End (N), and each corresponds to a right factor of c similar to a left factor of b. The result may be stated as

THEOREM 2.5. *Let R be a 2-fir, $b \in R^*$ and write γ for the cardinal of the eigenring of bR. Suppose that $m \in R^*$ has two right factors b, b' that are similar and left comaximal. Then the number of non-left-associated right factors of m that are similar to a left factor of b is at least γ.*

In particular, if b is an atom in R, then the number of non-left-associated right factors of m similar to b is 0, 1 or at least γ. ∎

If R is an algebra over an infinite field k, any eigenring is an algebra over k and hence infinite. Thus we obtain the

COROLLARY. *Let R be a 2-fir which is also a k-algebra, where k is an infinite field. Then the number of non-left-associated right factors of $c \in R^*$ similar to a given atom is 0, 1 or infinite.* ∎

Finally we specialize to the case of principal ideal domains. In this case we can give a simple criterion for the distributivity of the factor lattice.

Let R be an integral domain, then an element c of R^* is said to be *right invariant* if cR is a 2-sided ideal, i.e. $Rc \subseteq cR$.

LEMMA 2.6. *In a 2-fir, any two similar elements that are right invariant are right associated.*

Proof. Let c be right invariant, then $Rc \subseteq cR$, hence cR annihilates R/cR. In fact cR is the precise annihilator, for if a annihilates R/cR, then $Ra \subseteq cR$ and so $a \in cR$.

Now let $c \sim c'$ and assume that both c, c' are right invariant, then $R/cR \cong R/c'R$ and equating annihilators, we find that $cR = c'R$; thus c and c' are right associated. ∎

THEOREM 2.7. *A principal left ideal domain has a distributive factor lattice if and only if every non-zero element is right invariant.*

Proof. Let R be a principal left ideal domain whose non-zero elements are right invariant. Then any two similar elements are right associated, by Lemma 2.6, and so by Theorem 2.4, R has a distributive factor lattice.

Conversely, if R has a distributive factor lattice, then by duality (Theorem 3.2.1), the principal right ideals containing a fixed non-zero element c

form a distributive lattice with DCC, hence by Proposition 1.5, each is mapped into itself by all endomorphisms, i.e. $ba R \subseteq aR$. Thus $Ra \subseteq aR$, so that a is right invariant. ∎

Exercises 4.2

1. Let R be an atomic 2-fir in which no two similar elements are right comaximal. Show that R has a distributive factor lattice and hence is a uniform 2-fir.

2. Show that any right invariant UFD is uniform.

3. Show that an atomic 2-fir in which any two atoms are either right associated or comaximally transposable is right invariant (and hence uniform).

4. Show that the complex-skew polynomial ring does not have a distributive factor lattice.

5. Show that a skew polynomial ring over a field has a distributive factor lattice if and only if it is commutative.

6. Show that a principal right ideal domain which is right uniform is right invariant.

7. Give a direct proof of Theorem 2.4.

8*. In a ring R with distributive factor lattice, if a, b, $c \in R$ are such that any two of aR, bR, cR have a non-zero intersection, show that $aR \cap bR \cap cR \neq 0$. Find a generalization to n terms.

9^0. In Exercise 8, is it enough to assume $aR \cap bR \neq 0$, $aR \cap cR \neq 0$?

10*. Show that a uniform UFD with a right large element has a right invariant element and hence is a right Ore ring.

11^0. Is the conclusion of Exercise 10 true for any 2-fir with distributive factor lattice?

12. Show that every rigid domain is uniform.

13. Show that every uniform UFD has elements that are not fully reducible.

14^0. In Theorem 2.4, does (c) ⇒ (b) ⇒ (a) hold generally?

15. (Bergman [67]) Let K/k be a Galois extension with group $G = \text{Gal}\,(K/k)$ and let M be any K-bimodule satisfying $\lambda x = x\lambda\,(x \in M,\ \lambda \in k)$. Show that $M = \Sigma \oplus M_\sigma\,(\sigma \in G)$, where $M_\sigma = \{x \in M \mid x\alpha = \alpha^\sigma x\}$.
Let R be a uniform UFD which is a k-algebra and contains K as subfield, but not in its centre. Defining R_σ as above, show that for any $x \in R_\sigma$, $x(x - \alpha) = (x - \alpha^\sigma)x$ is a comaximal relation, and deduce that for any $x \in R_\sigma\,(\sigma \neq 1)$, $1 - x$ is a unit. (Hint: For the last part replace x by $x\alpha$.)

16^0. Show that in a 2-fir with right ACC_1, any element similar to a right invariant element c is right associated to c, and hence is itself right invariant. Does this still hold without right ACC_1?

17^0. Determine the structure of non-commutative invariant principal ideal domains.

4.3 Conservative 2-firs

Let us return to the situation of Theorem 2.5, Corollary; we saw that if an element c in a 2-fir R over a field k has two factorizations $c = ab = a'b'$ with similar but not left-associated atomic right factors, then it has a whole 1-parameter family of such factorizations. We now examine what happens when we adjoin an indeterminate t to k. In general this will lead to factorizations of c in $R \otimes k(t)$ which cannot be lifted to R. Suppose however that R is such that every factorization in $R \otimes k(t)$ can be lifted to R; this holds e.g. if R is a free algebra. Then this situation cannot arise, and we conclude that similar right factors are necessarily associated, in other words, the ring R is then uniform. We shall see below how this property can be used to provide us with many examples of uniform 2-firs, and hence of rings with a distributive factor lattice. Throughout this section, k is a commutative field, and all tensor products are understood to be over k.

Definition. Let R be a k-algebra; if both R and $R \otimes k(t)$ are 2-firs and R is 1-inert in $R \otimes k(t)$, then R is said to be a *conservative 2-fir over k*.

For example, the free algebra $R = k\langle X \rangle$ is a conservative 2-fir over k; for $R \otimes k(t)$ is the free algebra over $k(t)$, which is again a 2-fir, and the inertia follows from

PROPOSITION 3.1. *Let R be a k-algebra which is an integral domain and remains one under all field extensions. Then $R[t] = R \otimes k[t]$ is 1-inert in $R \otimes k(t)$, and hence so is R.*

Proof. Let $a_\lambda, b_\mu \in R \otimes k(t)$ be two families such that $a_\lambda b_\mu = c_{\lambda\mu} \in R[t]$ for all λ, μ. For each λ we can find $f_\lambda \in k[t]$ such that $a_\lambda' = f_\lambda a_\lambda \in R[t]$ and such that f_λ has least degree subject to this condition; this ensures that if in $\bar{k}$, the algebraic closure of k, we specialize t to a zero t_0 of f_λ then $a_\lambda'(t_0) \neq 0$ in $R \otimes \bar{k}$. Further let $f \in k[t]$ be the polynomial of highest degree dividing each a_λ' and put $a_\lambda' = a_\lambda'' f$, so that $a_\lambda'' \in R[t]$. Now put $b_\mu' = b_\mu f$ and let $g_\mu \in k[t]$ be the polynomial of least degree such that $b_\mu'' = b_\mu' g_\mu \in R[t]$. Then $a_\lambda'', b_\mu'' \in R[t]$ and $b_\mu'' = b_\mu' g_\mu = b_\mu f g_\mu$, hence

$$f_\lambda g_\mu c_{\lambda\mu} = f_\lambda g_\mu a_\lambda b_\mu = a_\lambda' g_\mu b_\mu = a_\lambda'' b_\mu'' \quad \text{for all} \quad \lambda, \mu. \tag{1}$$

If some g_μ has positive degree, specialize t to a zero of such a g_μ. Then the left-hand side of (1) becomes zero, but b_μ'' stays different from 0, by the definition of g_μ. Hence each a_λ'' becomes 0 and so the a_λ'' have a common factor in $k[t]$, against the choice of f. Therefore $g_\mu = 1$ for all μ, i.e.

$b_\mu' \in R[t]$. By repeating the argument with the a's and b's interchanged we find $g \in k[t]$ such that $b_\mu f g^{-1}$ and $a_\lambda g f^{-1}$ all lie in $R[t]$; thus $R[t]$ is 1-inert in $R \otimes k(t)$.

Now R is 1-inert in $R[t]$ because R is an integral domain, hence by transitivity R is 1-inert in $R \otimes k(t)$. ∎

COROLLARY. *Let R be a k-algebra. If R is a 2-fir and* (i) *R remains an integral domain under all field extensions and* (ii) *R remains a 2-fir under purely transcendental field extensions, then R is a conservative 2-fir.* ∎

Before we come to the main result we need another lemma. For any element $a \in R[t]$, we shall indicate the value of a obtained by specializing t to 0 by a suffix: $a_0 = a|_{t=0}$.

LEMMA 3.2. *Let R be a 2-fir which is a k-algebra, and*

$$ab = cd \tag{2}$$

an equation holding in $R[t]$ such that b, d are left comaximal in $R \otimes k(t)$, b_0, d_0 are not both 0, and a_0, c_0 right comaximal in R. Then b_0, d_0 are left comaximal in R.

Proof. Since a_0, c_0 are right comaximal and $a_0 b_0 = c_0 d_0$, this product is the least common left multiple of b_0, d_0 in R (it was only to get this conclusion that we had to assume R to be a 2-fir).

Now b, d are left comaximal in $R \otimes k(t)$, therefore we have an equation

$$pb - qd = f \qquad f \in k[t]^*, \tag{3}$$

in $R[t]$. If $f_0 = 0$, then $p_0 b_0 = q_0 d_0$ and this is a left multiple of $a_0 b_0 = c_0 d_0$. Hence by subtracting a suitable left multiple of (2) from (3) we can modify p, q so that both become divisible by t. We can then divide p, q, f all by t and obtain an equation of the same form as (3). If this process is continued, we reduce the degree of f and so eventually reach a case where $f_0 \neq 0$. Taking the constant term of f as 1, we therefore get from (3)

$$p_0 b_0 - q_0 d_0 = 1. \quad ∎$$

We can show that every conservative 2-fir satisfies the condition (a) of Theorem 2.4. Thus let R be a conservative 2-fir over k and suppose that $ba' c \in aR$, where a is similar to a'. Say

$$ba' c = ad, \qquad c \neq 0,$$

and let

$$au' = ua'$$

be a comaximal relation between a and a'. Then in $R[t]$ we have

$$a(dt + u'c) = (bt + u)a'c. \tag{4}$$

Now any common right factor (in $R \otimes k(t)$) of $dt + u'c$ and $a'c$ can by inertia be taken in R. Hence it must divide d, $u'c$ and $a'c$, i.e. generate a left ideal containing $Rd + Ru'c + Ra'c = Rd + Rc = R$. Thus (4) is in fact left comaximal in $R \otimes k(t)$, further the constant terms of the right factors are not both 0 and those of the left factors are right comaximal in R. Hence by the lemma, $u'c$ and $a'c$ are left comaximal, i.e. c is a unit and $ba' \in aR$. This proves the main result of this section:

THEOREM 3.3. *Any conservative 2-fir satisfies conditions* (a)–(f) *of Theorem* 2.4, *in particular it is* (*left and right*) *uniform and has a distributive factor lattice.* ∎

For example, this shows the free associative algebra $k\langle X \rangle$ to have a distributive factor lattice. Further, by Theorem 2.7, we have the

COROLLARY. *If a principal ideal domain is a conservative 2-fir over a field, it is invariant.* ∎

Of course the converse need not hold, since a commutative principal ideal domain is not necessarily conservative (*cf.* 4.6). But there is a partial converse to Theorem 3.3.

PROPOSITION 3.4. *Let R be a 2-fir with a distributive factor lattice, and assume that R is an algebra over an infinite field k, then R is 1-inert in $R \otimes k(t)$.*

Proof. Let a_λ, b_μ be families in $R \otimes k(t)$ such that $a_\lambda b_\mu \in R$, and suppose that a_λ, $b_\mu \in R^*$ for one pair of suffixes λ', μ' say. Then $a_{\lambda'} b_\mu \in R$ and hence $b_\mu \in R$ for all μ; similarly $a_\lambda b_{\mu'} \in R$ and so $a_\lambda \in R$ for all λ. Thus it will be enough to show that a single factorization can be lifted to R.

Let $a, b \in R \otimes k(t)$, and $ab = c \in R$. Then in $R[t]$ we have, on modifying a and b each by a factor in $k[t]$:

$$ab = cf, \quad a, b \in R[t], \quad c \in R, \quad f \in k[t]. \tag{5}$$

Denote the degrees of a, b, f in t by p, q, r respectively.

Let $\mathfrak{a}$ be the right ideal of R generated by the coefficients of $a = a(t)$, considered as a polynomial in t. Then for any $p + 1$ distinct elements

$\alpha_0, \ldots, \alpha_p \in k$ we can express all the coefficients of $a(t)$ as k-linear combinations of $a(\alpha_0), \ldots, a(\alpha_p)$, hence these $p + 1$ elements generate $\mathfrak{a}$ as right ideal. If we choose $p + 1$ such values so as to avoid the zeros of f (at most r in number) then the $a(\alpha_i)$ will be left factors of c, hence the $a(\alpha_i)R$ will belong to $\mathbf{L}(cR, R)$ and so will their sum $\mathfrak{a}$; therefore $\mathfrak{a} = dR$ for some $d \in R$. Similarly, using the factorial duality, we find that the intersection of any $q + 1$ of these ideals will be a fixed member of $\mathbf{L}(cR, R)$, which we write as eR. Clearly $eR \subseteq dR$; we assert that equality holds.

For if $eR \neq dR$, then the principal right ideals between eR and dR form a non-trivial distributive lattice, and hence (Appendix 1) there is a homomorphism of this lattice onto the 2-element lattice $\mathbf{2} = \{0, 1\}$ such that $eR \mapsto 0$, $dR \mapsto 1$. Since the sum of the ideals $a(\alpha)R$, as α ranges over any $p + 1$ values of k, avoiding the zeros of f, is the maximal element dR, which maps to 1, it follows that $a(\alpha)R$ can map to 0 for at most p values of α. But the intersection of the $a(\alpha)R$, as α ranges over any $q + 1$ values as before, is eR, which maps to 0, so $a(\alpha)R$ maps to 1 for at most q values. Since k is infinite, we have a contradiction, which shows that $eR = dR$. Now d is a left factor of $a(t)$ in $R \otimes k(t)$ and e a left multiple, hence $a(t)$ is associated to d in $R \otimes k(t)$, and our factorization of c can be lifted to R. ∎

We note however that in the situation of this Proposition $R \otimes k(t)$ may not be a 2-fir, e.g. take $R = k(s)[x]$, where s, x commute.

To obtain a good supply of conservative 2-firs we shall now prove that a filtered ring with 2-term weak algorithm is a conservative 2-fir provided that the elements of degree 0 lie in the centre.

THEOREM 3.5. *Let R be a filtered algebra over an infinite commutative field k, such that every element of degree 0 lies in k. If R satisfies 2-term weak algorithm, then R is a conservative 2-fir.*

Proof. We must show that (i) $R \otimes k(t)$ is a 2-fir and (ii) R is 1-inert in $R \otimes k(t)$.

Let $A = \operatorname{gr} R$ be the graded ring associated with R, then A satisfies the 2-term weak algorithm as graded ring. Moreover,

$$\operatorname{gr} R \otimes k(t) = A \otimes k(t),$$

where $A \otimes k(t)$ is the graded ring with ith component $A_i \otimes k(t)$. So to prove (i) it will be enough to show that $A \otimes k(t)$ satisfies 2-term weak algorithm. Let

$$ab' = ba' \qquad (6)$$

be a non-trivial relation between a, $b \in A \otimes k(t)$ and suppose that $v(a) \geqslant v(b)$ say; we have to find $q \in A \otimes k(t)$ such that $a = bq$. On clearing denominators in (6) we may assume that a, b, a', $b' \in A \otimes k[t] = A[t]$, and that no two of these elements that occur on opposite sides of (6) have a common factor in $k[t]$.

On equating the terms independent of t in (6) we obtain a relation in A:

$$a_0 b_0' = b_0 a_0'.$$

If one of these elements is zero, say $a_0 = 0$, then $a_0' = 0$ or $b_0 = 0$ and so either a, a' or a, b have a factor t in common, contrary to hypothesis. By the 2-term weak algorithm in A, there exists $q_0 \in A$ with $a_0 = b_0 q_0$. If we replace a by $a - b q_0$ and a' by $a' - q_0 b'$ in (6), we obtain an equation of the same form, in which a, a' are divisible by t. On dividing by t and using induction on the degree in t, we find that $a - b q_0 = bq$, whence $a = b(q + q_0)$ as claimed.

Secondly let $c \in R$ have a factorization $c = ab$ in $R \otimes k(t)$. By multiplying up we get an equation

$$fc = ab, \qquad a, b \in R[t], \qquad f \in k[t]. \tag{7}$$

We may assume that $f(0) \neq 0$, for otherwise, on equating the terms independent of t, we find $a_0 b_0 = 0$, and so either a or b is divisible by t, and we can cancel this factor t from (7). Hence $f(0) \neq 0$, and we may assume without loss of generality that $f(0) = 1$. Now equate the terms independent of t in (7): we get $c = a_0 b_0$, and substituting in (7) we find

$$f a_0 b_0 = ab \qquad v(a_0) = v(a),$$

hence by the weak algorithm in $R \otimes k(t)$, $a = a_0 q$, where $v(q) = 0$ and hence q is a unit. Thus aq^{-1}, $f^{-1}qb \in R$. Next take two families a_λ, b_μ such that $a_\lambda b_\mu \in R$ for all λ, μ. By what has been shown, we may assume that for a particular λ, μ, say $\lambda = \mu = 1$, a_1, $b_1 \in R$. Then $a_1 b_\mu \in R$, hence on specializing t to a value t_0 for which the denominator in b_μ does not vanish, we get $a_1 b_\mu = a_1 b_\mu(t_0)$, whence $b_\mu = b_\mu(t_0)$ is independent of t. Similarly for a_λ, and it follows that R is 1-inert in $R \otimes k(t)$. ∎

Exercises 4.3

1. If a conservative 2-fir satisfies the right Ore condition, show that any two similar elements are right associated.

2. Let R be a conservative 2-fir over a field k, and a, $a' \in R$. If a, a' are similar in $R \otimes k(t)$, show that they are similar in R.

3. Let R be an n-fir over an algebraically closed field k. Show that $R[t]$ is n-inert in $R \otimes k(t)$. (Hint: The transforming matrix can be taken to lie in the subgroup generated by $GL_n(R[t])$ and the diagonal matrices over $k(t)$.)

4*. Let R be a $2n$-fir over a commutative field k and assume that R remains an n-fir under all extensions and a $2n$-fir under all purely transcendental extensions of k. Let c be a matrix in R_n with no zero-divisor as factor, such that any factor of c in $R[t]_n$ can be reduced to one in R_n on multiplying by an element of $\mathbf{GL}_n(R[t])$. Show that the corresponding module R_n/cR_n is distributive.

Taking $n = 2$ and p, q dissimilar atoms in R, show that $c = [p, q]$ satisfies the above hypotheses, but not $[p, p]$. Verify that when $c = [p, p]$, R_2/cR_2 is not distributive. ($[p, q]$ stands for $\mathrm{diag}\,(p, q)$ as in 2.6.)

5. In the complex-skew polynomial ring $R = \mathbf{C}[x;^-]$ show that $x^2 - 1$ has the factorizations $x^2 - 1 = (x - u)\,(x + \bar{u})$, where u ranges over the unit circle. Obtain the corresponding factorizations over $\mathbf{C}(t)[x;^-]$ with $u = (t + i)\,(t - i)^{-1}$ and show that these cannot be lifted to R.

6. Let R be a k-algebra, M an R-module, P a submodule of M with inclusion map $i:P \to M$ and $f:P \to M$ a homomorphism. Assume that the image of the homomorphism $i + tf$ of $P \otimes k(t)$ into $M \otimes k(t)$ is of the form $N \otimes k(t)$ for some $N \subseteq M$. Show that $fP \subseteq P$ and hence deduce another proof of Theorem 3.3.

7^0. Investigate Proposition 3.4 when the field k is finite.

8^0. Is every right invariant left Ore domain invariant?

9. In a free algebra, show that a non-unit c cannot satisfy an equation of the form $pcq + rcs = 1$.

10^0. In a free algebra, does a non-unit necessarily generate a proper ideal? (*cf.* Exercise 7.6.11.)

11*. In a free algebra, if $ab = bc^2$, show that a is a square.

4.4 Finite distributive lattices

In an atomic 2-fir with distributive factor lattice, the left factors of a given non-zero element form a distributive lattice of finite length; for a closer study of this lattice we shall in this section describe its structure in terms of partially ordered sets.

Let us denote by Pos the category of finite partially ordered sets, with isotone (i.e. order-preserving) mappings as morphisms. By DL we shall denote the category of all distributive lattices of finite length, with lattice–homomorphisms, i.e. mappings preserving meet, join, 0 and 1, as morphisms. In each of these categories $\mathbf{2} = \{0, 1\}$ denotes the chain of length 1.

Take $P \in \mathrm{Pos}$ and consider $P^* = \mathrm{Hom}_{\mathrm{Pos}}(P, \mathbf{2})$; this set P^* may be regarded as a finite distributive lattice, namely a subset of $\mathbf{2}^P$. The elements of P^*—isotone mappings from P to $\mathbf{2}$—may also be described by the subsets of P mapped to 0. They are precisely the *lower segments* of P, i.e. subsets X with the property:

$$a \in X, b \leqslant a \quad \text{implies} \quad b \in X.$$

A lower segment of the form $\lambda_a = \{x \in P | x \leqslant a\}$ is said to be *principal*; we observe that the partially ordered set of all principal lower segments of P is isomorphic to P, as member of Pos.

Clearly each $\alpha \in P^*$ is completely determined by the lower segment mapped to 0, and every lower segment defines such a mapping. Hence P^* may also be identified with the set of lower segments of P.

Next take $L \in DL$ and write $L^* = \mathrm{Hom}_{DL}(L, 2)$. Here we can regard L^* as a partially ordered set, writing $f \leqslant g$ if and only if $xf \leqslant xg$ for all $x \in L$. The elements of L^* may be characterized by the subsets they map to 1; given $f \in L^*$, let $a \in L$ be the meet of all $x \in L$ satisfying $xf = 1$. Then $af = 1$ and a is the unique minimal element with this property. Clearly $a > 0$ and if $a = x \vee y$, then $1 = xf \vee yf$, hence $xf = 1$ or $yf = 1$, i.e. $x \geqslant a$ or $y \geqslant a$. Thus a is *join-irreducible*:

$$a \neq 0 \quad \text{and} \quad a = x \vee y \quad \text{implies} \quad x = a \quad \text{or} \quad y = a.$$

Conversely, any join-irreducible element a gives rise to an $f \in L^*$, defined by the rule $xf = 1$ if and only if $x \geqslant a$. We may thus identify L^* with the partially ordered set of its join-irreducible elements.

LEMMA 4.1. *In a distributive lattice of finite length, every element is the join of the join-irreducible elements below it.*

Proof. Let $L \in DL$; if the lemma is false, take $a \in L$ such that a is not a join of join-irreducible elements, and lowest in L subject to this condition, i.e. such that the length of $[0, a]$ is minimal. Then a cannot be join-irreducible, so $a = b \vee v$, $b < a$, $c < a$, and by the minimality b, c are joins of join-irreducible elements; hence so is $a = b \vee c$. ∎

THEOREM 4.2. *The categories* Pos *and* DL *are dual to each other, via the contravariant functors*

$P \mapsto P^* = $ *lattice of lower segments of P,*

$L \mapsto L^* = $ *set of join-irreducible elements of L.*

Moreover, if P and L correspond, then the length of L equals card(P).

Proof. It is clear that two contravariant functors are defined between these categories by means of $\mathrm{Hom}\,(-, 2)$; it only remains to show that $P^{**} \cong P$, $L^{**} \cong L$.

Let $P \in$ Pos, then P^* consists of all lower segments of P. If $\alpha \in P^*$ and $a_1, \ldots, a_r$ are the different maximal elements of α, then $x \in \alpha$ if and only if $x \leqslant a_1$ or ... or $x \leqslant a_r$. Hence $\alpha = \lambda_{a_1} \vee \ldots \vee \lambda_{a_r}$, where λ_c is the principal lower segment corresponding to c. This shows α to be join-irreducible if and only if it is principal, and so P^{**}, the set of join-irreducible lower segments, is just the set of principal lower segments, which we know is $\cong P$.

Secondly, given $L \in DL$, consider L^*, the partially ordered set of its join-irreducible elements. This set determines L, by Lemma 4.1: each $a \in L$ can be represented by the set of join-irreducible elements $\leqslant a$, and the sets of join-irreducible elements occurring are just the lower segments, thus $L^{**} \cong L$.

Let L and P correspond under this duality, and suppose P has n elements. Then we can form a chain in L by picking a minimal element $a_1 \in P$, next a minimal element a_2 in $P \backslash \{a_1\}$, etc. therefore every chain has n elements. ∎

Clearly every P^* is finite, as subset of 2^P, hence we obtain

COROLLARY. *Any distributive lattice of finite length n is finite, of at most 2^n elements.* ∎

There is another way of describing the correspondence of Theorem 4.2 which is of importance for us in what follows.

In a distributive lattice of finite length, every link is projective to exactly one link with join-irreducible upper end-point. For let $[a, b]$ be a link in which b is not join-irreducible, say $b = x \vee y$, $x < b$, $y < b$. Then each of $x \vee a$, $y \vee a$ lies in $[a, b]$, i.e. is a or b, and not both can be a, say $x \vee a = b$. Then $[a, b]$ is perspective to a 'lower' link, namely $[x \wedge a, x]$ Hence the lowest link in the projectivity class will have a join-irreducible upper end-point. But such a lowest link is unique in a distributive lattice: Let $[a, b]$ be in perspective with lower links $[a', b']$ and $[a'', b'']$, then we have

a cube diagram as shown. $b = a \vee b' = a \vee b''$, $a \wedge b' = a'$, $a \wedge b'' = a''$; put $b_0 = b' \wedge b''$, then $b_0 \wedge a' = b' \wedge b'' \wedge a' = b' \wedge b'' \wedge a = a_0$ say, and by symmetry, $b_0 \wedge a'' = a_0$, while $a' \vee b_0 = (a \wedge b') \vee (b'' \wedge b') = (a \vee b'') \wedge b' = b \wedge b' = b'$; similarly $a'' \vee b_0 = b''$. This shows that in each projectivity class of links there is a unique lowest link, whose upper end-point is necessarily join-irreducible.

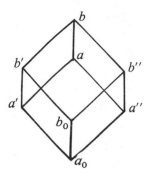

It follows that the join-irreducible elements correspond to the projectivity classes of links. If our lattice L has length n, there are just n join-irreducible elements and hence n projectivity classes of links. Since each class has a representative in each chain (by the Jordan–Hölder theorem), there is exactly one representative from each class in each chain.

It remains to describe the partial ordering in terms of these classes. Let $a, b \in P$ and let α, β be the projectivity classes of links corresponding to a, b respectively. If a, b are incomparable in P, then we can form chains in P^* in which the representative of α lies lower than that of β, and chains in which it lies higher, depending on whether we choose a before b or b before a in forming the chain. But if $a < b$, in P, then we must choose a before b and hence in every chain the representative of α lies lower than that of β. This expresses the partial ordering in terms of the projectivity classes.

Exercises 4.4

1. Show that a modular lattice has finite length if and only if every chain in it is finite. Give examples of (i) an infinite modular lattice of finite length and (ii) a general lattice of infinite length, all of whose chains are finite.

2. Show that a finite distributive lattice is complemented if and only if the corresponding partially ordered set is totally unordered. (Hint: In a Boolean algebra, the join-irreducible elements are precisely the minimal non-zero elements.)

3. Examine how the correspondence of Theorem 4.2 is affected if we take (i) upper instead of lower segments, (ii) meet- instead of join-irreducible elements, and (iii) make both these changes.

4. Let L be a modular lattice of finite length in which any two projective intervals are perspective. Show that L must be distributive. Determine all such lattices, using Theorem 4.2.

4.5 The factor lattice

Let R be an atomic 2-fir with a distributive factor lattice. For each $c \in R^*$, $L_c = \mathbf{L}(cR, R)$ is a distributive lattice of finite length. We shall write

P_c for the corresponding partially ordered set L_c^*. Each atomic factorization of c:

$$c = p_1 p_2 \cdots p_n$$

corresponds to a chain in L_c; if $c = q_1 \ldots q_n$ is another atomic factorization of c, then q_i is said to be *equivalent* to p_j if we can pass from q_i to p_j by a series of comaximal transpositions. Since comaximal transpositions correspond to perspectivities in L_c, the equivalence classes of atomic factors correspond to projectivity classes of links in L_c, and thus to the elements of P_c. We shall refer to an equivalence class of atomic factors of c as an *abstract atomic factor* of c. Each abstract atomic factor has just one representative in each complete factorization of c, and of two abstract atomic factors, p and q say, p precedes q if it occurs on the left of q in every complete factorization of c; on the other hand when p, q are incomparable then they may be comaximally transposed whenever they occur next to each other in a complete factorization. Every complete factorization is completely determined by the order in which the abstract factors occur; in particular an element with n factors cannot have more than $n!$ complete factorizations.

Any expression of c as a product $c = ab$ corresponds to a decomposition of P_c into a lower and a complementary upper segment, which may be identified with P_a, P_b respectively. Given two factorizations

$$c = ab' = ba', \tag{1}$$

we see that the highest common left factor and least common right multiple of a, b will correspond to the intersection and union of P_a, P_b respectively. In particular, a comaximal relation (1) for c corresponds to an expression of P_c as a union of two disjoint lower segments, which means a partition of its diagram into two disconnected components. We note also that in this case $L_c = L_a \times L_b$, in agreement with Theorem 3.5.5, Corollary.

Recalling that projective links in L_c correspond to similar factors, we see that with every element of $P_c = L_c^*$ we can associate a similarity class of atoms in R. Abstract factors corresponding to the same similarity class cannot be incomparable in P_c because similar atoms cannot be comaximally transposed in R (by Theorem 2.4), hence every similarity class corresponds to part of a chain within P_c. It follows that the only automorphism of P_c preserving similarity classes is the identity; hence the same holds for L_c. Thus for any similar elements c and c' the isomorphism between L_c and $L_{c'}$ (and between P_c and $P_{c'}$) is unique. We state this as

PROPOSITION 5.1. *Let c, c' be similar elements of an atomic 2-fir R with distributive factor lattice. Then there is a unique isomorphism $L_c \to L_{c'}$ between the factor lattices.* ∎

If $f:R \rightarrow R'$ is a homomorphism of atomic 2-firs with distributive factor lattice, then for any $c \in R$ such that $c \notin \ker f$, we get a lattice homomorphism from L_c to L_{cf}: the obvious mapping preserves HCLF's because it preserves comaximality and it preserves LCRM's by the factorial duality. By Theorem 4.2, a homomorphism in the opposite direction is induced from P_{cf} to P_c.

In a commutative principal ideal domain, or indeed in any commutative UFD, two atoms are coprimely transposable if and only if they are non-associated. It follows that the only possible structures for the sets P_c in this case are disjoint unions of finite chains. E.g. in Z, $720 = 2^4 . 3^2 . 5$, hence P_{720} consists of three chains, of lengths 3, 1 and 0. By contrast, in the non-commutative case, all possible structures for P_c are realized:

THEOREM 5.2. *Let* $A_n = k\langle x_1, \dots, x_n \rangle$ *be the free algebra of rank* n. *Given any partially ordered set* P *of* n *elements, there exists* $c \in A_n$ *with* $P_c \cong P$.

Proof. The case $n = 0$ is clear, so assume $n > 0$ and let α be any element of P. Assume inductively that we have found $c_0 \in A_{n-1}$ such that $P_{c_0} \cong P_0 = P \backslash \{\alpha\}$.

Write $P = U \cup V \cup W$, where U is the set of elements $< \alpha$ in P, V the set of elements incomparable with α and W the set of elements $> \alpha$. Clearly U, $U \cup V$ and $U \cup V \cup W$ are lower segments of P_0; they correspond to left factors a, ab, and $abd = c_0$ of c_0. Put $c = a(bx_n + 1)bd$, then we assert that $P_c \cong P$.

In the first place, $bx_n + 1$ is an atom, since it is linear in x_n and in any factorization the term independent of x_n must divide 1.

We can now identify P_c with P by letting the factors of a, b and d correspond as in the identification of P_{c_0} and $P \backslash \{\alpha\}$, and letting the abstract atomic factor to which $bx_n + 1$ belongs correspond to α. This identification is well-defined, because the automorphism of P_b induced by the comaximal relation $(bx_n + 1)b = b(x_n b + 1)$ preserves similarity classes and so is the identity. It remains to check that the partial ordering of P_c agrees with that of P.

Since ab is a left factor of c and bd is a right factor, the orderings on the corresponding subsets of P_c will agree with those on P_{ab} and P_{bd}, as required. The new abstract factor is incomparable with the factors of b, due to the comaximal relation $(bx_n + 1)b = b(x_n b + 1)$. Now the partial ordering will be completely determined if we show that the factor corresponding to $bx_n + 1$ lies above all factors of a and below all factors of d. By symmetry, it suffices to prove the first statement.

Suppose the contrary; then for some non-unit right factor e of a, we would have a comaximal relation $e(bx_n + 1) = fe'$. Now we obtain a ring homomorphism of A_n into A_{n-1} by putting $x_n = 0$; this will preserve co-maximal relations and hence maps f to an element similar to $b \cdot 0 + 1 = 1$, i.e. a unit. But in A_n f is similar to the non-unit $bx_n + 1$, hence it must involve x_n. But then the product $e(bx_n + 1) = fe'$ will involve monomial terms in which x_n occurs, but is not the last factor (since e' is a non-unit). This is a contradiction, and it shows that every factor of a lies below $bx_n + 1$. ∎

In fact all these partially ordered sets may already be realized in A_2. This will follow by showing that A_n (for any n) can be embedded as a 1-inert subring in A_2; the next result shows more generally how the free algebra of countable rank can be embedded 1-inertly in A_2.

THEOREM 5.3. *The free algebra of countable rank can be embedded* 1-*inertly in the free algebra of rank* 2.

Proof. Consider the free algebra $A_2 = k\langle x, y \rangle$; it has the usual grading by total degree in x and y. We shall also grade it by associating with each monomial p the function $i(p) = d_x(p) - d_y(p)$, the degree in x minus the degree in y, which we shall call the *imbalance* of p. The *balanced* elements of A_2, i.e. elements that are homogeneous of i-degree 0, form a subalgebra which we shall denote by B.

Let U be the class of monomials $u \neq 1$ such that $i(u) = 0$, but $i(u'') > 0$ for every right factor $u'' \neq 1$, u. Note that every such u must begin with a factor x. Now A_2 has an automorphism $^-: u \to \bar{u}$ of order 2 which consists in interchanging x and y, and this automorphism maps B into itself; we claim that $U' = \{u, \bar{u} | u \in U\}$ is a free generating set of B, as free k-algebra. This is clear since every balanced monomial can be decomposed in a unique manner into factors in U' by writing it out as a product of x's and y's, and putting a dividing mark at every point having the same number of x's as y's to the right of it.

It follows that the subalgebra C generated by the elements $u + \bar{u}$ $(u \in U)$ is also free. Clearly it has infinite rank; we shall prove that it is 1-inert in A_2. To do this we need only show that for any $c \in C$, $c = ab$ with $a, b \in A_2$ implies $a, b \in C$. For if we are given two families a_λ, b_μ with $a_\lambda b_\mu \in C$, it follows that each a_λ and b_μ ⁚ in C. This simplification arises because every unit of A_2 already lies in C.

We shall take the monomials of A_2 to be ordered by total degree and those of a given degree lexicographically, then the leading term in A_2 of any element $(u_1 + \bar{u}_1) \dots (u_n + \bar{u}_n)$ $(u_i \in U)$ will be $u_1 \dots u_n$ (because u is the leading term of $u + \bar{u}$). It follows that the leading term in every non-zero element of C is of this form, and that every element in $B \backslash C$ can be written as the sum of a term in C and an element whose leading term contains factors $\bar{u}$ $(u \in U)$ in its decomposition.

Now let $c = ab$, where $c \in C$, a, $b \in A_2$. Since $c \in B$, it is i-homogeneous, of imbalance 0, hence a and b are also i-homogeneous, of imbalance m and $-m$ respectively. It follows that every monomial term of maximal degree occurring in c has the property that if we split it into a left factor of degree $d(a)$ and a right factor of degree $d(b)$, the former will have imbalance m and the latter imbalance $-m$. But combining this with the fact that $\bar{c} = c$, we see that m must be 0. Hence a and b lie at least in B.

Note that the leading term of c is the product of the leading terms of a and b, hence neither of these leading terms can involve a $\bar{u}$ in its factorization into elements of U and $\bar{U}$. Let us write $a = a' + f$, $b = b' + g$, where a', $b' \in C$ and f, g are elements of B which are 0 or have leading terms involving at least one factor $\bar{u}$. Our earlier remark shows that $d(f) < d(a')$, $d(g) < d(b')$. Now C will contain $c - a'b' = (a' + f)(b' + g) - a'b' = a'g + fb' + fg$. We note that the leading term of $a'g$, if non-zero, will involve a $\bar{u}$ in its right subsegment of degree $d(g)$ but not to the left thereof, and we can make parallel observations about the leading terms of fb' and fg. We deduce that no two of these leading terms, if they exist, are equal, and hence that the leading term of our sum is one of these unless $f = g = 0$. But, being in C, the leading term of our sum cannot involve a $\bar{u}$, so we must have $f = g = 0$, i.e. a, $b \in C$. ∎

Exercises 4.5

1. Show that a factorization $c = a_1 \ldots a_n$ in a uniform UFD corresponds to an isotone map of P_c into the ordered set of n elements.

2. A subset X of a partially ordered set is called *convex* if $x, y \in X$, $x < a < y$ implies $a \in X$. Show that a subset X of P_c is convex if and only if c has a factorization $c = aub$, where $P_u \simeq X$, and if $c = a'u'b'$ is another such factorization, then u' is obtainable from u by a series of comaximal transpositions.

3. Let R be a uniform UFD and let $c = a_1b_1 = a_2b_2$. If each similarity class of atoms contributes at least as many terms to a factorization of a_1 as it does to a factorization of a_2, show that $a_1 \in a_2R$.

4. Find elements in $k\langle x, y \rangle$ with factor lattices corresponding to the following partially ordered sets:

(i) ∘ ∘ ∘   (ii) ⋀   (iii) ⋁   (iv) ⋁⋁   (v) ⋎

5. In $k\langle x, y, z \rangle$ show that the factor lattice of
$$(x^2 + x)[(y(x^2 + x) + 1)z + 1](y(x^2 + x) + 1)$$
is the free distributive lattice on 3 free generators.

6^0. Is the embedding in the proof of Theorem 5.3 totally inert?

4.6 Eigenrings

Let R be any ring; we have seen in 0.4 that the eigenring of an element $c \in R^*$ is essentially the endomorphism ring of R/cR. If R has a distributive factor lattice, this will naturally have consequences for the eigenrings of elements, some of which we shall study here. As before our rings will be algebras over a commutative field k; hence the eigenring of an element will also be an algebra over k. If the eigenring of c is k itself, we shall say that c has *scalar eigenring*.

From Theorem 1.3 we obtain the following result on the structure of eigenrings:

PROPOSITION 6.1. *Let R be an atomic 2-fir with distributive factor lattice. Then for any $c \in R^*$, the eigenring of c is a direct product of a finite number of completely primary rings.* ∎

Beyond this rather general fact it seems difficult to apply the results on distributive modules to the study of arbitrary rings with distributive factor lattice. In what follows we shall therefore put further restrictions on the rings; most of these will be satisfied by free algebras.

In 3.2 we saw that in a 2-fir over a field k, the eigenrings of atoms tend to be algebraic over k if k is large. This makes it plausible that the same conclusion will hold if R stays a 2-fir under purely transcendental field extensions. In fact more is true, but we need to make some observations on $R \otimes k(t)$ to provide us with a substitute for Proposition 3.1.

We shall write $R(t) = R \otimes k(t)$ in what follows. Let us call an element of $R(t)$ *essentially monic* if it can be written as pf^{-1}, where $p \in R[t]$ has a unit as highest coefficient and $f \in k[t]$. Clearly any left or right factor of an essentially monic element is again essentially monic. Further, if p, $q \in R[t]$ and p is essentially monic, then we can express each coefficient of q as a left linear combination of the coefficients of pq (over $R[t]$). In particular, if pq has some $a \in R$ as right factor, then q also has a as right factor. Of course this property still holds for $R(t)$; in particular, if $p \in R(t)$ is essentially monic and has a right multiple in R: $pq = a \in R$, then $q = q'a$, hence $pq' = 1$ and so p is then invertible. It follows that an essentially monic element of $R(t)$ and any non-zero element of R will be left coprime in $R(t)$ provided they are right commensurable.

With these preparations we can prove

PROPOSITION 6.2. *Let R be a 2-fir over a field k, such that $R(t) = R \otimes k(t)$ is again a 2-fir. Then the eigenring of any non-zero element of R is algebraic over k.*

Proof. Let $e \in R^*$ and consider its eigenring $\mathbf{E}(eR) = \mathbf{I}(eR)/eR$. Take $y \in \mathbf{I}(eR)$, say $ye = ey'$. Then in $R(t)$ we have

$$(t - y)e = e(t - y').$$

By the preceding observations, $t - y$ and e are left coprime, and hence right comaximal (because $R(t)$ is a 2-fir). Writing down an equation of comaximality and clearing denominators, we get

$$(t - y)p + eq = f \quad \text{where} \quad p, q \in R(t), \quad f \in k[t]^*.$$

In this equation write all powers of t on the left and substitute y for t; put another way, we reduce each term modulo $(t - y)R[t]$ to an element of R. The first term goes to 0, the second to a member of eR, because $t^i e = y^i e = ey'^i$. Hence $f(y) \in eR$, i.e. the image of y in $\mathbf{E}(eR)$ is a zero of $f \in k[t]$, and so $\mathbf{E}(eR)$ is algebraic over k, as asserted. ∎

COROLLARY 1. *Let R be a 2-fir over an algebraically closed field, and assume that R remains a 2-fir under all pure transcendental field extensions. Then every atom of R has a scalar eigenring.*

For we know that the eigenring of an atom is a field, by Schur's lemma (Proposition 3.2.4), and the only algebraic (skew) field extension of k is k itself. ∎

In the commutative case the eigenring of an ideal $\mathfrak{a}$ is just the residue class ring modulo $\mathfrak{a}$, and we obtain

COROLLARY 2. *Let R be a commutative Bezout domain which is an algebra over k and remains a Bezout domain under all purely transcendental field extensions. Then for every ideal $\mathfrak{a}$ of R, $R/\mathfrak{a}$ is algebraic over k.* ∎

Proposition 6.2 can be applied to free algebras to show that eigenrings are algebraic; but in this case there is a stronger result. In the proof we shall need the notion of a retract of a ring; we briefly recall the relevant facts. Let R be any ring; a subring K of R is called a *retract* of R if there is a homomorphism $\lambda: R \to K$ which reduces to the identity on K. Then the functor

$$F: M \mapsto M \otimes_K R, \quad \text{where } M \text{ is any right } K\text{-module},$$

provides a category-equivalence between all right K-modules and all right R-modules of the form $M^F = M \otimes R$. For F has the inverse $G: P \to P \otimes_R K$, where K is defined as left R-module by the rule $r \cdot k = \lambda(r) \cdot k$. It follows that for any right K-module M,

$$\mathrm{End}_R (M \otimes R) \cong \mathrm{End}_K (M). \tag{1}$$

We can now prove the following result:

PROPOSITION 6.3. *Let $R = k\langle X \rangle$ be a free associative algebra over k, then the eigenring of any finitely generated non-zero right ideal of R is finite-dimensional over k.*

Proof. Let $\mathfrak{a} = \Sigma e_i R$ be the given right ideal, then $\mathbf{E}(\mathfrak{a}) = \mathbf{I}(\mathfrak{a})/\mathfrak{a} \cong \mathrm{End}_R (R/\mathfrak{a})$. Let $e_1, ..., e_r$ be a generating set of $\mathfrak{a}$; if we put all free generators not occurring in any e_i equal to 0, we obtain a retract K of R, which is a finitely generated free k-algebra. Clearly $R/\mathfrak{a} = M \otimes R$, where $M = K/\Sigma e_i K$, hence by (1) we may replace R by K, which is again free, but finitely generated. Now any $y \in K$ lies in $\mathbf{I}(\Sigma e_i K)$ precisely if

$$y e_j = \Sigma e_i y_{ij} \quad \text{for some} \quad y_{ij} \in K. \tag{2}$$

Let d be the maximum of the degrees of $e_1, ..., e_r$ in a free generating set of K, then by (2) and the weak algorithm in K,

$$y = \Sigma e_i u_i + y' \qquad \deg y' < d. \tag{3}$$

Without loss of generality the e_i may be taken to be free generators, then y' is uniquely determined by y, and $y \mapsto y'$ is a k-linear mapping from $\mathbf{E}(\mathfrak{a})$ to the space of elements of degree less than d in K. This space is finite-dimensional, and since the mapping is clearly injective, $\mathbf{E}(\mathfrak{a})$ is also finite-dimensional. ∎

COROLLARY. *The eigenring of any non-zero element of a free associative k-algebra is finite-dimensional over k.* ∎

It seems plausible that the endomorphism ring of any distributive module of finite length is right invariant, but we have not been able to establish this. However, in the special case of free algebras we can prove even more:

PROPOSITION 6.4. *Let R be a k-algebra which is an atomic 2-fir with distributive factor lattice; assume that each atom of R has a scalar eigenring. Then any non-zero element of R has a commutative eigenring.*

Proof. Let $M = R/cR$; by induction on the length we may assume that every proper $\mathscr{C}$-submodule of M has a commutative eigenring. Every $\alpha \in \mathbf{E}(cR)$ maps each submodule of M into itself, so if M is the sum of its proper submodules, we can embed $\mathbf{E}(cR)$ into the direct product of the corresponding endomorphism rings, hence $E(cR)$ is then commutative. The alternative is that the sum of all proper submodules of M is a unique maximal submodule M', say. By hypothesis every endomorphism of M/M' is induced by multiplication by an element of k, hence every endomorphism of M has the form $\lambda + \alpha$, where $\lambda \in k$ and $\alpha M \subseteq M'$. It is therefore enough to show that any two non-surjective endomorphisms of M commute. Let us take such endomorphisms α, β of M; by Proposition 1.4, Corollary, one of αM, βM is contained in the other, say $\alpha M \subseteq \beta M \subseteq M'$.

Now α maps M into $\beta M \cong M/\ker \beta$, hence by Proposition 1.5*, α is induced by an endomorphism α' of βM, i.e. $\alpha = \alpha' \beta$. Let β_1 be the restriction of β to βM, then we have

$$\alpha\beta = \alpha' \beta\beta = \alpha' \beta_1 \beta,$$

but on βM all endomorphisms commute, by the induction hypothesis. Hence $a' \beta_1 \beta = \beta_1 \alpha' \beta = \beta_1 \alpha = \beta\alpha$. Thus $\alpha\beta = \beta\alpha$. ∎

COROLLARY. *Let R be a k-algebra which is an atomic 2-fir and remains one under arbitrary field extension. Then the eigenring of any non-zero element is commutative.*

For let K be an algebraically closed field extension of k of cardinal greater than $\dim_k R$. Then all atoms in $R \otimes K$ have scalar eigenrings (Proposition 3.2.5) and $R \otimes K$ has distributive factor lattice by Theorem 3.3, hence we can apply the result just proved to reach the conclusion. ∎

This Corollary shows e.g. that all eigenrings of elements in a free associative algebra are commutative.

For examples of non-commutative eigenrings, consider the elements $x^2 + 1$ and $x^2 - 1$ in the complex-skew polynomial ring. Both are invariant (even central), hence their eigenrings are quotients of the whole ring by the two-sided ideals they generate. The element $x^2 + 1$ is an atom, so the quotient is a field (Proposition 3.2.4); clearly this is the field of quaternions. The element $x^2 - 1$ is an I-atom, a product of two atoms, hence the eigenring is a 2×2 matrix ring over $\mathbf{C}$.

Exercises 4.6

1. In an atomic 2-fir R, characterize the elements whose eigenring has zero radical.

2. In the free algebra $\mathbf{R}\langle x, y \rangle$, show that $a = xy^2x + xy + yx + x^2 + 1$ is an atom but does not remain one under extension to $\mathbf{C}$. Deduce that the eigenring of a is $\mathbf{C}$. Find an element of the idealizer mapping to i.

3. By taking the right ideal in Proposition 6.3 two-sided, show that the hypothesis that the right ideal be finitely generated cannot be omitted. Deduce also that a two-sided ideal in $k\langle X\rangle$ cannot be finitely generated as right ideal unless it has finite codimension.

4*. (Bass) Let R be a commutative principal ideal domain containing a field k. Show that $R \otimes k(t)$ is again a principal ideal domain if and only if every prime ideal in $R[t]$ that is not minimal among the non-zero primes (i.e. of height >1) meets $k[t]^*$. Deduce that the condition that $R/\mathfrak{p}$ be algebraic over k is sufficient as well as necessary for $R \otimes k(t)$ to be a principal ideal domain.

5. Apply Exercise 4 to test whether $R \otimes k(t)$ is a principal ideal domain in the following cases:

   (i) $R = k(x)[y]$,
   (ii) R is a principal ideal domain algebraic over a subfield k,
   (iii) $R = k[[t]]$.

6^0. Is the eigenring of a finitely generated non-zero right ideal in $k\langle X\rangle$ necessarily commutative?

7^0. Is the endomorphism ring of a distributive module of finite length right invariant?

8*. (Bergman) In the ring of integral quaternions, show that the eigenring of each atom is commutative, but that this need no longer hold for general (non-zero) elements.

Notes and comments on Chapter 4

Most of the results in this chapter are due to Bergman and the author and have not been published before. In particular, Theorem 2.4 was obtained by the author in 1964 (unpublished) and he conjectured that it applied to free algebras. This conjecture was proved by Bergman in 1966. Much of this chapter is contained in Bergman [67], especially the later version, and 4.4–5 follow this source rather closely. The material of 4.1 went through several versions and was improved as a result of discussions with G. M. Bergman and W. Stephenson; Proposition 1.1 is due to Roos [67]. Theorem 2.5 generalizes (and simplifies) a result in Noether–Schmeidler [20]; Proposition 6.2 was proved for commutative principal ideal domains by H. Bass (unpublished).

5. Linear algebra over firs and semifirs

The main topic of this chapter is the study of a class of modules over firs which forms a natural generalization of torsion modules over principal ideal domains (and to which they reduce in the commutative case). The modules are discussed in **5.**3–4 and the corresponding matrices in **5.**5–6. These 'torsion' modules also obey a duality, valid in a rather more general context (**5.**1–2). The remainder of the chapter deals with the intersection theorem for firs (**5.**7) and the interrelation of various chain conditions in **5.**8.

5.1 Bound and unbound modules

Let R be any ring and $\mathcal{T}$, $\mathcal{F}$ two classes of right R-modules such that

(i) $X \in \mathcal{T}$ *if and only if* $\mathrm{Hom}\,(X, Y) = 0$ *for all* $Y \in \mathcal{F}$,

(ii) $Y \in \mathcal{F}$ *if and only if* $\mathrm{Hom}\,(X, Y) = 0$ *for all* $X \in \mathcal{T}$.

If we regard Hom as a bifunctor (i.e. functor in two arguments) on the category of modules, (i) and (ii) express the fact that $\mathcal{T}$ and $\mathcal{F}$ are the annihilators of each other, and we shall sometimes write $\mathcal{T} = \mathrm{Ann}\,\mathcal{F}$ $\mathcal{F} = \mathrm{Ann}\,\mathcal{T}$. There is a certain parallel here with the concept of orthogonality in a metric linear space, but by contrast to that case, Hom is *not* symmetric in its two arguments. Any $\mathcal{T}$ and $\mathcal{F}$ satisfying (i) and (ii) are called a *torsion class* and its associated *torsion free class* respectively. Given any class $\mathcal{C}$ of right R-modules, we can obtain a torsion class containing $\mathcal{C}$, and its associated torsion free class, by setting $\mathcal{F} = \mathrm{Ann}\,\mathcal{C}$, $\mathcal{T} = \mathrm{Ann}\,\mathcal{F}$. Similarly one can form the least torsion free class containing $\mathcal{C}$. We shall be particularly interested in the least torsion free class containing R and its torsion class.

Definition. An R-module M is said to be *bound*, if

$$M^* = \mathrm{Hom}_R\,(M, R) = 0.$$

Clearly bound modules form a torsion class in the above sense. As an example consider a bound module M over a right fir R. If $f: M \to R$ is a non-zero homomorphism, the image, being free, must split M; thus M is bound if and only if it does not have a direct summand which is free of

175

positive rank, which is the case precisely when R is not a direct summand of M. A corresponding criterion holds for semifirs and n-firs, if we restrict M suitably. Thus we have

PROPOSITION 1.1. *A module over a right fir (or a finitely generated module over a semifir, or an n-generator module over an n-fir) R is bound if and only if it does not contain R as a direct summand.* ∎

Every torsion class is closed under certain operations; for the class of bound modules (over any ring) this is easily verified directly:

PROPOSITION 1.2. *Over any ring R, the class of bound modules is closed under the formation of homomorphic images, module extensions, direct limits and in particular, direct sums.* ∎

The modules in the corresponding torsion free class are called *unbound*. An unbound module can also be defined as a module with no non-zero bound submodules. For if N satisfies this condition and M is any bound module, then so is any homomorphic image of M, hence $\mathrm{Hom}\,(M, N) = 0$, and so N is unbound. Conversely, if N has a bound submodule $N' \neq 0$ then $\mathrm{Hom}\,(N', N) \neq 0$.

The above description for unbound modules can be rephrased thus: N is unbound precisely when every non-zero submodule N_1 has a non-zero linear functional: $N_1{}^* \neq 0$. Sometimes we wish to consider a wider class obtained as follows. Given $n \geqslant 1$, if every non-zero n-generator submodule of a module N has a non-zero linear functional, N is said to be *n-unbound*.

Clearly R is unbound, for any ring R, and we have the following analogue of Proposition 1.2.

PROPOSITION 1.3. *Let R be a ring and n a positive integer. Then the class $\mathscr{C}_n$ of n-unbound right R-modules contains all free modules and is closed under the formation of submodules and arbitrary direct products (and hence under arbitrary inverse limits and direct sums), and under module extensions. Likewise for the class of unbound modules.* ∎

Let M be any R-module; by Proposition 1.2, M has a unique maximal bound submodule M_b (viz. the sum of all bound submodules of M) and M/M_b has no non-zero bound submodules, i.e. is unbound. Dually, M_b may also be characterized as the least submodule of M with unbound quotient.

Over an integral domain the 1-unbound modules are just the torsion-free modules. This follows from the next result, which describes n-unbound modules over n-firs.

THEOREM 1.4. *Let R be an n-fir. Then an R-module M is n-unbound if and only if every n-generator submodule of M is free.*

Proof. Let M be n-unbound, $n > 0$, and assume the result for integers less than n. Any submodule N of M generated by n elements is a homomorphic image of ${}^nR$, say $N = f({}^nR)$. Let $g: N \to R$ be a non-zero linear functional, then $gf: {}^nR \to R$ is non-zero and by Theorem 1.1.1(f), im gf is free of some rank $r > 0$; thus im $gf = {}^rR$ and applying an appropriate automorphism of ${}^nR$ we find maps

$$ {}^rR \oplus {}^{n-r}R \to N \to {}^rR $$

such that the first map is still surjective, while the composition is the projection onto ${}^rR$. This induces the desired decomposition $N = {}^rR \oplus f({}^{n-r}R)$. By hypothesis $f({}^{n-r}R)$ is free and hence so is N. The converse is clear by Proposition 1.3. ∎

COROLLARY. *Let R be a ring, n a positive integer and M a non-zero R-module. Then the following are equivalent:*

(i) *every n-generator submodule of M is free of unique rank,*

(ii) *R is an n-fir and M is n-unbound.*

Proof. (ii) implies (i) by Theorem 1.4. To show that (i) implies (ii) we note that since $M \neq 0$, by (i) it contains a submodule isomorphic to R, and so R is an n-fir by Theorem 1.1.1(d). Now the rest follows by Theorem 1.4. ∎

Of particular interest for our purpose is the fact that by Proposition 1.3 all direct powers R^I are n-unbound, so that by Theorem 1.4, when R is an n-fir, any n-generator submodule of R^I is free.

Bound modules satisfy chain conditions under fairly mild restrictions. To state them let us recall that a *right hereditary* ring is a ring in which every right ideal is projective. Such rings are certainly weakly semihereditary, and so we can apply Theorem 0.2.9 which tells us that any projective module over a right hereditary ring is a direct sum of finitely generated modules. We apply this result to bound modules as follows:

THEOREM 1.5. *Let R be a right hereditary ring, and let M be a finitely related right R-module, i.e. a module with a presentation*

$$ {}^nR \to {}^IR \to M \to 0. $$

Then each submodule of M is a direct sum of a finitely generated module and a projective module.

Proof. By hypothesis, $M = F/N$, where F is free and N is finitely generated. Every submodule of M has the form P/N, where $N \subseteq P \subseteq F$ and here P is projective, because R is right hereditary. By Theorem 0.2.9, P is a direct sum of finitely generated modules, and so it contains a finitely generated direct summand P' containing N (because N was finitely generated). Write $P = P' \oplus P''$, then we have the exact sequence

$$0 \to P'/N \to P/N \to P/P' \cong P'' \to 0.$$

Since P'' is projective, the sequence splits and P'/N as image of P' is finitely generated. ∎

Since a bound module contains no non-zero projective submodule as a direct summand, we see that in the situation of Theorem 1.5 every bound submodule of M is finitely generated. Hence we have

COROLLARY 1. *A finitely related module over a right hereditary ring satisfies* ACC *on bound submodules.* ∎

Taking M itself to be bound, we obtain

COROLLARY 2. *Let M be a finitely related bound module over a right hereditary ring, then every bound submodule of M (in particular M itself) is finitely presented.* ∎

As a matter of fact it is easy to show that over any ring, a finitely related bound module is finitely presented.

Exercises 5.1

1. Show that over a right hereditary ring R the largest bound submodule M_b of a module M may also be defined as

$$M_b = \bigcap \{\ker h \mid h \in \text{Hom}\,(M, R)\}.$$

Deduce that M is unbound if and only if it can be embedded in a direct power R^I.

2. Show that a finitely related bound module over any ring is finitely presented.

3. Let R be a right hereditary ring and M a finitely related right R-module. Show that the projective submodules of M form an inductive system.

4. Let R be an integral domain and M an R-module. Show that a homomorphism $M \to R$ annihilates all torsion elements; deduce that any module consisting entirely of torsion elements is bound. Show that for finitely generated right modules over left Ore rings the converse holds.

5. Let R be a ring embeddable in a field K and assume that R contains two left R-linearly independent elements a, b. Verify that the submodule $a^{-1}R + b^{-1}R$

of K is bound, as right R-module. Deduce that K is not locally free as right R-module (*cf.* Ex. **0.6.3**).

6*. Find an example of an n-unbound module for all n, yet not unbound. (Hint: Over a semifir this requires a module M whose finitely generated submodules are free, but $M^* = 0$.)

7. Show that over a semifir, every finitely generated module is the direct sum of a free and a bound module.

8. Let R be an n-fir and $\mathscr{C}_n$ the class of all n-unbound R-modules with ACC_n. Show that $\mathscr{C}_n$ contains all free modules and is closed under (a) the formation of submodules, extensions and unions of ascending sequences with quotients in $\mathscr{C}_n$ (and hence direct sums), and (b) inverse limits (and hence direct products). Show that R must have right ACC_n for $\mathscr{C}_n$ to be non-trivial.

9. (Bergman [67]). (a) Let k be a field with an endomorphism α which is not surjective and define R' as the opposite skew polynomial ring in y over k, with the commutation rule $ya = a^\alpha y$; thus $R' = k^0[y; \alpha]^0$, where ^{0} denotes passage to the opposite ring. Show that $R'y$ is a two-sided ideal which is free as right ideal; if (u_i) is a right basis for k over k^α, show that $(u_i y)$ is a right basis for $R'y$ over R'.

(b) Let $R = k^\alpha + R'y$ be the subring of polynomials in y with constant term in k^α. Verify that R is a left Ore ring with the same field K of fractions as R', but R is not a right Ore ring (note that even R' is not right Ore).

(c) Take $x \in k$, $x \notin k^\alpha$ and let $f: {}^2R' \to R'$ be the linear functional

$$\begin{pmatrix} a \\ b \end{pmatrix} \mapsto yxa - yb.$$

Then f also defines a linear functional f_R (by restriction) on ${}^2R$. Show that $\ker f_R = (1, x) R'y = \Sigma(1, x) u_i y R'$; deduce that a submodule of an R-module of rank 1 can have infinite rank (for a suitable choice of k and α). Take a particular suffix $i = 1$ say and put $M = {}^2R/(1, x) u_1 y R'$. Verify that f_R extends to a mapping $f_M: M \to R$ and by showing that $\ker f_M$ is embeddable in ${}^2R$, show that M is unbound. Show also that $M \otimes K = {}^2K/(1, x) K$ and hence that M, although unbound, is not embeddable in $M \otimes K$.

10. Let R be a left Ore ring, M and N right R-modules and $f: N \to M$ a homomorphism. If M is finitely generated and $\operatorname{coker} f$ is bound show that $rk\ N \leqslant rk\ M$. Does this remain true without the condition that M be finitely generated? (Hint: Try $M = K$.)

5.2 Duality

We shall now establish a duality for bound modules over hereditary rings, from which it will follow that any finitely presented bound module satisfies DCC as well as ACC on bound submodules. We shall present the actual results so as to apply to arbitrary rings and then specialize at the end.

Let us call a module M over any ring R *special* if it is finitely presented and of projective dimension at most 1. Thus a special module M is a module with a presentation

$$0 \to Q \to P \to M \to 0.$$

where P, Q are finitely generated projective R-modules. If we dualize this exact sequence, we get the exact sequence of left R-modules

$$0 \to M^* \to P^* \to Q^* \to \text{Ext}\,(M, R) \to 0,$$

bearing in mind that $\text{Ext}\,(P, R) = 0$ because P is projective. Let us assume that M is bound, so that $M^* = 0$; Dualizing once more and writing $\hat{M} = \text{Ext}\,(M, R)$ for short, we obtain the following commutative diagram with exact rows:

$$0 \to Q \longrightarrow P \longrightarrow M \to 0$$
$$\qquad \downarrow\alpha \qquad \downarrow\beta \qquad \downarrow\gamma$$
$$0 \to (\hat{M})^* \to Q^{**} \to P^{**} \to \hat{\hat{M}} \to 0.$$

Here α and β are isomorphisms because P and Q are finitely generated projective. This allows us to define γ and to show that it is an isomorphism too (by the 5-lemma); likewise we conclude that $(\hat{M})^* = 0$. These remarks suggest the truth of

PROPOSITION 2.1. *In any ring R, there is a duality (i.e. anti-equivalence) "$\wedge$" between the categories of special bound right R-modules and special bound left R-modules, such that if $M = \text{coker}\,(Q \to P)$ where P, Q are finitely generated projective, then $\hat{M} = \text{coker}\,(P^* \to Q^*)$.*

Proof. Consider a mapping between two special bound right R-modules given by presentations of the above form

$$0 \longrightarrow Q \xrightarrow{i} P \xrightarrow{j} M \longrightarrow 0$$
$$\qquad \downarrow h \qquad \downarrow g \qquad \downarrow f$$
$$0 \longrightarrow Q' \xrightarrow{i'} P' \xrightarrow{j'} M' \longrightarrow 0$$

By projectivity of P, fj lifts to a map $g : P \to P'$ and $h = g|Q$ maps into Q'. The map f is completely determined by the pair (g, h), i.e. any such pair of maps giving a commutative diagram induces a map f of the cokernels. Moreover, two such pairs (g, h) and (g_1, h_1) give the same f if and only if we can find a map $e : P \to Q'$ such that $g_1 - g = i'e$, $h_1 - h = ei$ (i.e. (g, h) and (g_1, h_1) are 'homotopic'). Hence the category of special bound right R-modules is equivalent to the category of maps $Q \to P$, which are injective and have injective duals, under homotopy-equivalence, as defined above. *This* category is clearly dual to the corresponding category of projective left R-modules, and hence to the category of special bound left R-modules. ∎

We observe that the duality of Proposition 2.1 is given explicitly by the functor $\text{Ext}\,(-, R)$. In the special case $R = \mathbf{Z}$ it is just the familiar duality of abelian groups given by $A \mapsto \operatorname{Hom}(A, \mathbf{Q}/\mathbf{Z})$. We indicate briefly the conditions under which this simplification can be made.

A module M is said to be *strongly bound* if M and all its submodules are bound. For any ring R, let I be the injective hull of R, as right R-module, and put $K = \operatorname{coker}(R \to I)$, so that we have the exact sequence

$$0 \to R \to I \to K \to 0. \tag{1}$$

Clearly M is strongly bound if and only if $\operatorname{Hom}(M, I) = 0$. By (1) we have, since I is injective, the exact sequence

$$0 \to \operatorname{Hom}(M, R) \to \operatorname{Hom}(M, I) \to \operatorname{Hom}(M, K) \to \operatorname{Ext}(M, R) \to 0. \tag{2}$$

But $\operatorname{Hom}(M, I) = 0$ because M is strongly bound, and we have

PROPOSITION 2.2. *Let R be any ring and define I, K as above. Then for any strongly bound module M,*

$$\operatorname{Ext}(M, R) \cong \operatorname{Hom}(M, K). \ \blacksquare \tag{3}$$

Over a principal ideal domain (commutative or not) every bound module is strongly bound and we can therefore use Proposition 2.2 to express the duality in terms of Hom. Moreover, in this case I is just the field of fractions of R. This is an R-bimodule, hence K is also an R-bimodule and it follows that (for a right R-module M) (3) is actually a left R-module isomorphism. When R is a fir but not a principal ideal domain, there will generally be modules that are bound but not strongly bound. Even then we can use the exact sequence (2) to describe $\operatorname{Ext}(M, R)$ as the cokernel of the mapping $\operatorname{Hom}(M, I) \to \operatorname{Hom}(M, K)$. However, a closer analysis of I is necessary to reveal its R-bimodule structure. This is always possible but it is not clear that this would throw further light on $\operatorname{Ext}(M, R)$.

Returning to Theorem 2.1, let us apply the result to hereditary rings. In the first place, every finitely presented module is now special. Moreover, by Theorem 1.5, Corollary 2, every finitely presented bound module satisfies ACC on bound modules (necessarily finitely presented by Corollary 2) and applying the above duality, we find that the module satisfies DCC for bound submodules. Thus we obtain

THEOREM 2.3. *A finitely presented bound module over a (left and right) hereditary ring satisfies both chain conditions for bound submodules.* ■

Exercises 5.2

1. What becomes of the duality of Proposition 2.1 in the case where R is right self-injective (i.e. injective as right R-module)?

2. For this exercise only, let us call a module M *extra-special* if it has a presentation $0 \to P \to P \to M \to 0$, where P is finitely generated projective. Show that in the duality of Proposition 2.1, extra-special bound modules correspond to extra-special bound modules.

3. Let R be a right fir. If every bound right R-module is strongly bound, show that R is a principal right ideal domain.

4. Let R be a right Ore ring and K its field of fractions. Show that K is the injective hull of R. Does this remain true for more general rings that are embeddable in fields?

5*. Let R be a two-sided fir and I its injective hull as right R-module. Describe the R-bimodule structure of I and compare it with J, the injective hull of R as left R-module.

6. Let R be right hereditary and M a finitely related right R-module. If P is a maximal projective submodule of M, show that M/P is strongly bound. Deduce that for every finitely related module M there is an exact sequence

$$0 \to P \to M \to Q \to 0,$$

where P is projective and Q is finitely generated and strongly bound.

7*. Give an example of a strongly bound module over $k\langle x, y\rangle$ which is not finite dimensional over k.

8. Let R be an integral domain and $\mathfrak{a}$ a right ideal. Show that $R/\mathfrak{a}$ is strongly bound if and only if R is an essential extension of $\mathfrak{a}$ (i.e. every non-zero right ideal meets $\mathfrak{a}$ non-trivially).

9. Let R be a two-sided fir and I its injective hull as right R-module. Then for any finitely related right R-module M, show that there exists a set X such that $\operatorname{Hom}(M, I) \cong I^X$.
Let $R = k\langle x_1, x_2, \ldots\rangle$, $M = R/x_1 R$; show that $P = \Sigma x_i R/x_1 R$ is a maximal projective module. Find X in this case.

5.3 Torsion modules over firs and semifirs

In this section we shall use the results of 5.1–2 to study a class of modules over firs, which for principal ideal domains reduce to torsion modules; many properties of the latter will be found to carry over to the more general case. In any question where finiteness is not needed (or assumed separately), the results extend to semifirs, and we shall therefore state them for that case, when possible.

Let R be a semifir and M a right R-module. If M has a finite presentation

$$0 \to {}^m R \to {}^n R \to M \to 0, \tag{1}$$

then the *characteristic* $\chi(M)$ of M is defined by

$$\chi(M) = n - m. \tag{2}$$

This is independent of the choice of the presentation (1) by Schanuel's lemma (Appendix 2).

If M is finitely generated but not finitely related, we put $\chi(M) = -\infty$; if M is not finitely generated, we put $\chi(M) = \infty$. For a free module the characteristic is of course just the rank. Sometimes we may wish to extend the definition of the characteristic to matrix rings over semifirs. If $R = T_k$ say, where T is a semifir, and P is the minimal projective, we again have a presentation (1), with R replaced by P. The characteristic is defined correspondingly, taking $\chi(P) = 1/k$, so as to ensure that $\chi(R) = 1$, as before. Thus in this case the characteristic will be $+\infty$ or a rational number with denominator dividing k. Many of our results will have analogues for the matrix case, but these will usually be fairly obvious extensions, and their formulation will be left to the reader.

The characteristic of a module is related to that of a submodule by the following formula.

PROPOSITION 3.1. *For any short exact sequence of modules over a semifir* R,

$$0 \to M' \to M \to M'' \to 0,$$

we have

$$\chi(M) = \chi(M') + \chi(M''), \tag{3}$$

with the usual conventions about ∞, *together with the rules*:

(a) *if* $\chi(M'') = \infty$, *then* $\chi(M) = \infty$,

(b) *if* $\chi(M'') = -\infty$ *and* $\chi(M') = \infty$, *there is no conclusion*.

Proof. Let $M = F/H$, where F is free; then M'' can be written as homomorphic image of M, say $M'' = F/K$ and we obtain the following commutative diagram with exact rows and columns:

$$
\begin{array}{ccccccccc}
& & 0 & \to & 0 & & & & \\
& & \downarrow & & \downarrow & & & & \\
0 & \to & H & \to & H & \to & 0 & & \\
& & \downarrow & & \downarrow & & \downarrow & & \\
0 & \to & K & \to & F & \to & M'' & \to & 0 \\
& & \downarrow & & \downarrow & & \downarrow & & \\
0 & \to & M' & \to & M & \to & M'' & \to & 0 \\
& & \downarrow & & \downarrow & & \downarrow & & \\
& & 0 & & 0 & & 0 & &
\end{array}
$$

The rows are exact by construction, as are the second and third columns; hence the first column can be filled in so as to keep the diagram commutative, and the first column is then exact too, by the 3×3 lemma (Appendix 2).

If $\chi(M)$ and $\chi(M'')$ are finite, then F, H, K can all be taken to be finitely generated, and then (3) follows easily, using the first column for a presentation of M'. If $\chi(M'') = \infty$, then M'' is not finitely generated, so neither is M, whence (a) follows. Thus we may assume $\chi(M'') < \infty$. Further, if $\chi(M'') = -\infty$, then we need only look at the case $\chi(M') < \infty$, by (b). But then M', M'' are both finitely generated, hence so is M, therefore F has finite rank, while K is not finitely generated. Now the presentation for M' shows that H is not finitely generated, whence $\chi(M) = -\infty$. There remains the case where $\chi(M'')$ is finite but $\chi(M)$ is not. If $\chi(M) = \infty$, then M' cannot be finitely generated, so $\chi(M') = \infty$; if $\chi(M) = -\infty$, we can take F, K to be of finite rank, but H is not finitely generated, hence $\chi(M') = -\infty$. ∎

We note particularly that the equation (3) is meaningful whenever M' and M'' are finitely generated.

If in Proposition 3.1, M is finitely generated, by n elements say, then $\chi(M'')$ is finite precisely if K is finitely generated, and in that case, $\chi(M'') = n - rk(K)$; hence we obtain the

COROLLARY. *Let M be any module on n generators over a semifir. Then a submodule M' can be generated by $n - \chi(M/M')$ elements.* ∎

Consider for a moment the special case where R is a principal ideal domain. In that case every submodule of an n-generator module can be generated by n elements, so every module has non-negative characteristic, and $\chi(M) = 0$ if and only if M is a finitely generated torsion module.

Over a fir, the characteristic may assume any integer value, positive, negative or zero. E.g. if $R = k\langle x, y\rangle$ is the free algebra and $M = R/(xR + yR)$, then $\chi(M) = -1$, and it is not hard to find cyclic modules of any negative characteristic. Guided by these examples, we make the following

Definition. Let R be a semifir; an R-module is called a *torsion module* if

(i) $\chi(M) = 0$ and

(ii) for any submodule M' of M, $\chi(M') \geqslant 0$.

From Proposition 3.1 we see that (in the presence of (i)) condition (ii) of this definition is equivalent to

(ii)' for every quotient M'' of M, $\chi(M'') \leqslant 0$.

Every torsion module is finitely generated (by (i)), and in (ii) it is enough to test for finitely generated submodules of M. We also note that a cyclic torsion module is just a strictly cyclic module, as defined in **3.2**.

PROPOSITION 3.2. *In any short exact sequence of modules over a semifir,*

$$0 \to M' \to M \to M'' \to 0,$$

if two terms are torsion modules, then so is the third.

Proof. By hypothesis, two of the numbers in (3) are zero, hence cases (a) and (b) of Proposition 3.1 are ruled out and so (3) holds, i.e. the third term is also zero. It remains to show that submodules have non-negative characteristic.

(α) M' and M'' are torsion modules. Let N be a submodule of M, then

$$\chi(N) = \chi(N \cap M') + \chi(N/(N \cap M')). \tag{4}$$

Here the first term on the right is non-negative because $N \cap M'$ is a submodule of M' and likewise for the second term, because

$$N/(N \cap M') \cong (N + M')/M',$$

where the right-hand side is a submodule of $M/M' \cong M''$.

(β) M is a torsion module. Any submodule N of M' is also a submodule of M, hence $\chi(N) \geqslant 0$, thus M' satisfies (ii), and any quotient module Q of M'' is also a quotient of M and so $\chi(Q) \leqslant 0$, i.e. M'' satisfies (ii)'. ∎

We denote by $\mathcal{T}_R$ the category consisting of all right torsion modules over a semifir R, and all homomorphisms between them. Thus $\mathcal{T}_R$ is a full subcategory of $\mathcal{M}_R$, the category of all right R-modules and homomorphisms. Similarly the category of left torsion modules is written $_R\mathcal{T}$ By Proposition 3.2, the direct sum of two torsion modules is again a torsion module, so we have an additive category (Appendix 2). Moreover, in any homomorphism between torsion modules,

$$f: M \to N$$

ker f and coker f are torsion. To show this we observe that im $f \cong$ coim f; since N is torsion, $\chi(\text{im } f) \geqslant 0$, and since M is torsion, $\chi(\text{coim } f) \leqslant 0$, hence $\chi(\text{im } f) = 0$. Moreover, any submodule of im f is a submodule of N and so has non-negative characteristic. Hence im f is a torsion module; now it follows by Proposition 3.2 that coker $f = N/\text{im } f$ is torsion, and likewise for ker f. The fact that $\mathcal{T}_R$ is a full subcategory of $\mathcal{M}_R$ with finite direct sums, kernels and cokernels may be summed up thus:

THEOREM 3.3. *Let R be any semifir, then the category $\mathcal{T}_R$ of torsion modules over R is an abelian category.* ∎

It is clear that any torsion module is bound, in the sense of 5.1. For a non-zero homomorphism $M \to R$ would have as image a finitely generated free module, and this has positive characteristic. Moreover, a fir is a special case of a hereditary ring, hence by Theorem 2.3 we obtain

THEOREM 3.4. *A torsion module over a fir satisfies both chain conditions on torsion submodules.* ∎

This theorem is also expressed by saying that the category $\mathcal{T}_R$ is Noetherian and Artinian. We note explicitly that the functor

$$M \mapsto \mathrm{Ext}\,(M, R)$$

provides a duality between the categories $\mathcal{T}_R$ and $_R\mathcal{T}$.

Theorems 3.3-4 allow us to apply any result known for abelian categories to torsion modules. For example, by the Jordan–Hölder theorem, in a torsion module over a fir, any chain of torsion submodules can be refined to a finite chain of maximal length, i.e. a *composition series*, and any two such composition series have the same length, and isomorphic factors. This result may be interpreted in terms of factorizations of matrices over firs, and will be discussed in this context in 5.6 below.

We note again the analogue of Schur's lemma, proved as in the usual case:

PROPOSITION 3.5. *Let M be a simple object in $\mathcal{T}_R$ (i.e. a non-zero torsion module without non-zero torsion submodules), where R is a semifir. Then $\mathrm{End}_R\,(M)$ is a field.* ∎

So far we have confined ourselves to finitely generated modules, as that is the most interesting case (for us). But it is also possible to extend the notion of torsion module. For any semifir R we define the category $\mathcal{T}_R^\uparrow$ of *general torsion modules* as consisting of those modules M in which every finite subset is contained in a finitely generated torsion submodule. Then $\mathcal{T}_R^\uparrow$ (as full sub-category of $\mathcal{M}_R$) is again an abelian category; moreover it has exact direct limits and a generator, i.e. it is a *Grothendieck category* (*cf.* Cohn [70']), and may be obtained as the completion of $\mathcal{T}_R$. Dually one defines the category $\mathcal{T}_R^\downarrow$ of *protorsion modules* to consist of all inverse limits of finitely generated torsion modules and all continuous homomorphisms (relative to the natural topology on the inverse limit). Now the functor $M \mapsto \mathrm{Ext}\,(M, R)$ establishes a duality between the categories $\mathcal{T}_R^\uparrow$ and $_R\mathcal{T}^\downarrow$ (*cf.* Cohn [70']).

Exercises 5.3

1. Let R be a semifir and $n \geqslant 1$. How are the characteristics of modules related which correspond under the category equivalence of $\mathcal{M}_R$ and $\mathcal{M}_{R_n}$?

2. Over a semifir, show that every torsion submodule of an n-generator torsion module can be generated by n elements. Deduce that a torsion module over a fir satisfies ACC for torsion submodules, using Theorem 1.2.3.

3. Let R be a fir and $c \in R^*$, then a non-unit right factor b of c is called *inessential* if c has a non-unit right factor right coprime to b, otherwise *essential*. Show that the cyclic torsion modules defined by two elements c, c' have isomorphic injective hulls if and only if c and c' have an essential right factor in common.

4. Let R be a fir and $c \in R^*$. If the elements with c as essential right factor are pairwise left commensurable, show that the injective hull of R/cR can be expressed as a direct limit of cyclic modules. Give an example in $R = k\langle x, y \rangle$ where this fails.

5. Let $R = k\langle x, y, z \rangle$ and consider the torsion modules $M = R/xyR$, $N = R/xzR$. Verify that both have $P = R/xR$ as quotient. Let T be the pullback; show that T is not a torsion module, but that it has a unique torsion submodule. Show that the lattice of torsion submodules of T is finite and isomorphic to that of $R/xy(zy + 1) R$.

$$\begin{array}{ccc} T & \longrightarrow & M \\ \downarrow & & \downarrow \\ N & \longrightarrow & P \end{array}$$

6*. Let R be a semifir, $a, b \in R^*$ and $A = R/aR$, $B = R/bR$. Show that $\mathrm{Ext}^1(B, A) \cong A \otimes B^* \cong R/(Rb + aR)$. Deduce that every extension of A by B splits if and only if $Rb + aR = R$.
If $R = k\langle x, y \rangle$, take $a = x$, $b = y$, and find the non-split extensions apart from the strictly cyclic one.

7*. Let R be a semifir, $a, c \in R^*$, $A = A/aR$, $C = R/Rc$. Show that $\mathrm{Tor}_1(A, C) \cong (aR \cap Rc)/aRc$.

8⁰. Classify the bound modules of characteristic 0 over a semifir.

9⁰. (Bergman [67]). A module is said to be *polycyclic*, if it has a finite chain of submodules with cyclic quotients. Does there exist a semifir, not left or right Bezout, over which every torsion module is polycyclic?

10⁰. (Bergman [67]). Determine the class of semifirs over which every polycyclic torsion module can be written as a direct sum of cyclic torsion modules (observe that this includes all principal ideal domains).

5.4 Torsion modules over n-firs

It is clear that much of **5.3** can be generalized to n-firs. Thus, just as strictly cyclic modules (i.e. 1-generator torsion modules) may be studied over 2-firs, so n-generator torsion modules can be studied over $2n$-firs. In order to obtain finiteness conditions we now have to assume ACC_n explicitly, and the category of torsion modules over a $2n$-fir need no longer be abelian, but we shall see below the modified assertion which takes its place. Although most of this could be proved as in **5.3** (or even be treated together), we shall use a slightly different approach here.

PROPOSITION 4.1. *Let R be a $2n$-fir and P a free R-module of rank n with a free submodule Q of rank n, such that no module between Q and P can be generated by less than n elements (equivalently: is free of rank less than n).*

Then the submodules of P containing Q and of rank exactly n form a sublattice of the lattice of all submodules between P and Q.

Proof. Given modules A and B between P and Q, each free of rank n, the exact sequence of Proposition **1.1.2** shows that $rk(A + B) + rk(A \cap B) = 2n$; now the hypothesis ensures that $rk(A + B) = rk(A \cap B) = n$. ∎

We shall call a module M over a ring R an *n-generator torsion module*, if R is a $2n$-fir and M has a presentation

$$0 \to Q \xrightarrow{f} P \xrightarrow{g} M \to 0, \tag{1}$$

where $P \cong Q \cong {}^n R$, such that P has no submodule containing $f(Q)$, of rank less than n. Clearly this reduces to the definition of a torsion module (on n generators) in the sense of **5.3**, when R is a semifir. We shall call the presentation (1) of M a *torsion presentation*, and shall always take f to be an inclusion unless the contrary is stated. For brevity we usually omit the zeros at the ends when writing (1). Note that an n-generator torsion module is also an n'-generator torsion module for all $n' > n$ such that R is still a $2n'$-fir: we need only add ${}^{n'-n} R$ to P and Q in (1). The module M will be called a *torsion module* if it is an n-generator torsion module for some n.

As in Proposition 3.2 we see that in a short exact sequence

$$0 \to M' \to M \to M'' \to 0,$$

if two terms are n-generator torsion modules and the third can be generated by n elements, then it is also an n-generator torsion module. More precisely, if M and one of M', M'' are n-generator torsion modules, then so is the remaining one. This follows because M'' as homomorphic image of M, can be generated by n elements, while M' can be generated by $n - \chi(M'') = n$ elements, by Proposition 3.1, Corollary.

In **5.3** we relied on Schanuel's lemma to show that the characteristic of a module is independent of the presentation. The next result gives an alternative proof of this fact for n-generator torsion modules:

PROPOSITION 4.2. *Let R be a 2n-fir, M a torsion module over R and $g: F \to M$ a surjection from a free module F of rank n to M; then*

$$\ker g \to F \to M$$

is a torsion presentation of M.

Proof. Let $Q \to P \to M$ be a torsion presentation of M, with P and Q of rank m. Then R is a $2m$-fir as well as a $2n$-fir, and hence an $(m + n)$-fir. We can find maps h, k to give a commutative diagram with exact rows

$$
\begin{array}{ccccccccc}
0 & \longrightarrow & \ker g & \longrightarrow & F & \overset{g}{\longrightarrow} & M & \longrightarrow & 0 \\
 & & \downarrow{k} & & \downarrow{h} & & \downarrow{1} & & \\
0 & \longrightarrow & Q & \longrightarrow & P & \longrightarrow & M & \longrightarrow & 0
\end{array}
$$

By Theorem 1.1.1(f), $\ker h$ is a direct summand in F, whose complement is free of rank at most n. It is clearly contained in $\ker g$, and hence a direct summand thereof; splitting off this summand, we see that it suffices to prove our contention for the presentation of M given by the complements. Thus we may assume that h is injective; for convenience we take h to be an inclusion.

Since g is surjective, $P = F + Q$. Hence $\ker g = F \cap Q$ is free of rank $rk\, F + rk\, Q - rk\, P = n$, by the dimension formula (Proposition 1.1.2). Now suppose we had $\ker g \subseteq U \subseteq F$, where U is free of rank less than n. Since $U \cap Q = \ker g$, another application of the dimension formula gives $rk(U + Q) = rk\, U + rk\, Q - rk\, \ker g = rk\, U + m - n < m$, which contradicts the fact that no module between Q and P has rank less than m. ∎

COROLLARY 1. *If an n-generator torsion module can be generated by m elements, where $m < n$, then it is an m-generator torsion module.* ∎

If $Q \to P \to M$ is an n-generator torsion presentation, then the submodules of M that are torsion modules correspond to the submodules of rank n between P and Q; hence, applying Proposition 5.1, we obtain

COROLLARY 2. *The torsion submodules of a torsion module M form a sublattice of the lattice* Lat (M) *of all submodules of M.* ∎

PROPOSITION 4.3. *The image, kernel and cokernel of a homomorphism between torsion modules over a ring R are again torsion modules.*

Proof. Given a homomorphism $f: M \to N$ and torsion presentations of M and $N: Q \to P \to M$, $Q' \to P' \to N$, which may both be taken to be n-generator presentations, we can again lift f to a map of the presentations:

$$
\begin{array}{ccccccc}
0 \to & Q & \to & P & \to & M & \to 0 \\
 & \downarrow{h} & & \downarrow{g} & & \downarrow{f} & \\
0 \to & Q' & \to & P' & \to & N & \to 0
\end{array}
$$

Consider the map $\alpha: Q' + P \to Q' + g(P)$ defined by $(x, y) \mapsto x + g(y)$; its kernel is $g^{-1}(Q') = \{y \in P | g(y) \in Q'\}$. So we have the exact sequence

$$0 \to g^{-1}(Q') \to Q' + P \to Q' + g(P) \to 0,$$

and since $Q' + g(P)$ is a $2n$-generator submodule of P', it is free and the sequence splits, whence

$$rk[g^{-1}(Q')] + rk[Q' + g(P)] = rk\, Q' + rk\, P = 2n. \tag{2}$$

Now both $S = g^{-1}(Q')$ and $S' = Q' + g(P)$ lie between the terms of an n-generator torsion presentation and hence have rank at least n; by (2), $rk\, S = rk\, S' = n$. Now it follows that $Q \to S \to \ker f$, $S \to P \to f(M)$, $Q' \to S' \to f(M)$, $S' \to P' \to \operatorname{coker} f$ are all torsion presentations. ∎

Over a semifir, this proposition shows again that torsion modules form an abelian category, but over a $2n$-fir, all we get is what might be called a "truncated filtered abelian category"; it is filtered by the minimal-number-of-generators function, which is subadditive for extensions, and non-increasing for subobjects and quotients (since these can be written as kernels and cokernels); but we are only assured of the existence of the direct sum of two objects in the category when their number of generators does not exceed n. Of course the minimal number of generators in a direct sum may well be less than the sum of the numbers for the summands: consider $\mathbf{Z}/2\mathbf{Z} \oplus \mathbf{Z}/3\mathbf{Z}$, or more generally, $R/aR \oplus R/bR$, where a, b are left comaximal elements of a 2-fir. To obtain a criterion in the general case, we observe that any n-generator torsion module M is given as the cokernel of a certain mapping ${}^nR \to {}^nR$, and this mapping is completely specified by an $n \times n$ matrix over R; thus we may speak of the matrix defining M. Of course the matrix will not be unique, but it determines the module up to isomorphism (cf. **5.6**). The next result generalizes Lemma **4.2.1**.

PROPOSITION 4.4. *Let M, N be any two n-generator torsion modules; then the following assertions are equivalent:*

(a) *$M \oplus N$ can be generated by n elements,*

(b) *$M \oplus N$ is an n-generator torsion module,*

(c) *the $n \times n$ matrices defining M and N can be chosen to be left comaximal in R_n.*

Proof. (a) $\Rightarrow$ (b). Clearly $M \oplus N$ is a torsion module, which is n-generated by hypothesis, so the assertion follows by Proposition 4.1, Corollary 1.

(b) $\Rightarrow$ (c). By passing to the matrix ring $T = R_n$ we may assume M, N cyclic. Let $M \oplus N = T/\mu T$ and suppose that M, N correspond to $\beta T/\mu T$, $\alpha T/\mu T$ respectively in the isomorphism, where $\alpha T \cap \beta T = \mu T$. Then

$$\alpha\beta' = \beta\alpha' = \mu, \tag{3}$$

where α', β' are right coprime. Hence the exact sequence

$$0 \to T\alpha' \cap T\beta' = T\mu \to T\alpha' \oplus T\beta' \to T\alpha' + T\beta' \to 0$$

shows that $rk(T\alpha' + T\beta') = 1$. Thus $T\alpha' + T\beta' = T\delta$, but since α' and β' are right coprime, δ is a unit, so α' and β' are left comaximal and

$$M = \beta T/\mu T = T/\alpha' \, T, \qquad N = \alpha T/\mu T = T/\beta' \, T. \tag{4}$$

(c) $\Rightarrow$ (a). When M, N are defined by α', β' respectively, which are left comaximal, then a comaximal relation (3) holds, and hence

$$T/\mu T = (\alpha T + \beta T)/\mu T = \alpha T/\mu T \oplus \beta T/\mu T$$
$$= T/\beta' \, T \oplus T/\alpha' \, T,$$

where the sum is direct, since $\alpha T \cap \beta T = \mu T$. ∎

For a $2n$-fir we thus have the category $\mathscr{T}_R{}^n$ of n-generator torsion modules, and it is clear from Proposition 2.1 that Ext $(-, R)$ provides a duality between $\mathscr{T}_R{}^n$ and $_R\mathscr{T}^n$, the corresponding category of left modules.

If we now specialize by imposing the right Ore condition, we get

PROPOSITION 4.5. *The torsion modules over a right Bezout domain are precisely the finitely presented modules consisting of torsion elements. Moreover, in an exact sequence,*

$$0 \to M' \to M \to M'' \to 0, \tag{5}$$

if M is a torsion module and M' is finitely generated, then M', M'' are torsion modules.

Proof. A torsion module is certainly finitely presented and has rank 0, whence all its elements are torsion elements. Conversely, if M is finitely presented and consists of torsion elements, its rank is 0. Moreover, for a Bezout domain the rank is always non-negative, by Proposition 0.6.3. This remark also shows that $rk \, M' = rk \, M'' = 0$ in (5), and when M' is finitely generated, both M' and M'' are finitely presented. ∎

COROLLARY. *Any n-generator torsion module over a left or right Bezout domain R has a chain of torsion submodules of length n, whose quotients are cyclic torsion modules.*

For right Bezout domains this is immediate, from the preceding result, and for left Bezout domains it follows by duality (Proposition 2.1). ∎

Exercises 5.4

1. Examine which results of 5.4 remain valid if we allow n-firs in defining n-generator torsion modules.

2. Use Theorem 3.3 to prove Proposition 4.2, Corollary 2 in the case of semifirs. (Hint: If M_1, M_2 are submodules of M, express $M_1 \cap M_2$ as the kernel of a map from M to another torsion module.)

3. Use Proposition 4.2, Corollary 2 to show that the factorizations of two similar elements correspond to each other bijectively.

4. Show that the extension of one torsion R-module by another is again a torsion module provided that it can be generated by n elements and R is a $2n$-fir.

5. Show that the category of special bound modules is not closed under kernels.

6. Let R be a 2-fir, but not a right Ore ring. Show that an element of a cyclic torsion module need not generate a torsion module.

Given an example of an n-generator torsion module with n elements which do not generate a torsion submodule.

7^0. (Bergman [67]). Develop a theory of "truncated filtered abelian categories". In particular, show that every such category gives rise to a universal filtered abelian category. Is it possible (and practicable) to use such a theory to derive the results of 5.4 from those of 5.3?

5.5 The ranks of matrices

The rank of a matrix is a numerical invariant which is certainly familiar to the reader, at least for matrices over a field or a commutative ring. We shall need an analogue for matrices over more general rings; there is then more than one invariant that can lay a claim to generalize the usual rank. This is not surprising since even in the commutative case one defines row and column rank separately and *then* proves them equal. We shall introduce three different notions of rank; we describe their properties and their relation to each other, this will make it clear which to use in any particular situation. For a smoother development we begin with some remarks on the decompositions of free modules over n-firs.

Let M be a module over any ring and N a submodule, then we define the *closure* $\bar{N}$ of N in M as the intersection of the kernels of all linear functionals on M which vanish on N. If $\bar{N} = M$, we say N is *dense* in M; this is the case precisely when M/N is bound in the sense of 5.1. If $\bar{N} = N$, we say that N is *closed*; this is so if and only if N is the kernel of a mapping into a direct power of the ring. Hence N is closed in M if and only if M/N is unbound.

Let R be an n-fir, then the submodules of ^nR which are annihilators of finite families of linear functionals are direct summands of ^nR, by Theorem 1.1.1(f). Moreover, any direct sum decomposition of ^nR has up to isomorphism the form $^nR = {}^rR \oplus {}^sR$ for unique r and s satisfying $r + s = n$. It follows (by an easy induction argument) that any descending chain of direct summands is finite, of length at most n. But any closed submodule of ^nR is the intersection of a descending chain of direct summands and so is itself a direct summand. This proves

PROPOSITION 5.1. *For any n-fir R, the closed submodules of ^nR are precisely the direct summands. They form a lattice in which every maximal chain has length n, and the height of any member of this lattice is its rank as a free module. Moreover, if M is any closed submodule of ^nR, then the linear functionals vanishing on M form a free module of rank $n - \mathrm{rk}\, M$.* ∎

We give a couple of examples to show that the relation between a module and its closure is not very close in general (for other examples, see the exercises).

Let R be any integral domain; if $x_1, ..., x_n$ are *right* linearly independent elements of R, then the elements $(x_1, 0, ..., 0)^T, ..., (x_n, 0, ..., 0)^T$ of ^mR generate a free submodule of rank n, but its closure $(R, 0, ..., 0)^T$ has rank 1. Secondly, let $y_1, ..., y_n$ be *left* linearly independent, then it is clear that no non-zero linear functional of ^nR vanishes on $(y_1, ..., y_n)^T$. So the submodule generated by this element is dense in ^nR. Here we have a submodule of rank 1, whose closure has rank n.

Let us now turn to matrices. An $m \times n$ matrix $\alpha = (\alpha_{ij})$ over a ring R may be interpreted in different ways, either as

(a) a right R-module homomorphism of columns $^nR \to {}^mR$, or

(b) a left R-module homomorphism of rows $R^m \to R^n$, or

(c) an element of the additive group $^mR \otimes R^n$, or

(d) a bilinear mapping $R^m \times {}^nR \to R$.

For our first result we shall specialize the ring.

PROPOSITION 5.2. *Let α be an $m \times n$ matrix over an n-fir R. Then the following numbers are equal and do not exceed n:*

   (i) *the rank of the submodule of ^mR spanned by the n columns of α, i.e. $\mathrm{rk}\,\mathrm{im}\,\alpha$ under interpretation* (a),

   (ii) $n - \mathrm{rk}\,\mathrm{ker}\,\alpha$ *under interpretation* (a),

   (iii) *the rank of the closed submodule generated by the rows of α under interpretation* (b).

Proof. The equivalence of (i) and (ii) is clear, as well as the fact that the common value cannot exceed n. To prove the equivalence of (i) and (iii) suppose that the closure of the image of the mapping $\alpha: R^m \to R^n$ has rank

r, the number described in (iii). Then we can factor α as follows:

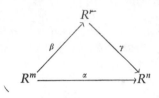

here β has a dense image (its range is the closure of its image), while γ is the inclusion of a direct summand. Dualising, we obtain maps represented by the same matrices and still factoring α:

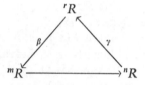

where γ is the projection onto a summand and β will be injective. Hence the image of α in ^mR is isomorphic to ^rR, where r is the number defined in (i). ∎

The common value of these numbers is called the *column rank* of α. This definition makes it clear that the column rank of a matrix is independent of the choice of bases. In particular the column rank is unchanged by elementary transformations. However, we have had to restrict the class of rings (to n-firs) to ensure that the image is a free module.

Over an m-fir we can similarly define the *row rank* of α. Now case (i) of Proposition 5.2 (and its left-hand analogue) relates these numbers to the usual row and column ranks of matrices, over a commutative ring, say. In the general case, an $m \times n$ matrix over a max (m, n)-fir will have both kinds of ranks and in general they will be unrelated. This is clear from the examples given earlier, if we note that the row rank of α equals the rank of $R^m\alpha$, while the column rank equals the *closure* thereof.

Denoting the row and column ranks by ρ_r, ρ_c respectively, we have for a product of matrices:

$$\rho_r(\alpha\beta) \leqslant \rho_r(\alpha), \qquad \rho_c(\alpha\beta) \leqslant \rho_c(\beta).$$

The reader may also formulate characterizations of these two types of rank in terms of interpretations (c) and (d) of the matrix α, but these are less simple. Various versions of the definitions of row and column rank may also be formulated for matrices over rings other than n-firs, but they are not in general equivalent. However, there is a third notion, which is left–right symmetric and is naturally defined for any ring:

PROPOSITION 5.3. *Let R be any ring and α an $m \times n$ matrix over R. Then the following four numbers are equal and do not exceed $\min(m, n)$:*

(i) *the least r such that α is the product of an $m \times r$ and an $r \times n$ matrix, i.e. such that the map α under interpretation* (a) *or* (b) *can be factored through R^r,*

(ii) *the least r such that α can be written $\Sigma_1^r\, a_i \otimes b_i$ (under interpretation (c)),*

(iii) *the least r such that the image of α in R^n is contained in a submodule generated by r elements (interpretation* (b)*),*

(iv) *the least r such that the image of α in $^m R$ is contained in a submodule contained by r elements (interpretation* (a)*).*

Further, this number does not exceed $\rho_r(\alpha)$ if this is defined, and does not exceed $\rho_c(\alpha)$ when this is defined.

The proof is straightforward. ∎

The number described in this theorem is called the *inner rank* of α, and is denoted by $\rho(\alpha)$. Like the other ranks, it does not depend on the choice of bases and is unaffected by elementary transformations. We next examine when two of the ranks are equal.

PROPOSITION 5.4. *If R is a right Bezout domain, the inner rank of any matrix equals the column rank.*

Proof. We know that in any case $\rho(\alpha) \leq \rho_c(\alpha)$, so it will suffice to prove the reverse inequality. Let $\rho(\alpha) = r$, then using interpretation (a), we can factor α by $^r R$; the image in $^r R$ is a finitely generated submodule of $^r R$, hence (by Proposition 0.6.3) of rank at most r, therefore the image in $^m R$ also has rank at most r, i.e. $\rho_c(\alpha) \leq r$. ∎

In a left and right Bezout domain, by symmetry, the row rank, column rank and inner rank of any matrix are equal.

Over any ring R let us define a matrix to be *full* if it is square, say $n \times n$, and of inner rank n. From Proposition 5.2 we see that an $m \times n$ matrix α is a left zero-divisor if and only if $\rho_c(\alpha) < n$, hence we obtain the

COROLLARY. *A square matrix over a right Bezout domain is a left non-zerodivisor if and only if it is full.* ∎

Unfortunately these results do not hold generally, even in commutative rings. Thus let R be the commutative ring generated over a field k by four elements a, b, c, d subject to the relation $ad = bc$. This is an integral domain, but not a Bezout domain; it is easily seen that the matrix $\begin{pmatrix} a & b \\ c & d \end{pmatrix}$ has inner rank 2, but is a zero-divisor. An even simpler example is obtained by looking at 1×1 matrices. In any ring the non-zero elements are full, as 1×1 matrices, but they may be zero-divisors.

Turning to non-commutative rings, let us take an integral domain R which is neither a left nor a right Ore ring. This means that R contains a pair of left linearly independent elements a, b and a pair of right linearly independent elements c, d. The matrix

$$\binom{a}{b} (c \ d) = \begin{pmatrix} ac & ad \\ bc & bd \end{pmatrix}$$

has row and column ranks equal to 2 (when they are defined), but inner rank 1.

For inner ranks we have the following analogue of Sylvester's law of nullity:

PROPOSITION 5.5. *Let α be an $m \times n$ matrix and β an $n \times p$ matrix over a 2n-fir R. Then*

$$\rho(\alpha) + \rho(\beta) - n \leq \rho(\alpha\beta) \leq \min \{\rho(\alpha), \rho(\beta)\}. \tag{1}$$

Proof. The right-hand inequality is clear: if α (or β) can be factored through ^rR, then so can $\alpha\beta$. Now suppose that $\alpha: X \to Y$, $\beta: Y \to Z$ are given and suppose that $\alpha\beta: X \to Z$ can be factored through W, giving a factorization $\alpha\beta = \gamma\delta$. Then the maps

$$X \xrightarrow{\ \alpha + \gamma\ } Y \oplus W \xrightarrow{\ \beta - \delta\ } Z$$

compose to zero. Hence they can be factored through the kernel U and the image V of $\beta - \delta$:

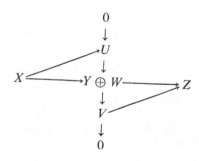

By composition with the projection of $Y \oplus W$ onto Y and the inclusion of Y in $Y \oplus W$ respectively we see that α and β can be factored through U and V respectively, and $Y \oplus W$ is an extension of U by V. Now W may be chosen to have rank $r = \rho(\alpha)$, where $r \leqslant \rho(\alpha) \leqslant n$, and Y has rank n, hence $Y \oplus W$ is free of rank at most $2n$. Therefore its image V in the free module Z is itself free, hence the vertical sequence in the diagram splits, U is free, and

$$rk\ U + rk\ V = r + n.$$

By definition, $\rho(\alpha) \leqslant rk\ U$, $\rho(\beta) \leqslant rk\ V$, hence $\rho(\alpha) + \rho(\beta) - n \leqslant r$ and this proves the left-hand inequality in (1). ∎

Taking α and β to be full, we find the

COROLLARY. *The product of two full $n \times n$ matrices over a $2n$-fir is again full.* ∎

Exercises 5.5

1. Let R be an n-fir; show that a mapping $^nR \to {}^nR$ defines a torsion module if and only if it has inner rank n.

2. Let R be any ring with a field of fractions K (i.e. a field containing R and generated by R as field); define the *rank* of any right R-module M as the dimension over K of $M \otimes K$.
   If R is a left Ore ring with field of fractions K and M is a finitely generated right R-module, show that $rk\ M = rk\ \mathrm{Hom}\ (M, R)$. Show that $rk\ M$ is the maximum length of chains of closed submodules in M.

3*. (Klein [72]). Let $n \geqslant 1$ and let R be a ring such that any chain of closed submodules of ^nR has length at most n; show that any nilpotent $n \times n$ matrix α over R satisfies $\alpha^n = 0$. If R is an integral domain, prove the converse.

4. Let R be a non-zero ring. Show that of the following properties:
   (a) R is weakly n-finite,
   (b) the unit-matrix of order n is full,
   (c) R^n has unique rank,
each implies the next (*cf.* Cohn [66], where it is shown that these properties are distinct).

5. Let $R = \Sigma_1^n\ Rx_i$; show that $(x_1, ..., x_n)^T$ generates a direct summand of ^nR and hence a closed submodule which is free of rank 1.

6*. (Bergman [67]). Let R be an integral domain and x, y left linearly independent elements of R. Show that each of $(x, y, 1)^T$, $(0, 0, 1)^T$ generate a closed submodule of ^3R, but their sum is dense in ^3R. If R is a 3-fir, show that the lattice of closed submodules of ^3R is not modular.

7. (Bergman [67]). Given rings $R \subseteq K$ and an R-module M, a submodule M' is said to be K-*closed* if it is the zero-set of a family of maps $M \to K$. Thus 'R-closed' is the same as 'closed'. Show that every closed submodule is K-closed.

If R is a left Ore ring, K its field of fractions and M a finitely generated right R-module, show that every K-closed submodule is closed.

If R is a right Ore ring, K its field of fractions and M a right R-module, then there is a natural bijection between the set of K-closed submodules of M and submodules of $M \otimes K$. Deduce that if R is a left and right Ore ring with field of fractions K, and M a finitely generated right R-module, then the lattice of closed submodules of M is isomorphic to $\mathrm{Lat}_K(M \otimes K)$, and hence is modular.

8. Let R be a semifir and M a right R-module with generators $e_1, \ldots, e_n$ and defining relations $\Sigma e_i a_{iv} = 0$, where $A = (a_{iv})$ is a given matrix. Show that M is bound if $\rho_r(A) = n$. When $R = k\langle x, y\rangle$, give an example of a bound R-module with positive characteristic.

9. Let R and S be any rings and A_R, B_S, $_RC_S$ any modules with the actions shown. Define a natural mapping

$$A \otimes_R \mathrm{Hom}_S(B, C) \to \mathrm{Hom}_S(B, A \otimes_R C), \tag{1}$$

and show that (1) is an isomorphism when $S = R$ is a left Ore ring, $C = R$, $A = K$ its field of fractions and B a finitely generated R-module. Deduce that in this case for any finitely generated right R-module M, $\mathrm{Hom}_R(M, K) \cong K \otimes M^*$.

10. Let R be a right Ore ring. Show that an $r \times s$ matrix, where $r < s$, annihilates a non-zero column. Define the notions of row and column rank in this case, and show that over a left and right Ore ring, the row and column ranks of an $r \times s$ matrix are equal (their common value is called the *outer rank*).

11. Show that a commutative integral domain for which the inner and outer ranks coincide has the property: $a|b_1b_2$ implies $a = a_1a_2$ and $a_i|b_i$. (Hint: consider 2×2 matrices. Integrally closed domains with this property are called *Schreier rings*, cf. Cohn [68].)

12^0. Characterize commutative integral domains for which the inner and outer ranks coincide.

5.6 The factorization of matrices over semifirs

The factorization theory of 2-firs in 3.2 was based on the interpretation of factorizations as chains of strictly cyclic modules. Similarly we shall find that a study of torsion modules leads to a factorization theory of matrix rings over semifirs, or what comes to the same thing, factorization of matrices over semifirs.

Over a semifir R every finitely presented module is completely specified by a matrix. Thus if M is a right R-module, with the presentation

$$0 \to {}^nR \overset{\lambda}{\to} {}^mR \overset{\mu}{\to} M \to 0, \tag{1}$$

then λ corresponds to left multiplication by an $m \times n$ matrix α say over R, which determines M completely: $M \cong {}^mR/\alpha({}^nR)$, where $\alpha({}^nR)$ stands for the submodule of ${}^mR$ spanned by the columns of α. The fact that λ is injective means that α is a left non-zerodivisor; from the duality of 5.2 (or by a direct argument) it follows that α is a left and right non-zerodivisor precisely when M is bound. Of course a given module may be defined by more than one matrix: in fact an $m \times n$ matrix α and an $r \times s$ matrix α' define isomorphic modules if and only if there exists an $m \times r$ matrix β such that

$$\alpha(^nR) + \beta(^rR) = {}^mR, \qquad \alpha'(^sR) = \{x \in {}^rR | \beta x \in \alpha(^nR)\}.$$

This is proved in the same way as Proposition 3.3.1.

To describe annihilators in general we need a lemma.

LEMMA 6.1. *Let R be a right semihereditary ring and α an $m \times n$ matrix over R. Then the right annihilator of α has the form eR_n where e is an idempotent $n \times n$ matrix over R.*

Proof. Denote the right annihilator of α in R_n by $\mathfrak{a}$, then $R_n/\mathfrak{a} \cong \alpha R_n$. Now αR_n, the right ideal generated by the rows of α, is a finitely generated right ideal of R_n and hence is projective, so $R_n = \mathfrak{a} \oplus \mathfrak{a}'$, where $\mathfrak{a}'$ is a right ideal. If $1 = e + f$ is the decomposition of 1 in this direct sum, then e is idempotent and $\mathfrak{a} = eR_n$. ∎

We can now give a description of torsion modules in terms of matrices.

THEOREM 6.2. *Let R be a semifir. Then any finitely presented module M over R has a presentation (1) where the map λ may be realized as left multiplication by an $m \times n$ matrix α over R which is a left non-zerodivisor. Moreover, M is a torsion module if and only if $m = n$ and α is a full matrix.*

Proof. The first sentence has already been proved, so let M be a torsion module, then $\chi(M) = 0$, hence $m = n$ and α is then a square matrix. By passing to the matrix ring R_n we may assume that M is cyclic with generator u and defining relation $u\alpha = 0$ (Theorem 0.1.2). Now let $\alpha = \beta\gamma$ by any factorization in R_n. This corresponds to an extension, described by a short exact sequence

$$0 \to M' \to M \to M'' \to 0,$$

where M'' is generated by u with definining relation $u\beta = 0$ and M' is generated by $v(= u\beta)$ with defining relation $v\gamma = 0$. In particular, $\chi(M')$, $\chi(M'')$ are both finite, and hence

$$\chi(M'') = -\chi(M') \leqslant 0. \tag{2}$$

We first show that β is a left non-zerodivisor in R_n. By Lemma 6.1, the right annihilator of β has the form $(1 - e)R_n$, where e is an idempotent in R_n. Then $\beta = \beta e$ and since R is projective-free, e is conjugate to

$$E_r = \begin{pmatrix} I_r & 0 \\ 0 & 0 \end{pmatrix}$$

(Proposition 0.2.6). Hence $\beta = \beta e$ may be regarded as an $n \times r$ matrix (by omitting the last $n - r$ columns, which consist of zeros only). It follows that $\chi(M'') = n - r \geqslant 0$, and this contradicts (2) unless $r = n$. Thus $e = 1$ and β is a left non-zerodivisor. This shows that no left factor of α can be a left zero-divisor; clearly this shows that α is full.

To prove the converse, assume that M is not a torsion module. Then either $m \neq n$, or $m = n$ but M has a submodule N such that $\chi(N) = -r < 0$. In the former case the conclusion follows, so take M to be presented as in (1) with $m = n$. Then $\mu^{-1}(N)$ is a submodule of ${}^n R$ and so is a free module containing im λ, with quotient isomorphic to N. Hence

$$\chi\big(\mu^{-1}(N)\big) = \chi(N) + n = n - r,$$

which shows that $\mu^{-1}(N) = {}^{n-r}R$. Since im $\lambda = \mu^{-1}(N)$, we can factor λ through ${}^{n-r}R$, obtaining

$$0 \longrightarrow {}^n R \xrightarrow{\ \lambda\ } {}^n R$$
$$\searrow \qquad \nearrow$$
$$\qquad {}^{n-r}R$$

This shows that α has inner rank at most $n - r$, and hence α is not full. ∎

As an example take R to be the free associative algebra over a field, on $x_1, x_2, \ldots$ and consider the right R-module M with generators e_1, e_2, e_3 and defining relations $e_1 x_1 + e_2 x_2 + e_3 x_3 = 0$, $e_3 x_4 = 0$, $e_3 x_5 = 0$. This module satisfies $\chi(M) = 0$, but the submodule M' generated by e_3 satisfies $\chi(M') = -1$. In terms of matrices M can be described by the matrix

$$\begin{pmatrix} x_1 & 0 & 0 \\ x_2 & 0 & 0 \\ x_3 & x_4 & x_5 \end{pmatrix} = \begin{pmatrix} x_1 & 0 \\ x_2 & 0 \\ x_3 & 1 \end{pmatrix} \begin{pmatrix} 1 & 0 & 0 \\ 0 & x_4 & x_5 \end{pmatrix}$$

and the factorization on the right shows that the matrix is not full (although it is a non-zerodivisor).

The following test for fullness of matrices over a semifir is sometimes useful.

PROPOSITION 6.3. *Let α be an $n \times n$ matrix over a semifir R. Then α is a left non-zerodivisor precisely if $\chi(\alpha R_n) = 1$, and α is full if and only if*

$$\chi(\alpha R_n + \mathfrak{a}) \geqslant 1 \qquad \textit{for every right ideal } \mathfrak{a} \textit{ of } R_n. \qquad (3)$$

Proof. The first assertion is clear. Let us write $T = R_n$, then by Theorem 6.2, α is full if and only if $\chi(T/\alpha T) = 0$ and every homomorphic image of $T/\alpha T$ has non-positive characteristic. If α is full, then $\chi(\alpha T) = \chi(T) = 1$.

and for any right ideal $\mathfrak{a}$ of T, $T/(\alpha T + \mathfrak{a})$ is a homomorphic image of $T/\alpha T$, whence $\chi(\alpha T + \mathfrak{a}) \geqslant 1$, i.e. (3). Conversely, if (3) holds, then for any homomorphic image of $T/\alpha T$, say $T/(\alpha T + \mathfrak{a})$,

$$\chi(T/(\alpha T + \mathfrak{a})) = 1 - \chi(\alpha T + \mathfrak{a}) \leqslant 0,$$

therefore $T/\alpha T$ is a torsion module, and hence α is full. ∎

We can now state the main result on the factorization of matrices over firs:

THEOREM 6.4. *Let R be a fir and k any positive integer. Then the total matrix ring R_k is a ring with unique factorization of all full matrices, in the sense that every full matrix is either a unit or a product of atoms, and any two such atomic factorizations of a given full matrix are isomorphic.*

Proof. Write $T = R_k$ and let c be any element of T corresponding to a full matrix. Any factorization

$$c = a_1 \ldots a_r, \tag{4}$$

of c corresponds to a chain of right ideals of T,

$$T \supseteq a_1 T \supseteq \ldots \supseteq cT, \tag{5}$$

with quotients that are torsion modules: $T/a_1 T$, $T/a_2 T$, ..., $T/a_r T$.

Conversely, let $\mathfrak{a}/cT$ be a torsion submodule of T/cT, then $\chi(\mathfrak{a}) = 1$ hence $\mathfrak{a}$ is principal, say $\mathfrak{a} = aT$, where $aT \supseteq cT$, hence $c = ab$. It follows that a is full. Thus there is a bijection between the factorizations (4) of c and the chains (5) of principal right ideals from T to cT with torsion quotients. The result now follows from the Jordan–Hölder theorem, since $\mathscr{T}_R$ is an abelian category with both chain conditions. ∎

A similar result holds for matrices over semifirs (or even $2n$-firs) with left and right ACC_n; its precise formulation and proof may be left to the reader.

In order to obtain a theory of finitely generated modules over firs which will specialize to finitely generated modules over principal ideal domains, we need to be able to handle arbitrary finitely presented modules over firs, or equivalently, the factorization of rectangular matrices over firs. Although some of the above results can be extended to cover this case, there are difficulties in the way of a satisfactory theory; we shall discuss the extension, and the difficulties, briefly below.

Let us say that an $m \times n$ matrix α is *right non-singular* if its right annihilator (in R_n) is zero, and define *left non-singular* matrices similarly. Thus bound submodules correspond to non-singular matrices.

An $m \times n$ matrix α is said to be a *unit*, if there is an $n \times m$ matrix β such that $\alpha\beta = I_m$, $\beta\alpha = I_n$. Of course, when α is non-singular, either of these conditions implies the other. Over a ring with invariant basis number (IBN) all units are square; in fact this condition is sufficient as well as necessary.

By a *proper* factorization of a matrix α we mean a representation of α as a product

$$\alpha = \alpha_1 \ldots \alpha_r$$

of $r > 1$ non-singular non-unit matrices α_i (of any size, subject to the product being defined). A non-singular non-unit which has no proper factorization is called *unfactorable*. In particular, an element of R which is unfactorable in this sense, is an atom, but the converse need not hold. E.g. over the free associative algebra on $x_1, x_2, \ldots$, we have the element

$$c = x_1 x_{r+1} + x_2 x_{r+2} + \ldots + x_r x_{2r}. \tag{6}$$

This is an atom if $r > 1$, but not unfactorable, because it can be written as a proper product of a row and a column; clearly this was possible only because we had rectangular matrices that are non-singular. By contrast, over a commutative ring, all non-singular matrices must be square. E.g. if we allow the x's to commute and consider the element (6) over the commutative polynomial ring in the x's, we find that it is unfactorable, because now there is no proper factorization: the row and column used before are singular.

We can use the chain conditions of Theorem 2.3 to obtain a factorization of matrices:

THEOREM 6.5. *Every non-singular matrix over a fir is either a unit or a proper product of a finite number of unfactorable matrices.*

Proof. Any proper factorization $c = a_1 a_2 \ldots a_t$, where a_v is an $r_{v-1} \times r_v$ matrix $(r_0 = m, r_t = n)$ corresponds to a series of submodules

$${}^m R \supset a_1({}^{r_1}R) \supset a_1 a_2({}^{r_2}R) \supset \ldots \supset c({}^n R).$$

The inclusions are proper, since no a_v is a unit, and the quotients are bound modules. Thus the factorization corresponds to a chain of bound submodules of ${}^m R / c({}^n R)$, and by Theorem 2.3, any such chain has a maximal refinement, which is still finite. ∎

Theorem 6.5 naturally leads one to ask how the various factorizations of a given matrix are related. We shall see that apart from the rather special case of strongly bound modules, there is no very close relationship between different factorizations.

The strongly bound modules (over any ring) form an abelian category, and over a hereditary ring the finitely presented bound modules have finite composition length, so the Jordan–Hölder theorem holds in this case. The examples which follow illustrate the failure of the Jordan–Hölder theorem for bound modules over the free associative k-algebra $R = k\langle x, y, z, u, v, \ldots \rangle$, where k is a field.

(i) Let M be generated by e, f with defining relations $ex = fy = 0$, $ez = fz$. This module is bound, as well as its submodules eR, fR, but their intersection $eR \cap fR = ez\,R$ is free. M has the following maximal chains of bound submodules: $0 \subset eR \subset M, 0 \subset fR \subset M, 0 \subset (e - f)\,R \subset M$, whose quotients are defined by the following matrices: $x, (y, z); y, (x, z); z, (x, y)$.

(ii) Let M be generated by e, f with defining relations $ex = fy = 0$, $ez = fu$. Here we have two maximal chains: $0 \subset eR \subset M, 0 \subset fR \subset M$, whose quotients have the matrices $x, (y, u)$ and $y, (x, z)$.

(iii) Let M be generated by e, f with defining relations $ex = fy = 0$, $ez = fuv$. Here we have the following maximal chains: $0 \subset eR \subset eR + fuR \subset M$ and $0 \subset fR \subset M$. The quotients have matrices $x, v, (u, y)$ and $y, (x, z)$. This example shows that not even the number of terms in a chain need be constant.

Exercises 5.6

1. Let R be the $n \times n$ matrix ring over a $2n$-fir and let a, b be elements of R with a full common right multiple. Show that $aR \cap bR$ and $aR + bR$ are both principal.

2. Let R be a $2n$-fir with left and right ACC$_n$. Show that the ring R_n has unique factorization of full elements (cf. Theorem 6.4).

3. Show that an unfactorable $n \times n$ matrix over any ring R is an atom in R_n and that the converse holds when R is commutative.

4^0. Develop a theory of unique factorization rings that allows for factorization of zero-divisors (taking e.g. the factorization of matrices over a principal ideal domain as a model, cf. **8.1**).

5.7 The intersection theorem for firs

In this section we shall apply the ACC for bound submodules of finitely related modules over a fir to show that the intersection of the powers of an ideal in a fir is zero. We begin by considering an arbitrary ring R. If $\mathfrak{a}$ is any (left, right or two-sided) ideal in R, we write

$$\mathfrak{a}^{\omega} = \bigcap \mathfrak{a}^n.$$

LEMMA 7.1. *Let $\mathfrak{a}$ be a free left ideal in a ring R, with basis (e_λ) say, and let $a \in \mathfrak{a}^\omega$. Then*

$$a = \Sigma a_\lambda e_\lambda \qquad \text{where} \qquad a_\lambda \in \mathfrak{a}^\omega. \tag{1}$$

Proof. For any $n \geqslant 0$ we can express a as $\Sigma u_1 u_2 \ldots u_n$ $(u_i \in \mathfrak{a})$ and hence

$$a = \Sigma c_{\lambda_1 \ldots \lambda_n} e_{\lambda_1} \ldots e_{\lambda_n}.$$

In particular, there is an expression of the form (1) and since the e_λ are free, we have

$$a_\lambda = \Sigma c_{\lambda_1 \ldots \lambda_{n-1}\lambda} e_{\lambda_1} \ldots e_{\lambda_{n-1}} \in \mathfrak{a}^{n-1}.$$

Since n was arbitrary, we see that $a_\lambda \in \mathfrak{a}^\omega$. ∎

LEMMA 7.2. *Let R be an integral domain and $\mathfrak{a}$ a free left ideal of R which is proper, with basis (e_λ) say. Given any elements $a_i \in \mathfrak{a}$ $(i = 1, \ldots, r)$ such that the a_i are right linearly independent over R, write*

$$a_i = \Sigma b_{i\lambda} e_\lambda,$$

then

$$\Sigma b_{j\lambda} R \supset \Sigma a_i R, \tag{2}$$

where the summation in (2) is over all suffixes.

Proof. Clearly $\Sigma a_i R \subseteq \Sigma b_{j\lambda} R$. If equality holds, we have

$$b_{j\lambda} = \Sigma a_i c_{ij\lambda},$$

hence $a_j = \Sigma b_{j\lambda} e_\lambda = \Sigma a_i c_{ij\lambda} e_\lambda$. Since the a_i were right linearly independent, we find that $1 - \Sigma c_{ii\lambda} e_\lambda = 0$, which contradicts the fact that $\mathfrak{a}$ is proper. ∎

We can now prove the intersection theorem:

THEOREM 7.3. *Let $\mathfrak{a}$ be a proper two-sided ideal in a right fir R, which is free as a left ideal, then $\mathfrak{a}^\omega = 0$.*

Proof. Suppose that $\mathfrak{a}^\omega \neq 0$. Given any elements $a_1, \ldots, a_r \in \mathfrak{a}^\omega$, not all zero, on replacing them by a basis of the right ideal $\Sigma a_i R$, we obtain a right linearly independent set in $\mathfrak{a}^\omega$. Thus we may take $a_1, \ldots, a_r \in \mathfrak{a}^\omega$ to be right linearly independent. Consider the module $M = R/\Sigma a_i R$; since $\Sigma a_i R \neq 0$, $x a_i = 0$ for any $x \in R$ and all i implies $x = 0$. This shows that M is bound. Now write

$$a_i = \Sigma b_{i\lambda} e_\lambda, \tag{3}$$

where (e_λ) is a basis of $\mathfrak{a}$ as left R-module. By Lemma 7.1, $b_{i\lambda} \in \mathfrak{a}^\omega$ and by Lemma 7.2, $\Sigma a_i R \subset \Sigma b_{j\lambda} R$. Hence M contains the nonzero submodule $M_1 = \Sigma b_{j\lambda} R / \Sigma a_i R$. This is again a bound module, for if $\lambda = 1, 2, ..., t$ is the precise range for which non-zero coefficients occur in (3), then a relation matrix for M_1 is

$$\begin{pmatrix} e & & & \\ & e & & 0 \\ & & \ddots & \\ 0 & & & \ddots \\ & & & e \end{pmatrix} \qquad \text{where} \qquad e = \begin{pmatrix} e_1 \\ \vdots \\ e_t \end{pmatrix}$$

this is non-singular, and hence M_1 is bound. Hence the quotient $M/M_1 = R/\Sigma b_{i\lambda} R$ is also bound, with $b_{i\lambda} \in \mathfrak{a}^\omega$. We can therefore continue the process and get an ascending chain

$$M_1 \subset M_2 \subset ... \subset M$$

of modules with bound quotients; hence all the M_i are bound and by Theorem 2.3, the series must break off, which is a contradiction. ∎

COROLLARY. *In a two-sided fir, the intersection of the powers of any proper two-sided ideal is zero.* ∎

Exercises 5.7

1. Give a direct proof of Theorem 7.3, Corollary for principal ideal domains.

2. Let R be an integral domain with left ACC_1. Show that any proper two-sided ideal $\mathfrak{a}$ that is principal as right ideal satisfies $\mathfrak{a}^\omega = 0$.

3. Show that the conclusion of Theorem 7.3, Corollary does not hold for the one-sided fir constructed in 2.9.

5.8 Ascending chain conditions

We have already seen, in 1.2, that firs satisfy right and left ACC_n for all n. This result will be rederived here in a wider context; we shall introduce a chain condition which entails ACC_n and holds in all firs.

A module M is said to satisfy ACC_{ds} if every chain of finitely generated submodules of M with bound quotients (or equivalently: dense inclusions) must break off. If a ring R satisfies ACC_{ds} as right (or left) module over itself, we shall say that R satisfies *right* (or *left*) ACC_{ds}.

We first prove a result which elucidates the connexion between ACC_{ds} and ACC_n. Let us call two ascending sequences of partially ordered sets *cofinal*, if each element of either sequence precedes some element of the other sequence.

LEMMA 8.1. *In an n-fir R, let M be a module with a chain*

$$M_1 \subseteq M_2 \subseteq \dots \tag{1}$$

of submodules, each free of rank at most n, and assume that (1) *is not cofinal with any chain of submodules all of rank less than n. Then M_i/M_{i-1} is bound for all large i.*

Proof. Any M_i/M_{i-1} that is not bound will have R as a direct summand, by Proposition 1.1, and its complement corresponds to a submodule M_i' such that $M_{i-1} \subset M_i' \subset M_i$, where M_i' has smaller rank than M_i, and hence is of rank less than n. If this happens for infinitely many i, then the sequence of M_i' is cofinal, but of rank less than n. ∎

Suppose that R is an n-fir with right ACC_{ds}. Given any ascending chain of n-generator right ideals in R, let $m \leqslant n$ be the least integer for which our chain is cofinal with a chain of right ideals that are free of rank at most m. Applying Lemma 8.1 to this chain, we see that it has bound quotients and hence terminates. This proves the

COROLLARY. *For n-firs, right ACC_{ds} implies right ACC_n. In particular, a semifir with right ACC_{ds} satisfies right ACC_n for all n.* ∎

The next result tells us when a semifir satisfies ACC_{ds}.

THEOREM 8.2. *A semifir is a right $\aleph_0$-fir if and only if it satisfies right ACC_{ds}.*

Proof. ⇒. Let $\mathfrak{a}_1 \subseteq \mathfrak{a}_2 \subseteq \dots$ be a chain of dense inclusions of finitely generated right ideals of a right $\aleph_0$-fir R. The union $\mathfrak{a}$ of the $\mathfrak{a}_i$ is countably generated and hence free. Since $\mathfrak{a}_1$ is a finitely generated submodule of $\mathfrak{a}$, it involves only finitely many members of a basis of $\mathfrak{a}$ and so is contained in a finitely generated direct summand $\mathfrak{b}$ of $\mathfrak{a}$, say $\mathfrak{a} = \mathfrak{b} \oplus \mathfrak{b}'$. Let $p: \mathfrak{a} \to \mathfrak{b}' \subseteq R$ be the projection onto $\mathfrak{b}'$. This is a linear functional, zero on $\mathfrak{a}_1$, and hence by density, zero on each $\mathfrak{a}_i$, and hence is zero on $\mathfrak{a}$, i.e. $\mathfrak{a} = \mathfrak{b}$. Thus $\mathfrak{a}$ is finitely generated and our chain must terminate.

⇐. Let $\mathfrak{c}$ be a right ideal of the semifir R generated by countably many elements $x_1, x_2, \dots$ and suppose that R has right ACC_{ds}. Put $\mathfrak{a}_0 = 0$ and for each $i > 0$, let us inductively construct $\mathfrak{a}_i$ as a maximal finitely generated subideal of $\mathfrak{c}$ in which $\mathfrak{a}_{i-1} + x_i R$ is dense; this exists by ACC_{ds}. We claim that each $\mathfrak{a}_i$ is a direct summand in $\mathfrak{a}_{i+1}$. Indeed, let $\mathfrak{a}_i'$ be a direct summand of $\mathfrak{a}_{i+1}$ containing $\mathfrak{a}_i$, which is free of least possible rank. Any linear functional on $\mathfrak{a}_i'$ which is zero on $\mathfrak{a}_i$ will have for kernel a direct summand of $\mathfrak{a}_i'$ (and hence of $\mathfrak{a}_{i+1}$), which contains $\mathfrak{a}_i$ and is free of smaller rank than $\mathfrak{a}_i'$ (by Theorem 1.1.1(f)), unless the functional is zero. But $\mathfrak{a}_i'$ was of minimal rank, hence the functional must be zero and $\mathfrak{a}_i$ is dense in $\mathfrak{a}_i'$. Hence by construction of $\mathfrak{a}_i$ we have $\mathfrak{a}_i' = \mathfrak{a}_i$.

We thus have $\mathfrak{a}_{i+1} = \mathfrak{a}_i \oplus \mathfrak{b}_{i+1}$ say, where $\mathfrak{b}_{i+1}$, being finitely generated, is free. Hence $\mathfrak{c} = \bigcup \mathfrak{a}_i = \mathfrak{b}_1 \oplus \mathfrak{b}_2 \oplus \dots$ is also free. ∎

As a corollary we obtain another proof of Theorem 1.2.3, that any right $\aleph_0$-fir satisfies right ACC_n for all n. Moreover, we have

COROLLARY 1. *A right fir satisfies right* ACC_{ds} *and hence right* ACC_n *for all* n. ∎

If we look at the proof of the sufficiency in Theorem 8.2, we see that a countably generated module M over a semifir R will be free provided that M has ACC_{ds} and every finitely generated submodule of M is free. In view of Theorem 1.4 this may be stated as

COROLLARY 2. *Let R be a semifir. Then any countably generated R-module which is n-unbound for all n and has* ACC_{ds}, *is free.* ∎

If R is a left Bezout domain, we can replace ACC_{ds} by ACC_n, by the following result:

PROPOSITION 8.3. *Over a left Bezout domain R, any torsion free ($=1$-unbound) right R-module with* ACC_n *for all n has* ACC_{ds}.

Proof. Let M be a torsion free right R-module and take a chain

$$M_1 \subseteq M_2 \subseteq \dots \tag{2}$$

of finitely generated submodules with bound quotients. We recall that any finitely generated torsion free right R-module is free (by Exercise 1.1.11). If for some i, $rk\, M_{i-1} < rk\, M_i$, then the induced map $M_{i-1} \otimes K \to M_i \otimes K$ is not surjective (where K is the field of fractions of R), hence there is a non-zero K-linear functional on $M_i \otimes K$ which vanishes on the image of $M_{i-1} \otimes K$. By left multiplication with the appropriate element of R we may assume that the induced map takes M_i into R. Thus we have a linear functional on M_i which is zero on M_{i-1} without vanishing, and this contradicts the fact that M_i/M_{i-1} is bound. Thus all the ranks in (2) must be equal, to n say, and the sequence terminates by ACC_n. ∎

Proposition 8.3 cannot be simplified by taking ACC_n for a fixed n only, since even over $\mathbf{Z}$ the conditions ACC_n are independent (*cf.* Exercise 9). If we combine Proposition 8.3 with Theorem 8.2, Corollary 2, we obtain a somewhat surprising conclusion:

COROLLARY. *Let R be a left Bezout domain with right ACC_n for all n. Then any countably generated right R-module embedded in a direct power R^I is free.* ■

This shows e.g. that every countably generated subgroup of $\mathbf{Z}^I$ is free, although of course $\mathbf{Z}^I$ itself is not free, unless I is finite.

We conclude this section with a result on what may be called "α-complete" direct limits of α-firs. We shall call a partially ordered set α-*directed* if every subset of cardinal at most α has an upper bound. A directed system over an α-directed set is also called an α-*directed* system.

PROPOSITION 8.4. *Let α be any cardinal greater than 1. Then the direct limit of any α-directed system of α-firs is again an α-fir.*

Proof. Let $\{R_i\}$ be the given system of α-firs, with maps $f_{ij}: R_i \to R_j$, and write $R = \lim_{\to} R_i$. When α is finite, the assertion is simply that the direct limit of any directed system of n-firs is an n-fir; this follows easily from the characterization of n-firs given in Theorem 1.1.1(a) (*cf.* Exercise 1.1.3). Thus we may take α to be infinite. Further, by the finite case, R will be a semifir, so it remains to show that given any set $X \subseteq R$, of cardinal at most α, XR will be a free right ideal.

Since our system is α-directed we can find a ring R_{i_0} such that each $x \in X$ has an inverse image x' in R_{i_0}. Write X' for the set of all these x'. For each finite subset Y of X' and each $i \geqslant i_0$ consider the rank of the right ideal $(Yf_{i_0i})R_i$. For fixed Y this rank is non-increasing in i, so it ultimately equals a minimum which it attains for some i depending on Y. Since there are no more than α finite subsets Y of X', we can find $i_1 \geqslant i_0$ such that all the ranks $rk(Yf_{i_0i})R_i$ have their minimum value for $i \geqslant i_1$. Put $X'' = X'f_{i_0i_1}$ and let B be a basis for the right ideal $X''R_{i_1}$ of the α-fir R_{i_1}. We assert that for all $i \geqslant i_1$ and each finite $C \subseteq B$, Cf_{i_1i} is right linearly independent in R_i. It will follow that the image of C in R is right linearly independent, hence the image of B will be right linearly independent and therefore form a basis for XR.

Thus assume that Cf_{i_1i} is linearly independent. Pick a finite subset Y of X'' such that $C \subseteq YR_{i_1}$; then CR_{i_1}, being a direct summand in $BR_{i_1} \supseteq YR_{i_1}$, will be a direct summand in YR_{i_1}. Hence C can be extended to a basis C' for YR_{i_1}. But $C'f_{i_1i}$ is linearly dependent, hence $rk(Yf_{i_1i}R_i) < \text{card}(C') = rk(YR_{i_1})$, which contradicts our choice of i_1. ■

Exercises 5.8

1. (Continuation of Exercise 1.8). If R is a semifir and $\mathscr{C}$ the class of R-modules that are n-unbound for all n and have ACC_{ds}, show that $\mathscr{C}$ is closed under the operations listed in (a), but not (b). Show that R must have right ACC_{ds} for $\mathscr{C}$ to be non-trivial.

If R is left Bezout, show that $\mathscr{C} = \bigcap \mathscr{C}_n$ and hence in this case $\mathscr{C}$ admits (b) too.

2. Let R be a left Ore ring. Show that if a torsion free right R-module satisfies, for each n, ACC on submodules of rank $\leqslant n$, then it also satisfies ACC_{ds} (use Proposition 8.3).

3*. (Bergman [67]). Let R be an integral domain which is not a left Ore ring. Show that R^N as right R-module does not have ACC_{ds}. (Hint: Let $a, b \in R$ be left linearly independent and define $e_i, f_i \in R^N$ by $\pi_j(e_i) = \delta_{ij}$, $\pi_j(f_i) = ba^{j-i}$, where $a^r = 0$ for $r < 0$ and π_j is the projection on the jth factor. Verify that f_iR is dense in $e_iR + f_{i+1}R$ and deduce that the $M_i = e_1R + \dots + e_iR + f_{i+1}R$ form a strictly ascending chain of dense inclusions.)

4. Use the last example to construct a right but not left fir. (Hint: Try $k[[x; \alpha]]$, where α is a non-surjective endomorphism.)

5. Let R be a semifir and $\mathfrak{a} = \bigcup \mathfrak{a}_n$, where $\mathfrak{a}_n$ is a finitely generated right ideal properly containing $\mathfrak{a}_{n-1}$ as dense submodule. Show that $\mathfrak{a}$ is countably generated but not free.

6. Let $R = k\langle\!\langle x, y \rangle\!\rangle$ and $v = \Sigma_0^\infty x^i yxy^i$. Show that v satisfies the equation $v = xvy + yx$, and that this equation determines v uniquely. Let $\mathfrak{a}_n$ be the right ideal generated by $x^i y$ ($i = 0, 1, \dots, n$) and $x^{n+1}v$, Verify that $\mathfrak{a}_n$ is free on these generators, and that $\mathfrak{a}_n/\mathfrak{a}_{n-1}$ is bound but non-zero. Deduce that R is not a right $\aleph_0$-fir.

7*. Find an ACC (of the type of the ACC_{ds}) such that every semifir satisfying this condition on the right is a right fir (cf. Ex. 1.1.8).

8°. In a right Ore ring, does right ACC_n, for some n, or right ACC_{ds} imply the corresponding condition for free modules?

9*. (Bergman [67]). (a) For any prime number p denote by $\mathbf{Z}_p$ the ring of p-adic integers and by $\mathbf{Q}_p$ the field of p-adic numbers, thus $\mathbf{Q}_p = \mathbf{Z}_p[p^{-1}]$ and $\mathbf{Z}[p^{-1}] \cap \mathbf{Z}_p = \mathbf{Z}$ (taking $\mathbf{Z}$ to be embedded in $\mathbf{Z}_p$). Let $1, x_1, \dots, x_n$ be any $\mathbf{Z}$-linearly independent elements of $\mathbf{Z}_p$ and define a subgroup G of $\mathbf{Q}_p^{n+1}$ by the equation

$$G = (\mathbf{Z}_p^{n+1} + (1, x_1, \dots, x_n)\,\mathbf{Q}_p) \cap \mathbf{Z}[p^{-1}]^{n+1}.$$

Verify that any finitely generated subgroup of G can be generated by $n + 1$ elements. Show further, that for any $k > 0$, there exists $a = (1, a_1, \dots, a_n) \in \mathbf{Z}^{n+1}$ such that $p^{-k}a \in G$ and deduce that G is not finitely generated. Hence show that G does not satisfy ACC_{n+1}.

(b) Let G be defined as in (a) and suppose that C is a union of n-generator subgroups of G. Show that C is annihilated by a non-zero $\mathbf{Q}$-linear functional λ with coefficients in $\mathbf{Z}$. Further, show that C is contained in $p^{-k}\mathbf{Z}^{n+1}$, where p^k is the highest power of p dividing $\lambda((1, x_1, \dots, x_n))$. Deduce that C is finitely generated and that G has ACC_n. This shows that the conditions ACC_n ($n = 1, 2, \dots$) are independent, even over $\mathbf{Z}$.

Notes and comments on Chapter 5

The central part of this chapter, 5.3–4, arose in an attempt to obtain a factorization theory for matrices over firs (itself presented in 5.5–6). This is developed in Cohn [67] on which 5.3 is based. Bergman [67] gives a treatment applicable to n-firs, from a slightly different point of view, which is followed in 5.4.

Sections 5.1–2, taken from Cohn [72'], describes the background to 5.3–4 in a more general setting. The notion of torsion class used in 5.1 has been developed by many writers, frequently in the context of category theory. We need only the most elementary properties, all of which can be found in Dickson [66]; see also Stenström [71] for a comprehensive survey.

Much of 5.5, as well as 5.8 follows Bergman [67]. Theorem 8.2, Corollary 2 generalizes Pontrjagin's theorem: "A countably generated torsion free abelian group with ACC$_n$ for all n is free" (Pontrjagin [39], p. 168). In the case of $\mathbf{Z}$, related results have been obtained by Specker [50] who shows e.g. that in the subgroup B of $\mathbf{Z}^I$ consisting of all bounded sequences, every subgroup of cardinality at most $\aleph_1$ is free. Recently Nöbeling [68] has shown that B itself is free, and independently Bergman [72″] has given a very brief proof of this fact. The intersection theorem in 5.7 is taken from Cohn [70″], where it is used to construct a class of rings with inverse weak algorithm. For some far-reaching generalizations on infinite products of ideals, see Bergman [72].

6. Central elements and subrings

This chapter studies subrings of firs and semifirs which satisfy some finiteness or commutativity condition. The main results describe the form taken by the centre: A commutative ring can be the centre of a 2-fir if and only if it is an integrally closed integral domain, and it can be the centre of a principal ideal domain if and only if it is a Krull domain (**6**.1–3). Further, the centre of a fir that is not principal is a field (**6**.4). Sections **6**.2 and **6**.5 are devoted to a study of invariant elements in 2-firs and their factors, and **6**.6–8 examine subalgebras and centralizers in free algebras. The chapter ends with a fundamental result on free algebras: Bergman's centralizer theorem (Theorem 8.5).

6.1 Ore subrings of 2-firs

Just as 2-firs that are Ore rings and in particular, commutative 2-firs, have a rather special form, it is also possible to say more about Ore subrings of general 2-firs. We recall that in any ring R a left (right) ideal is said to be *large* if it has a non-zero intersection with every non-zero left (right) ideal of R.

Our basic aim will be to show that certain subrings of 2-firs are 'integrally closed' in a sense to be explained, and we begin with a result on the extension of modules. Let B be an integral domain, I a large left ideal of B and A a subring of B containing I. Then I meets every non-zero left A-submodule of B (not merely every non-zero left ideal). For given $b \in B^*$, we can find $b' \in B^*$ such that $b'b \in I$ and $b'' \in B^*$ such that $b''b' \in I \subseteq A$. Then $b''b'b \in Ab$ and also $\in BI \subseteq I$, hence $I \cap Ab \neq 0$.

PROPOSITION 1.1. *Let $B \supseteq A$ be integral domains, I a large left ideal of B contained in A and let M be a torsion free left A-module. Then there is a unique left B-module structure on IM extending the A-module structure, and the construction is functorial in M.*

Proof. We first note that if the structure of any torsion free left A-module N extends to a B-module structure, the extension is unique. For, given $b \in B$, $x \in N$, we choose $a \in A^*$ such that $ab \in I \subseteq A$. Then the product $y = bx$

211

in our extended structure must satisfy $ay = (ab)\,x \in IN$. So if y exists it is unique, by torsion-freeness.

Given $y = \Sigma a_i x_i$ $(a_i \in I,\ x_i \in M)$ and $b \in B$, let us define

$$b \cdot y = \Sigma(ba_i)x_i. \tag{1}$$

This again lies in IM and is independent of the representation chosen for y. For if $\Sigma a_i x_i = \Sigma a_j' x_j'$ are two such expressions, take $c \in A^*$ such that $cb \in I \subseteq A$. On multiplying by cb we get $\Sigma(cba_i)\,x_i = \Sigma(cba_j')x_j'$. Now

$$\Sigma(cba_i)\,x_i = \Sigma\big(c(ba_i)\big)\,x_i = \Sigma c\big((ba_i)x_i\big);$$

rewriting our sum accordingly and cancelling c (by torsion-freeness) we find $\Sigma(ba_i)x_i = \Sigma(ba_j')x_j'$ as desired. It is easily verified that (1) gives the required module structure and that the construction is functorial. ∎

If e is a non-zerodivisor in a ring R, then it is clear that the set of pairs $(x, y) \in R^2$ such that $xe = ey$ forms a subring of $R^2 = R \times R$ which is the graph of an isomorphism α_e between the idealizers of eR and Re in R respectively:

$$\mathbf{I}(eR) = \{x \in R \,|\, xe \in eR\} \qquad \text{and} \qquad \mathbf{I}(Re) = \{y \in R \,|\, ey \in Re\}.$$

A pair of mappings f, g of a set X into R will be called *conjugate* if $f = \alpha_e \cdot g$ for some non-zerodivisor e of R, i.e. if

$$g(x)\,e = ef(x) \qquad \text{for all} \qquad x \in X.$$

More precisely, f is said to be *right conjugate* to g and g is *left conjugate* to f.

PROPOSITION 1.2. *Let R be a 2-fir and B an integral domain, and let A be a right Ore ring which is a subring of R and B. Further, let I be a large left ideal of B contained in A, such that IR is principal in R. Then there exists an injection $f: B \to R$ such that $f|A$ is right conjugate to the inclusion of A in R, i.e. there exists $e \in R^*$ such that*

$$xe = ef(x) \qquad \text{for all} \qquad x \in A.$$

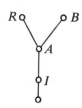

Proof. Consider R as left A-module; by Proposition 1.1 we get a left B-module structure on IR extending the A-module structure. Let us write this action as $b . x$ ($b \in B$, $x \in IR$). By functoriality it will respect right multiplication by elements of R.

By hypothesis, IR is principal as right ideal, say $IR = eR$, where clearly $e \neq 0$. Given $b \in B$, we have $b . e \in IR = eR$, say $b . e = ef(b)$. It is easily checked that the map $f: B \to R$ is a ring homomorphism and if $b \in A$, $ef(b) = b . e = be$, so that $f|A$ is indeed right conjugate to the inclusion of A in R. This shows in particular, that $\ker f \cap A = 0$, hence $\ker f \cap I = 0$ and so $\ker f = 0$, i.e. f is injective, as claimed. ∎

We observe that the right ideal IR of R is principal under either of the following conditions:

(a) IA is a finitely generated right ideal of A, or

(b) R has right ACC_1.

For since A is a right Ore ring, any two non-zero elements of I have a common non-zero right multiple in A and hence in R. So any finite subset of I generates a principal right ideal of R. When R has right ACC_1, IR is therefore principal, i.e. (b). Now suppose that IA is finitely generated, by $i_1 c_1, ..., i_n c_n$ say, $i_v \in I$, $c_v \in A$, and let $\Sigma i_v R = eR$. Then $IR = IAR = \Sigma i_v AR = eR$, whence IR is principal also in case (a).

An integral extension of a commutative domain can be defined in a number of ways; the next result shows that some of these carry over to the general case.

PROPOSITION 1.3. *Let B be an integral domain and A a subring. Then for any $y \in B$ the following are equivalent:*

(i) *the left A-module generated by the powers of y is finitely generated over A,*

(ii) *there is a monic polynomial $f = \Sigma c_i t^i$ with coefficients in A (written on the left) such that $\Sigma c_i y^i = 0$,*

(iii) *there is a monic polynomial in $A[t]$ which when considered over B, has $t - y$ as right factor.*

The proof is as in the commutative case and may be left to the reader. ∎

We express the fact that the conditions of Proposition 1.3 hold by saying that y is *left integral* over A. In the commutative case this reduces to the usual definition, but unlike that case, the elements of B that are left integral over A need not form a subring and in fact if y is left integral over A, the left A-module $\Sigma A y^i$ need not even be a ring. We therefore supplement the above definition as follows.

Let B be an integral domain and A a subring, then B is said to be *left integral* over A, if for every finite subset X of B, the left A-module generated by all the monomials in X is finitely generated. In particular, if $B = \Sigma_1^r A u_i$, we say that B is *finite left integral* over A. It is clear that every element of a left integral extension of A is itself left integral over A.

Now let A be a right Ore ring, K its field of fractions, and B a ring between A and K. Suppose that B is finite left integral over A, then $B = \Sigma A u_i$ say. Writing $u_i = a_i c^{-1}$ with a common denominator c, we find that

$$Bc \subseteq A. \tag{2}$$

This means that the *conductor* of A in B, defined as the set

$$I = \{c \in A | Bc \subseteq A\}$$

is different from 0. Clearly this agrees with the definition of the conductor in the commutative case. As in that case we see that generally I is a left ideal in B and a two-sided ideal in A. In fact I can be characterized as the largest ideal in A which is also a left ideal in B.

If A is a left as well as right Ore ring, then I is a large left ideal in B; more precisely, any non-zero principal left ideal of B which is contained in I is large in B. For let $b \in B^*$, $c \in I^*$; we must show that $Bc \cap Bb \neq 0$. By hypothesis, $Bc \subseteq A$, hence $b = ac^{-1}$ for some $a \in A$. Since A is left Ore, there exist $a_1, c_1 \in A^*$ such that $c_1 a = a_1 c^2$, hence $c_1 b = a_1 c \in Bc \cap Bb$.

THEOREM 1.4. *Let R be a 2-fir, A a left and right Ore subring of R and B a finite left integral extension of A in its field of fractions. If either* (i) *R has right* ACC_1 *or* (ii) *A is right invariant (i.e. every non-zero element of R is right invariant), then there exists an embedding $f: B \to R$ such that $f | A$ is right conjugate by a non-zero element e to the inclusion of A in R, i.e.*

$$xe = ef(x) \quad \text{for all} \quad x \in A. \tag{3}$$

Proof. By the remarks made earlier, the conductor I of A in B is non-zero and (i) follows by Proposition 1.2. To prove (ii) let $B = \Sigma A u_i = \Sigma A a_i c^{-1}$, i.e. $Bc = \Sigma A a_i$. Since c is in the conductor, $I = Bc$ is a large left ideal; moreover, $IA = BcA = \Sigma A a_i A = \Sigma a_i A$ is finitely generated as right ideal in A. Hence IR is principal and Proposition 1.2 applies again. ∎

COROLLARY. *The centre of a 2-fir is integrally closed in its field of fractions.*

We recall that an integral domain A is said to be *integrally closed* in its field of fractions K if every element of K integral over A already lies in A.

To prove the corollary, let C be the centre and K its field of fractions. We apply the theorem with $A = C$ and $B = C[f]$, for any element $f \in K$ and integral over C. Clearly f has the form $f = ab^{-1}$, where $a, b \in C$. Now for any $x \in R$, $xa = ax$, i.e. $xfb = xa = ax = fbx = fxb$, hence $(xf - fx)b = 0$, but $b \neq 0$, so $xf = fx$ which shows that $f \in C$, i.e. C is integrally closed. ∎

In conservative 2-firs (*cf.* 4.3) we can strengthen these results.

PROPOSITION 1.5. *Let R be a conservative 2-fir with right* ACC$_1$. *Then any maximal left and right Ore subring of R is left integrally closed in its field of fractions.*

The same holds without ACC$_1$ *if the subring is right invariant.*

Proof. Let A be a maximal Ore subring of R and B a finite left integral extension of A; we must show that $B = A$.

We can apply Theorem 1.4 to obtain an embedding $f : B \to R$ and $e \in R^*$ such that

$$xe = ef(x) \qquad (x \in A).$$

Take any $b \in B$ and write $b = c^{-1}a$, where $a, c \in A$. Then $a = cb$, hence

$$ae = ef(a) = ef(c)f(b) = cef(b).$$

Thus e and $ef(b)$ are left commensurable. By Theorem 4.2.4, $ef(b) = b_1 e$ for some $b_1 \in R$. Therefore $ae = cef(b) = cb_1 e$, i.e. $b_1 = c^{-1}a \in R$. This holds for any $b \in B$, hence $f(B)$ is right conjugate to a ring B_1 isomorphic to B and containing A. By maximality, $B_1 = A$, whence $B = A$. ∎

A similar argument can be applied to a maximal commutative subring of R:

COROLLARY. *In a conservative atomic 2-fir, any maximal commutative subring is integrally closed.* ∎

Exercises 6.1

1°. Is the relation of being a left integral extension transitive?

2. If B is a finite left and right integral extension of A, show that there exist $u_1, \ldots, u_n \in B$ such that $B = \Sigma Au_i = \Sigma u_i A$.

3. Obtain the conclusion of Theorem 1.4 under the following hypothesis: R is a 2-fir, A is a left and right Ore subring of R and B is a finite left and right integral extension of A in its field of fractions.

4. Let $B = k\langle x, y \rangle$ and A the subalgebra generated by x^2 and y^2. Find the set $\bar{A}$ of elements of B left integral over A. Is this set closed under addition or multiplication? Is ΣAa^i a subring for every $a \in \bar{A}$?

5. Show that a maximal invariant subring of a conservative 2-fir is integrally closed.

6.2 Invariant elements in 2-firs

In this section we turn to another method of studying the centres of 2-firs, which provides additional information in the case of atomic 2-firs. This is done by considering the set $I = \mathbf{I}(R)$ of invariant elements of R, i.e. the elements $c \in R^*$ such that $cR = Rc$. We begin by noting conditions under which invariants element occur in general rings:

PROPOSITION 2.1. (i) *In any ring R, a non-zerodivisor c is invariant if and only if the left and right ideals generated by c are both two-sided,*

(ii) *If R is an integral domain, then any non-zero ideal that is principal both as left and as right ideal has an invariant generator,*

(iii) *If R is a 2-fir, then the ideals with invariant generators form a sublattice of the lattice of all ideals.*

Proof. (i) is clear; for (ii) take $aR = Ra'$ and let $a = ua'$, $a' = av$, then $a' = ua'v$. Now $a'v \in Ra'$, say $a'v = wa'$, so $a' = uwa'$, i.e. $uw = 1$, u is a unit, and so $aR = Ra$.

(iii) Let $aR = Ra$, $bR = Rb$ be two non-zero ideals in a 2-fir R, then $aR \cap bR \neq 0$, hence $aR + bR = dR = Ra + Rb = Rd'$ and $aR \cap bR = mR = Ra \cap Rb = Rm'$. By (ii) d and m are invariant, as claimed. ∎

Proposition 2.1 shows that in any 2-fir R, the semigroup I of invariant elements is lattice-ordered by divisibility. Thus any two invariant elements have an HCF and this is the same whether calculated in I or in R; similarly for the LCM.

The quotient of a 2-fir by an ideal with an invariant generator has a rather special form, which is described in

THEOREM 2.2. *Let R be a 2-fir and c a non-unit invariant element. Then R/cR is a Bezout ring (not necessarily a domain). If R is moreover, atomic, R/cR is an Artinian principal ideal ring, which is a field if and only if c is an atom in R.*

Proof. The first assertion is that every finitely generated left or right ideal in $R cR$ is principal. Now cR meets every non-zero right ideal non-trivially, hence every principal right ideal of R/cR comes from a principal right ideal of R containing cR, and the sum of any two (and hence any finite number of) such principal right ideals is again principal. If R is also atomic, every chain of principal (left or) right ideals is bounded by the length of c. Hence R/cR is then left and right Artinian. Finally an Artinian ring is a field if and only if it is an integral domain, and R/cR is an integral domain precisely if c is an atom. ∎

When R is moreover atomic, I is clearly also atomic and we can apply Theorem 3.1.3. Let us recall that an *I-atom* is an invariant element which is an atom within I. Then we can state the result as

THEOREM 2.3. *Let R be an atomic 2-fir. Then R has unique factorization of invariant elements, i.e. every non-unit invariant element c can be written as a product of I-atoms*

$$c = a_1 \dots a_r \tag{1}$$

and if

$$c = b_1 \dots b_s \tag{2}$$

is any other factorization of c into I-atoms then $r = s$ and there is a permutation $i \mapsto i'$ of $1, \dots, r$ such that $b_{i'}$ is associated to a_i. Moreover, every order of the factors in (1) can be realized. ∎

In terms of ideals Theorem 2.3 states that in an atomic 2-fir the ideals with invariant generator form a free commutative semigroup under ideal multiplication. Of course this is merely a reflection of the fact that a lattice-ordered semigroup with ascending chain condition is necessarily free commutative.

Let A be a commutative integral domain and K its field of fractions, then any homomorphism $v: A^* \to \Gamma_+$ into the positive cone of a totally ordered group Γ (together with the convention $v(0) = \infty$), such that

$$v(a - b) \geqslant \min \{v(a), v(b)\},$$

is called a *general valuation*. Such a valuation can always be extended in just one way to a valuation of K, again written v, and the set $K_v = \{x \in K | v(x) \geqslant 0\}$ is a local Bezout domain containing A (called the *valuation ring* of v). Now the ring $\bar{A} = \bigcap K_v$ where v ranges over all general valuations on A consists of those elements of K that are integral over A. This ring $\bar{A}$ is called the *integral closure* of A in K; in fact A is integrally closed precisely if $\bar{A} = A$ (Lang [58], p. 12).

There is a related construction which we shall need here. This arises if instead of general valuations we limit our attention to $\mathbf{Z}$-valued valuations. Let A be a commutative integral domain and K its field of fractions, as before. If there is a family V of $\mathbf{Z}$-valued valuations on K such that (i) for any $x \in A^*$, $v(x) \geqslant 0$, for all $v \in V$, with equality for almost all v, and (ii) $A = \bigcap K_v$, then A is said to be a *Krull domain*. More generally, suppose there is a family V of $\mathbf{Z}$-valued valuations on A satisfying (i), then $\bar{A} = \bigcap K_v$ is clearly a Krull domain containing A.

From our earlier remarks it is clear that every Krull domain is integrally closed. Every commutative UFD is a Krull domain: we take V to be the family of valuations associated with the atoms of A. Likewise every Noetherian integrally closed domain is a Krull domain; here V is the class of valuations associated with the minimal prime ideals of A (*cf.* Bourbaki [65]).

We now have the following analogue of the result obtained in **6.1**.

THEOREM 2.4. *The centre of an atomic 2-fir is a Krull domain.*

Proof. Let R be an atomic 2-fir, C its centre and K the field of fractions of C. By Theorem 2.3 each element of C^* has a decomposition into I-atoms

$$a = u \Pi p^{\alpha_p} \qquad (u \in \mathbf{U}(R),\ \alpha_p \geqslant 0).$$

Fix p and consider the function v_p defined on C by

$$v_p(a) = \alpha_p.$$

Clearly this is $\mathbf{Z}$-valued, in fact it is non-negative on C, and $v_p(a) = 0$ for almost all p. Hence $\bar{C} = \bigcap K_v$ is a Krull domain containing C; we claim that $\bar{C} = C$. Let $f \in \bar{C}$, say $f = ab^{-1}$, where $a = u \Pi p^{\alpha_p}$, $b = v \Pi p^{\beta_p}$, u, $v \in \mathbf{U}(R)$ and $\alpha_p \geqslant \beta_p$. Then $f = w \Pi p^{\alpha_p - \beta_p}$, where $w \in \mathbf{U}(R)$; for we can divide by each power p^{β_p} in turn; each time the unit factor will change appropriately; we conclude that $f \in R$. Now as in the proof of Theorem 1.4 Corollary we find that $f \in C$, i.e. $\bar{C} = C$. ■

As we shall see in **6.3**, this theorem is best possible, in the sense that any Krull domain can occur as the centre of an atomic 2-fir.

Exercises 6.2

1. Let I be the set of right invariant elements in an integral domain R and let S be the set of all left factors of elements of I. Show that S is a right denominator set.

2. Let α be an automorphism of a field k such that no power of α is inner. Show that the monic invariant elements of $k[x; \alpha]$ are the powers of x. If α^r is inner, but no lower power, find all monic invariant elements (*cf.* **8.3**).

3. Show that a principal ideal domain is simple if and only if it has no non-unit invariant element.

4. Let k be a field of characteristic 0 and D an outer derivation of k. Show that $R = k[x; 1, D]$ is simple and hence has no non-unit invariant elements.

5. Let R be an atomic 2-fir and c a non-unit invariant element. Show that the Artinian ring $A = R/cR$ has the property that every left (or right) ideal is the annihilator of its annihilator in A (i.e. A is a *quasi-Frobenius ring*).

6. Let R be a 2-fir; show that the elements $\alpha\beta^{-1}$ ($\alpha, \beta \in R$, $\beta \neq 0$) satisfying $\alpha x\beta = \beta x\alpha$ for all $x \in R$ form an integrally closed ring.

7*. Find a ring in which the multiplication of invariant principal ideals is non-commutative (*cf.* Ex. **1.1.8**).

6.3 The centres of 2-firs

This section is devoted to proving the converse of Theorem 2.4, in the following strong form: Every integrally closed domain (resp. Krull domain) occurs as the centre of some Bezout domain (resp. principal ideal domain). Since every Bezout domain is a 2-fir, and every principal ideal domain an atomic 2-fir, this, with Theorem 2.4, completely characterizes the centres of (atomic) 2-firs.

The proof proceeds in two stages:

(i) given an integrally closed commutative domain C, we construct a commutative Bezout domain with an automorphism of infinite order whose fixed ring is C; moreover, when C is a Krull domain, the Bezout domain can actually be chosen to be a principal ideal domain.

(ii) given a commutative Bezout domain A with an automorphism α of infinite order, we construct a Bezout domain containing A, whose centre is precisely the fixed ring of α. Moreover, when A is a principal ideal domain, the ring containing it can be chosen to be principal.

It is convenient to begin with (ii). The two cases to be considered require rather different treatment, and we therefore take them separately.

PROPOSITION 3.1. *Let A be a commutative Bezout domain with an automorphism α of infinite order, and let C be the fixed ring of α acting on A. Then there exists a Bezout domain containing A, with centre C.*

Proof. Let K be the field of fractions of A; then α extends in a unique way to an automorphism of K, again denoted by α (*cf.* Exercise **0.5.13**). Now form the skew polynomial ring $K[x; \alpha]$ and let R be the subring of all polynomials with constant term in A. That R is a subring is clear; we claim that it has the desired properties. In the first place, if $f = \Sigma x^i a_i$ lies in the centre, then for any $b \in A$, $bf = fb$, whence on equating coefficients, $a_i b = a_i b^{\alpha^i}$. Now for each $i > 0$ there is a $b \in A$ such that $b^{\alpha^i} \neq b$, hence $a_i = 0$ for $i > 0$ and $f = a_0$. Further, the equation $fx = xf$ shows that $a_0{}^\alpha = a_0$, i.e. $f = a_0 \in C$, as claimed.

It remains to show that R is Bezout. Let $f, g \in R$, then we must show that $fR + gR$ is principal, and here we may assume that f, g have no common left factor of positive degree in x, for if $f = df_1$, $g = dg_1$ and $f_1 R + g_1 R = hR$ has been established, then $fR + gR = dhR$. By looking for the highest common left factor of f, g in $K[x; \alpha]$ we find polynomials u, v in the latter ring such that $fu - gv = 1$. On multiplying up by a suitable element of A we therefore obtain an equation

$$fu - gv = \gamma, \quad \text{where} \quad u, v \in R, \gamma \in A^*. \tag{1}$$

If the constant terms of f, g are α, β respectively, say $f = \alpha + f_1$, $g = \beta + g_1$, where f_1, g_1 have zero constant term, then $\alpha = f - \gamma(\gamma^{-1} f_1) \in fR + gR$ by (1), and similarly $\beta \in fR + gR$, hence the highest common factor δ say of α and β (in A) divides γ. It follows that δ is a left factor of f and g, and since $\delta \in \alpha R + \beta R$, we have $fR + gR = \delta R$. Thus any two elements generate a principal right ideal; by symmetry they also generate a principal left ideal and so R is a Bezout domain. ∎

By combining this result with Theorem 1.4, Corollary we get the

COROLLARY. *The fixed ring of an automorphism of infinite order acting on a commutative Bezout domain is integrally closed.* ∎

Of course this result can also be obtained directly.

PROPOSITION 3.2. *Let A be a commutative principal ideal domain with an automorphism α of infinite order, and let C be the fixed ring of α acting on A. Then there is a principal ideal domain containing A, with centre C.*

Proof. We form the ring of skew formal Laurent series $f = \Sigma x^i a_i$ where $a_i \in A$ and $a_i = 0$ for i less than some k depending on f, with the commutation rule $ax = xa^\alpha$. The verification that the centre of R is C is as before: the equations $bf = fb$ show that $f = a_0$ and now $xf = fx$ shows that $a_0 \in C$. It remains to show that R is a principal ideal domain.

Given $f = \Sigma x^i a_i \neq 0$, let a_k be the first non-zero coefficient. Then k is called the *order* and a_k the *leading coefficient* of f. Given f, $g \in R$ we must show that $fR + gR$ is principal. If f or g is 0 this is clear; otherwise, on multiplying f, g by suitable powers of x we may take them both to be of order 0. Denote their leading coefficients by a_0, b_0 respectively. If b_0 is a unit, then so is g, for $g = (1 + \Sigma_1^\infty x^i b_i b_0^{-1})b_0$ can be formally inverted. We may therefore use induction on the length of b_0 (as element of A). Let us put $a_0 A + b_0 A = d_0 A$, then there exist a_0', b_0', u, $v \in A$ such that

$$a_0' u - b_0' v = 1, \tag{2}$$

and $a_0 = d_0 a_0'$, $b_0 = d_0 b_0'$. On replacing f, g by $f^* = fb_0' - ga_0'$, $g^* = fu - gv$, we have $fR + gR = f^*R + g^*R$, because the matrix of transformation from f, g to f^*, g^* lies in A and has determinant 1, by (2). Moreover, the leading term of g^* has shorter length than b_0 unless $b_0|a_0$. By our induction hypothesis we may therefore assume the latter, say $a_0 = cb_0 (c \in A)$. Then $f^* = f - gc$ has higher order than f. If by suitable choice of $h \in R$, the order of $f - gh$ can be made arbitrarily large, then there is also an h such that $f - gh = 0$, and hence $fR + gR = gR$. Otherwise let us choose h so that $f_1 = f - gh$ has the largest possible order; then its leading term will not be divisible by b_0 and by repeating the process (of forming f^*, g^*) we can replace g by an element with leading term of shorter length. This shows that R is right Bezout, and by symmetry R is left Bezout. Now in any factorization of an element $f \neq 0$ the number of non-unit factors is bounded by the length of the leading term of f, hence R is atomic, and so is a principal ideal domain. ∎

Note that in this proof we had to use the fact that A is principal even to show that R is Bezout. In fact if we perform this construction with a Bezout domain A, we do not generally get a Bezout domain. Using Theorem 2.4 we again have a corollary (which as before can also be proved directly).

COROLLARY. *The fixed ring of an automorphism of infinite order acting on a commutative principal ideal domain is a Krull domain.* ∎

We now come to step (i) of our programme. This is in effect the converse of the corollaries to Propositions 3.1 and 3.2.

PROPOSITION 3.3. *Every integrally closed commutative integral domain is the fixed ring of an automorphism of infinite order acting on a commutative Bezout domain.*

Proof. We first give the basic construction which will in most cases produce the required ring, and then show how to modify it so as to get the result in all cases.

Let C be the given domain and K its field of fractions. By hypothesis, $C = \bigcap K_v$, where v ranges over the family V of all general valuations defined on C. We form the polynomial ring $K[t]$ in an indeterminate t. Each $v \in V$ can be extended to $K[t]$ by putting

$$v(\Sigma t^i a_i) = \min_i \{v(a_i)\}.$$

We assert that this is again a valuation. The rule $v(a - b) \geqslant \min \{v(a), v(b)\}$ is clear; to show that v is multiplicative, let $a = \Sigma t^i a_i$, $b = \Sigma t^j b_j$ and let a_r, b_s be the first coefficient attaining the minimum $v(a)$, $v(b)$ respectively. The product $c = ab$ has coefficients $c_k = \Sigma a_i b_{k-i}$ and $v(c_k) \geqslant v(a_r) + v(b_s)$, with equality holding for $k = r + s$, as is easily verified. Thus v is a valuation on $K[t]$ and in fact can be extended to a valuation, again denoted by v, of the quotient field $K(t)$. We now define

$$A = \bigcap K(t)_v, \qquad \text{where } v \text{ ranges over } V.$$

Thus A consists of the fractions f/g, where $f, g \in C[t]$ are such that $v(f) \geqslant v(g)$. A is sometimes called the *Kronecker functional ring*.

We claim that A is a Bezout domain: given two elements of A, on multiplying by a common denominator, we may take them to be $f, g \in C[t]$. Now take any n greater than the degree of f in t, and form $h = f + t^n g$. Then clearly $v(h) \leqslant v(f)$, $v(g)$, hence f/h, $g/h \in A$ and so $fA + gA = hA$.

Consider the map α of $K[t]$ defined by $f(t) \mapsto f(t + 1)$. This is an automorphism, clearly $v(f(t + 1)) \geqslant v(f(t))$ for any $v \in V$, and by writing down the same inequality with -1 in place of 1, we see that $v(f^\alpha) = v(f)$, hence α extends to an automorphism of A. We observe that α is of infinite order precisely when K has characteristic 0. In that case the fixed field of α acting on $K(t)$ is K, hence the fixed ring of α acting on A is $K \cap A = C$.

In the case of finite characteristic we modify our construction by starting, not with $K[t]$, but with $K[..., t_{-1}, t_0, t_1, t_2, ...]$. As before we get a Bezout domain, and for our automorphism α we can use instead of the translation $t \mapsto t + 1$ the substitution $t_n \mapsto t_{n+1}$. Clearly this is of infinite order and the fixed ring in the action on $K(..., t_{-1}, t_0, t_1, t_2, ...)$ is K, hence the fixed ring in A is C. Thus A is a Bezout domain with an automorphism of infinite order whose fixed ring is C. ∎

Suppose now that C is a Krull domain. Let K be its field of fractions and V the family of valuations defining C. We form A as in the proof of Proposition 3.3, using the family V instead of the family of all valuations.

Then it follows as before that A is Bezout with fixed ring C. We claim that A is in fact a principal ideal domain. Given $a \in A$, it is clear that $v(a) \neq 0$ for only finitely many $v \in V$, say $v_1, ..., v_r$. Now for any factor b of a, $0 \leqslant v(b) \leqslant v(a)$ for all $v \in V$, and if $v(b) = v(b')$ for all $v \in V$, then b and b' are associated. Hence there are only finitely many classes of factors of a, whence A is atomic and therefore a principal ideal domain. This proves

PROPOSITION 3.4. *Every Krull domain is the fixed ring of an automorphism of infinite order acting on a commutative principal ideal domain.* ∎

Putting all the results of this section together, we find

THEOREM 3.5. *Every integrally closed commutative integral domain occurs as the centre of a Bezout domain; every Krull domain occurs as the centre of a principal ideal domain.* ∎

Exercise 6.3

1. Give a direct proof (e.g. by valuation theory) that the fixed ring of an automorphism acting on a commutative Bezout domain R is integrally closed, and is a Krull domain when R is a principal ideal domain.

2*. Does Proposition 3.1 still hold when the automorphism has finite order?

3. Examine why the proof of Proposition 3.1 fails for principal ideal domains and that of Proposition 3.2 fails for Bezout domains.

4. Let R be a right hereditary local ring. Give a direct proof (using Theorem 3.4.7) that either the centre of R is a field or R is a right discrete valuation ring and its centre is a DVR.

5. In the proof of Proposition 3.3, if K has finite characteristic, complete the argument by taking in place of $K[t]$, the ring $K[t, t^{\frac{1}{2}}, t^{\frac{1}{4}}, ...]$, with automorphism $t \mapsto t^2$.

6. Verify that for any commutative field k, the polynomial ring $k[x_1, x_2, ...]$ is a Krull domain but not Noetherian. Deduce the existence of a Noetherian domain with a non-Noetherian centre.

7. Show that the centre of a 2-fir with right ACC_1 is a Krull domain.

6.4 The centre of a fir

The results of 6.2–3 give a complete description of the centre of a principal ideal domain or a Bezout domain, as well as some information on the centre of a 2-fir, but they leave open the question whether e.g. any Krull domain can occur as the centre of a genuine, i.e. non-Ore, fir. As we shall see, once we assume that our rings are non-Ore, the centre is very much more restricted. Thus the centre of a non-Ore fir is necessarily a field. More generally, the conclusion will hold for non-Ore 3-firs with right ACC_2.

We begin with a non-commutative analogue of a theorem of I. S. Cohen, which (in Kaplansky's formulation) states that any ideal that is maximal among the non-finitely generated ideals of a ring is prime.

LEMMA 4.1. *Let $\mathfrak{a}$ be a right ideal in a ring R which is maximal among the non-finitely generated right ideals of R, and let c be an invariant element of R which is not in $\mathfrak{a}$. Then $cR \cap \mathfrak{a} = \mathfrak{a}c$.*

Proof. From the invariance of c it follows that $cR \cap \mathfrak{a}$ can be written as $\mathfrak{a}'c$, where $\mathfrak{a}' = \{x \in R | xc \in \mathfrak{a}\}$ is a right ideal containing $\mathfrak{a}$. If the inclusion $\mathfrak{a}' \supseteq \mathfrak{a}$ were proper, then by the maximality of $\mathfrak{a}$, $\mathfrak{a}'$ would be finitely generated, hence also $\mathfrak{a}'c$. Again by the maximality of $\mathfrak{a}$, the right ideal $\mathfrak{a} + cR$ is finitely generated, and the exact sequence

$$0 \to cR \cap \mathfrak{a} \to cR \oplus \mathfrak{a} \to cR + \mathfrak{a} \to 0 \tag{1}$$

shows that $cR \oplus \mathfrak{a}$, and hence $\mathfrak{a}$ itself, is finitely generated, a contradiction. Therefore $\mathfrak{a}' = \mathfrak{a}$. ∎

The usefulness of Lemma 4.1 resides in the fact that in any ring, the set of non-finitely generated right ideals is inductive, so that maximal non-finitely generated right ideals always exist. Instead of taking $\mathfrak{a}$ "maximal non-finitely generated" we may take $\mathfrak{a}$ to be maximal in any class of right ideals which behaves appropriately under extensions. Here is another case which will be needed later.

LEMMA 4.2. *Let R be a 3-fir and let c be an invariant element of R. If $\mathfrak{a}$ is a right ideal which is maximal among the free right ideals of rank 2 of R, then $cR \cap \mathfrak{a} = \mathfrak{a}c$.*

Proof. Since c is invariant, cR is right large and hence any finitely generated right ideal containing cR must be principal (*cf.* Theorem 2.2), therefore $\mathfrak{a} + cR$ is principal, say $\mathfrak{a} + cR = dR$. Thus the sequence (1) takes the form

$$0 \to \mathfrak{a}'c \to cR \oplus \mathfrak{a} \to dR \to 0. \tag{2}$$

Since dR is free, this splits and since we are in a 3-fir, we can conclude that $\mathfrak{a}'c$ is free of rank 2, hence so is $\mathfrak{a}'$. But $\mathfrak{a}$ was maximal for this property, hence $\mathfrak{a}' = \mathfrak{a}$. ∎

We now come to the main result of this section.

THEOREM 4.3. *Let R be a 3-fir with right ACC_2, but not a right Ore ring. Then every invariant element of R is a unit.*

Proof. Since R is a domain but not right Ore, there exist $p, q \neq 0$ such that $pR + qR$ is direct. By ACC_2 we can find a right ideal $\mathfrak{a}$ which is maximal free of rank 2. Let c be any invariant element of R. By Lemma 4.2, $cR \cap \mathfrak{a} = \mathfrak{a}c$, and $cR + \mathfrak{a} = dR$ for some $d \in R$. Hence if $\mathfrak{a} = pR + qR$,

$$dR/cR \cong (pR + qR)/(pcR + qcR) \cong (R/cR)^2. \tag{3}$$

Now c is determined by R/cR up to associates (Lemma **4.2.6**); by (3), $(R/cR)^2$ is strictly cyclic, which can only happen when c is left comaximal with itself (Lemma **4.2.1**), hence c must be a unit. ∎

Recalling that a right fir satisfying the right Ore condition is a principal right ideal domain, we find the

COROLLARY. *If R is a right fir, then either R is a principal right ideal domain or every invariant element of R is a unit, and in particular, the centre of R is then a field.* ∎

It seems plausible that the conclusion of Theorem 4.3 is still true for 2-firs with right ACC_2 (but not right Ore). This has not been established so far, but there are examples showing it to be false for 2-firs with right ACC_1 (Bergman and Cohn [71]); see also Exercise 1.

Exercises 6.4

1. Show that the subalgebra of $Q\langle x, y \rangle$ consisting of elements in which the constant term has an odd denominator is a non-Ore 2-fir with non-unit central elements.

2. Let R be any ring. If a projective R-module P has a finitely generated dense submodule, show that P is itself finitely generated.
   Deduce that any right hereditary integral domain has right ACC_1.

3. Let R be a right hereditary integral domain, but not a right Ore ring. Show that R has no non-unit invariant elements.

4°. Determine the possible centres of (i) 2-firs with right ACC_2, and (ii) 2-firs with right ACC_1.

5°. What can be said about the right invariant or left invariant elements in a right fir?

6°. What can be said about the left large elements in a right fir? (For a right fir with right large non-units see **2.9**.)

6.5 Bounded elements in 2-firs

Let us return to consider invariant elements in a 2-fir R. We have seen that in an atomic 2-fir the decompositions into I-atoms play an important role. In general there is no reason to suppose that the I-atoms will be atoms, but at least we can factorize them into atoms, and this suggests that we look more closely at the factors of invariant elements. Such factors and the strictly cyclic modules they define are called *bounded*; however it will be convenient to begin with a general definition and specialize later.

Definition. A module M over an integral domain R is said to be *bounded* if there is a non-zero element c in R such that $Mc = 0$.

An equivalent definition is to require the annihilator of M in R to be non-zero. This annihilator is a two-sided ideal in R, called the *bound* of M. An element $a \in R$ is said to be *right bounded* if R/aR is bounded. Let $a = bc$, then for $d \in \mathrm{Ann}\,(R/aR)$, $Rd \subseteq aR = bcR$; in particular, $bd \in bcR$, so $d \in cR$. The same is true if instead of a we take any element similar to a. Thus

$$\mathrm{Ann}\,(R/aR) \subseteq \bigcap \{cR \mid c \text{ right factor of an element similar to } a\}. \tag{1}$$

In a 2-fir we can rewrite this as

$$\mathrm{Ann}\,(R/aR) \subseteq \bigcap \{cR \mid c \text{ similar to a right factor of } a\}. \tag{2}$$

In general equality need not hold here, but suppose now that a is right large. Given $x \in R^*$, we have $xR \cap aR \neq 0$, hence there is a right coprime relation

$$xa' = ax'.$$

Let d be the highest common left factor of a and x, say $a = da_0$, $x = dx_0$, then $x_0 a' = a_0 x'$ is a coprime relation, hence a' is similar to a_0 which is a right factor of a. Now choose any y in the right-hand side of (2). then in particular $y = a' y_1$ for some $y_1 \in R$, hence $xy = xa' y_1 \in aR$. Thus y annihilates $x \pmod{aR}$ and since x was arbitrary we see that $y \in \mathrm{Ann}\,(R/aR)$, i.e. equality holds in (2). Clearly any right bounded element is right large, for if $aR \supseteq \mathfrak{a}$, where $\mathfrak{a}$ is two-sided, then given $x \in R^*$, take $b \in \mathfrak{a}$, $b \neq 0$, then $xb \in \mathfrak{a} \subseteq aR$, so $xb = ay \neq 0$. The result can be summed up as

THEOREM 5.1. *Let R be a 2-fir. Then for any right large element a,*

$$\mathrm{Ann}\,(R/aR) = \bigcap \{cR \mid c \text{ similar to a right factor of } a\}. \tag{3}$$

In particular, this holds for any right bounded element. ∎

Suppose now that R is an atomic 2-fir and that a is right bounded. Then all the cR on the right of (3) contain a fixed non-zero element, d say, where $d \in \mathrm{Ann}\,(R/aR)$, and since the lattice $\mathbf{L}(dR, R)$ has finite length, it is complete and the intersection on the right of (3) is principal, say $\mathrm{Ann}\,(R/aR) = dR$. The element d, unique up to right associates, is called the *right bound* of a. Thus we have the

COROLLARY. *In an atomic 2-fir R, an element a is right bounded if and only if a is a left factor of a right invariant element d.*

In that case the right bound of a is a left factor of d.

Proof. We have seen that for a right bounded element a, $\mathrm{Ann}\,(R/aR) = dR$. Here $Rd \subseteq dR$, so d is right invariant, and $d \in aR$. Conversely, if $d = ab$ for a right invariant element d, then $Rd \subseteq dR \subseteq aR$, hence $d \in \mathrm{Ann}\,(R/aR)$, i.e. a is right bounded, with bound $d'R \supseteq dR$. ∎

An element a is said to be *bounded* if it is a left factor of an invariant element: $c = ab$; it is then also a right factor, for $cb = b'c = b'ab$, hence $c = b'a$. Thus the notion defined here is symmetric. A bounded element is clearly left and right bounded, and we have, by Theorem 2.2 and Proposition 2.1,

$$\mathrm{Ann}\,(R/aR) = \mathrm{Ann}\,(R/Ra). \tag{4}$$

Conversely, if a is left and right bounded and (4) holds, then $\mathrm{Ann}\,(R/aR) = dR = Rc$, by Theorem 5.1 and its left–right analogue, hence by Proposition 2.1, $\mathrm{Ann}\,(R/aR)$ has an invariant generator and a is then bounded.

Let a be bounded, then the invariant generator of $\mathrm{Ann}\,(R/aR)$, a^* say, is unique up to associates, and is called the *bound* of a. Clearly a^* depends only on the similarity class of a. By Theorem 5.1 a^* can also be defined by

$$a^* R = \bigcap \{cR \mid c \text{ similar to a right factor of } a\}. \tag{5}$$

This intersection can be taken to be a finite intersection, by the DCC in $R/a^* R$, and if we take it to be irredundant, we obtain a subdirect sum representation of $R/a^* R$, *qua* right R-module, by modules R/dR. This shows that every atomic factor of a^* is similar to a factor of a. By Theorem 5.1 this characterizes the atomic factors of a^* as the atoms similar to factors of a.

Let $a \in R$ have a bound a^*; if $a = bcd$, then c is again bounded, with bound dividing a^*. This follows by observing that R/cR is a quotient of a submodule of R/aR. Likewise the product of any bounded elements is bounded. These facts may be expressed by saying that the modules R/aR, *a* bounded, form the objects of a *dense* subcategory of the category $\mathscr{C}_R$ of strictly cyclic *R*-modules.

The next result gives the connexion between *I*-atoms and the atoms they bound.

PROPOSITION 5.2. *In an atomic 2-fir R, let p be an atom which is bounded. Then its bound p^* is an I-atom whose atomic factors are precisely the atoms similar to p. If K is the eigenring of pR and $n = l(p^*)$, then K is a field and*

$$R/p^*R \cong K_n.$$

Conversely, every I-atom of R occurs in this way.

Proof. Let *p* have bound $p^* = a^* b^*$, where a^*, b^* are non-unit invariant elements. Then *p* divides either a^* or b^*, and so has a smaller bound, a contradiction. Thus p^* is an *I*-atom. By Theorem 5.1, $p^*R = \bigcap p'R$, where p' runs over all elements similar to *p*, and in fact we can take a finite intersection. Thus R/p^*R is a submodule of $(R/pR)^N$ for some *N* (in fact p^* is fully reducible, *cf.* 3.6), and since R/pR is $\mathscr{C}$-simple, every submodule has the form $(R/pR)^n$ for some $n \leqslant N$. Hence

$$R/p^*R \cong (R/pR)^n \qquad (6)$$

as right *R*-modules. By comparing the lengths of composition series within $\mathscr{C}_R$, we see that $l(p^*) = n$, and comparing endomorphism rings in (6) we obtain the ring isomorphism

$$R/p^*R \cong K_n,$$

where *K*, the eigenring of *pR* is a field by Schur's Lemma.

By (5) an atom divides p^* if and only if it is similar to *p*, and to establish the final part, let p^* be any *I*-atom and *p* an atom dividing it. Then p^* is the bound of *p* and we can complete the proof by applying the first part. ∎

We next look at the direct decompositions of R/aR, where *a* is bounded. Our first task is to separate out the bounded components in such a decomposition; for this we need a definition and a lemma. Let us call two elements *a*, *b* of a 2-fir *R* *totally coprime* if no non-unit factor of *a* is similar to a factor of *b*, i.e. if R/aR and R/bR have no isomorphic factors apart from 0.

LEMMA 5.3. *A bounded element in a 2-fir can be comaximally transposed with any element totally coprime to it.*

Proof. Let a, b be totally coprime and let a be bounded, with bound $a^* = a_1 a$ say, then a^* and b are left coprime and hence right comaximal: $a^* u - bv = 1$. It follows that $u' a^* - bv = 1$ for some $u' \in R$, i.e. $u' a_1 a - bv = 1$. Hence by Lemma 3.4.3, a, b are comaximally transposable. By the symmetry of the situation, b, a are also comaximally transposable. ∎

Let a be any element in an atomic 2-fir R and take an atomic factorization

$$a = p_1 p_2 \cdots p_r. \tag{7}$$

Suppose that a bounded atom occurs in this factorization, say $p_{i_1}, \ldots, p_{i_k}$ are all bounded similar atoms, while the remaining atoms in (7) are not similar to p_{i_1}. By repeated application of Lemma 5.3 we can write a as

$$a = p_{i_1}' \cdots p_{i_k}' p_{j_1}' \cdots p_{j_h}',$$

where the p_i' but not the p_j' are similar to p_{i_1}. Bracketing the first k factors together, and the last h, we have $a = bc$, where b, c are totally coprime and b, like p_{i_1}, is bounded. Applying Lemma 5.3 again, we have $a = c'b'$, where b' is similar to b and c' similar to c. Therefore $a = bc = c'b'$ leads to a direct decomposition of a:

$$aR = bR \cap c' R.$$

If we repeat the process with a replaced by c', we eventually reach a direct decomposition of a into products of pairwise similar bounded atoms and a term containing no bounded non-unit factors. Let us call an element *totally unbounded* if it has no non-unit factor that is bounded. Then we can state our result as

THEOREM 5.4. *Any element a of an atomic 2-fir R has a direct decomposition*

$$aR = q_1 R \cap \ldots \cap q_k R \cap uR, \tag{8}$$

where each q_i is a product of similar bounded atoms, while atoms in different q's are dissimilar and u is totally unbounded. Moreover, the q_i and u are unique up to right associates.

Only the uniqueness remains to be proved. But (8) corresponds to a direct decomposition

$$R/aR \cong R/q_1 R \oplus \ldots \oplus R/q_k R \oplus R/uR.$$

Here the $R/q_i R$ are uniquely determined as the homogeneous components corresponding to a given $\mathscr{C}$-simple bounded isomorphism type, while R/uR contains all $\mathscr{C}$-simple submodules of unbounded isomorphism type. ∎

We note that neither u nor the q_i are in general indecomposable in the sense defined in 3.6.

For an invariant element c we are naturally interested in decompositions of R/cR within $I = \mathbf{I}(R)$. We therefore define an element $c \in I$, to be I-*decomposable* if it has a factorization

$$c = ab, \tag{9}$$

into non-unit invariant elements a, b which are left (and hence also right) coprime; otherwise c is I-*indecomposable*. There is a simple relationship between I-indecomposable and bounded elements that are indecomposable as defined in 3.6.

PROPOSITION 5.5. *Let R be an atomic 2-fir. Then*

(i) *an invariant element is I-indecomposable if and only if it is a power of an I-atom,*

(ii) *if $q \in R$ is bounded and indecomposable, then its bound is I-indecomposable, hence q is then a product of similar atoms.*

Proof. (i) follows by Theorem 2.3. To prove (ii) let q be bounded, with bound q^*. Since q^*R annihilates $M = R/qR$, we may regard M as Q-module, where $Q = R/q^*R$. If q^* is I-decomposable, Q is a non-trivial direct product and hence contains a central idempotent $e \neq 0, 1$. But then $M = Me + M(1 - e)$ is a direct decomposition of M into strictly cyclic modules, which contradicts the indecomposability of q. Hence q^* is indecomposable, and so is a power of an I-atom p^* say. Any atomic factor p of p^* has p^* as its bound and so p^* is a product of factors similar to p. The same holds for q^* and hence for q. ∎

Of course an *un*bounded indecomposable element need not be a product of similar atoms, as the example xy in the free algebra $k\langle x, y \rangle$ shows. Further, the converse of Proposition 5.5 (ii) is false, i.e. a decomposable bounded element need not have an I-decomposable bound; thus in Proposition 5.5 (ii), q^* itself may well be decomposable.

Let q^* be an I-indecomposable element in an atomic 2-fir R, say $q^* = p^{*e}$, $l(p^*) = h$. By Proposition 5.2, $R p^* R \cong K_h$, where the field K is the eigenring of an atomic factor of p^*. Now $R/p^* R \cong Q/\mathbf{J}(Q)$, where $Q = R/q^*R$ and $\mathbf{J}(Q)$ is the Jacobson radical of Q. Since Q is Artinian (Theorem 2.2), we can lift the matrix basis from R/p^*R to $R q^*R$ (cf. Jacobson [64], p. 56), whence $Q \cong L_h$, where $L/\mathbf{J}(L) \cong K$, i.e. Q is an Artinian matrix local ring over the scalar local ring L. Note that L, like Q, is Artinian, thus it is completely primary.

Now take a complete direct decomposition of R/q^*R as right R-module. The summands are necessarily strictly cyclic; thus

$$R/q^*R \cong R/q_1R \oplus \ldots \oplus R/q_kR. \tag{10}$$

But since $R/q^*R \cong L_h$ has a complete direct decomposition into h isomorphic right ideals, we see that $k = h$ and all the R/q_iR are isomorphic. Thus $R/q^*R \cong (R/qR)^h$, as right R-modules, for some indecomposable q. Since $l(q^*) = eh = hl(q)$, we see that $l(q) = e$. The result may be summed up as

THEOREM 5.6. *In an atomic 2-fir R, let q be bounded indecomposable with bound q^*. Express q^* as the power of an I-atom, say $q^* = p^{*e}$, then $l(q) = e$, and if $l(p^*) = h$, then*

$$R/q^*R \cong (R/qR)^h, \tag{11}$$

as right R-modules, while as ring, R/q^*R is a total matrix ring over a completely primary ring:

$$R/q^*R \cong L_h, \quad \text{where} \quad L = \text{End}_R(R/qR). \; \blacksquare \tag{12}$$

If q, q^* are as in Theorem 5.6, then q^*R is determined by the similarity class of q as the annihilator of R/qR, while R/qR is determined by q^*R as an indecomposable part of R/q^*R. Hence we have the

COROLLARY. *In an atomic 2-fir R, two bounded indecomposable elements have the same bound if and only if they are similar.* $\blacksquare$

Next we turn to the question of deciding when a given product of similar bounded atoms is indecomposable. Let p^* be any I-atom; its length $l(p^*)$ is called the *capacity* of p^*. For any integer $e \geqslant 0$ we have by Theorem 5.6, on decomposing p^{*e},

$$R/p^{*e}R \cong (R/q_eR)^h$$

for some indecomposable element q_e of length e. Thus q_e is a product of e atomic factors which are all similar.

Conversely, if p is a bounded atom with bound p^* and p_i is similar to p for $i = 1, \ldots, e$, then $q = p_1 \ldots p_e$ is bounded by p^{*e}, and if its exact bound is p^{*e}, then q is indecomposable. For if q could be decomposed it would have a smaller bound. This proves

PROPOSITION 5.7. *In an atomic 2-fir R, each bounded indecomposable element is a product of similar atoms. Let p be a bounded atom and p^* its bound. Then a product $q = p_1 \ldots p_e$ of atoms similar to p is indecomposable if and only* *if p^{*e} is the exact bound of q. Moreover, for any e, there is a bounded indecomposable element q_e of length e such that*

$$R/p^{*e} R \cong (R/q_e R)^h, \qquad \text{where} \qquad h = l(p^*). \quad \blacksquare$$

COROLLARY 1. *In an atomic 2-fir, if a has the bound p^{*e}, where p^* is an I-atom, then a direct decomposition of a has at most $l(p^*)$ terms.* $\blacksquare$

If an element a has bound p^{*e} and b has bound p^{*f}, then ab is bounded by p^{*e+f}. Applying this remark to a product of similar bounded atoms, we obtain

COROLLARY 2. *Any factor of a bounded indecomposable element is again bounded indecomposable.* $\blacksquare$

Now a product of two atoms $p_1 p_2$ is indecomposable if and only if they cannot be comaximally transposed. Hence any product $q = p_1 \dots p_n$ of bounded atoms has no indecomposable factors apart from atoms if and only if no pair of adjacent factors can be comaximally transposed, i.e. if q is rigid (*cf.* 3.4). By Proposition 5.7, two similar atoms p_1, p_2 with common bound q^* are comaximally transposable if and only if $p_1 p_2 \nmid p^*$. Thus we have

COROLLARY 3. *Let R be an atomic 2-fir and $q = p_1 \dots p_e$ a bounded product of atoms, then q is rigid if and only if the p_i have a common bound p^* and $p_{i-1} p_i \nmid p^* (i = 2, \dots, e)$. In particular, p^r is rigid if and only if $p^2 \nmid p^*$.* $\blacksquare$

There remains the problem of finding which elements are bounded. We shall confine ourselves to the case of an atomic 2-fir R. If p is an atom in R, $K = \text{End}_R (R/pR)$ is a field, and each element $a \in R$ defines a K-endomorphism of R/pR. We denote the natural mapping from R to R/pR by $x \mapsto \bar{x}$, then the K-endomorphism defined by a is

$$\rho_a : \bar{x} \to \overline{xa}.$$

We want to find an upper bound for the K-dimension of $\ker \rho_a$. Clearly if $a = a_1 \dots a_r$, then $\rho_a = \rho_{a_1} \dots \rho_{a_r}$ and by Sylvester's law of nullity (*cf.* 5.5),

$$\dim \ker \rho_a \leqslant \dim \ker \rho_{a_1} + \dots + \dim \ker \rho_{a_r}. \qquad (13)$$

We assert that ρ_a is injective when a has no factor similar to p. Thus assume $\ker \rho_a \neq 0$, then there exists $x \notin pR$ such that $xa \in pR$, say

$$xa = py \qquad (x \notin pR). \qquad (14)$$

Since p is an atom, (14) is left coprime, hence $a = p'a'$, where p' is similar to p. This shows that ρ_a is injective when a has no factor similar to p.

Suppose now that a is similar to p, then any element $x_1 \in R$ satisfying $x_1 a = p y_1$ for some $y_1 \in R$, while $x_1 \notin pR$, defines an isomorphism $R/aR \to R/pR$ and any two such isomorphisms differ by an endomorphism of R/pR, i.e. an element of K. Hence $\ker \rho_a$ is 1-dimensional in this case. Going back to (13) we see that $\ker \rho_a$ has a dimension at most equal to the number of factors of a that are similar to p. Thus we obtain

THEOREM 5.8. *Let R be an atomic 2-fir, p an atom and $K = \mathrm{End}_R (R/pR)$. Then for any $a \in R$, the mapping $\rho_a \colon \bar{x} \to \overline{xa}$ is a K-endomorphism of R/pR and*

$$\dim_K \ker \rho_a \leqslant m, \tag{15}$$

where m is the number of factors in a complete factorization of a that are similar to p. ∎

Suppose now that p is a right large atom. Then for any $x \notin pR$ there is a comaximal relation

$$xp' = px';$$

hence there exists p' similar to p and annihilating $\bar{x}$. Conversely, as we have seen, any p' similar to p annihilates some $\bar{x}$, hence when p is right large and q is any atom, then ρ_q is injective if and only if q is not similar to p. We derive two consequences.

Firstly let p be a right bounded atom, with right bound p^* of length m. Then p is certainly right large, and by (15),

$$\dim_K (R/pR) \leqslant m. \tag{16}$$

If this inequality were strict, we could find a product a of less than m factors similar to p which annihilates R/pR, and hence, by Theorem 5.1,

$$aR \subseteq \bigcap \{p' R \mid p' \text{ similar to } p\} = p^* R.$$

Thus $a \in p^* R$ and this contradicts the fact that $l(a) < m = l(p^*)$. Hence equality holds in (16) and we have proved

COROLLARY 1. *If p is a right bounded atom in an atomic 2-fir R with eigenring K and right bound p^*, then*

$$\dim_K (R/pR) = l(p^*). \quad \blacksquare$$

This then shows that for a right bounded atom p, R/pR is finite dimensional (and of course p is right large). Now let p be right large and

R/pR finite dimensional, then we can as before find a annihilating R/pR, hence p is right bounded and we obtain

COROLLARY. 2. *Let R be an atomic 2-fir; then an atom p in R with eigenring K is right bounded if and only if p is right large and* dim $_K(R/pR)$ *is finite.* ■

Exercises 6.5

1. Let R be an integral domain. Show that every bounded R-module is bound, but in general the converse is false.

2. Let R be a principal ideal domain and $a \in R^*$. If $\bigcap dR$, where d runs over all elements similar to a right factor of a, is non-zero, show that a is bounded. Show that this no longer holds for firs, by taking $R = k\langle x, y \rangle$, $a = x$.

3. (Jacobson [43]). Let K be a field of finite dimension over its centre and α an automorphism of K such that α^r is inner for some $r \geqslant 1$. Show that every non-zero element of the skew polynomial ring $K[x; \alpha]$ is bounded; illustrate this by the complex-skew polynomial ring.

Give an example of a product of similar bounded atoms which is decomposable in the sense of 3.6.

4. For a bounded element in an atomic 2-fir R, show that 'left indecomposable' = 'indecomposable'. If q is such an element, show that R/qR has only one composition series.

5. Let R be a ring. For any R-module M define the *tertiary radical* (of 0 in M) to be the set of all $x \in R$ which annihilate a large submodule of M. If R is an atomic 2-fir and c is bounded in R, find conditions for R/cR to have a tertiary radical of the form $bR \supseteq cR$. If c is indecomposable, show that $b = p^*$, where p^* is the I-atom corresponding to c.

6. Prove (4) for a bounded element in an atomic 2-fir.

6.6 Free semigroups

Before discussing subalgebras of free algebras, it is helpful to look at free semigroups, where we shall meet the same problems, though in a simplified form, since there is only one operation. To preserve the analogy we shall take all our semigroups to have a unit-element, denoted again by 1; then units and atoms can be defined as in rings. It is particularly instructive to see what becomes of the weak algorithm in semigroups; as we saw in 2.4, the weak algorithm may be used to characterize free algebras, and we shall obtain a corresponding result for semigroups.

Let $S = S_X$ be the free semigroup on a set X; a normal form for the elements of S_X is obtained by taking all finite sequences of elements of X—including the empty sequence to represent 1—and defining the multiplication in S_X by juxtaposition. Then it is clear that 1 is the only unit of S and X is the precise set of atoms. Thus the free generating set X of S can be characterized in terms of the semigroup structure of S alone and this shows that S has only one free generating set. It is this fact that accounts for the simplicity of the theory.

An arbitrary semigroup S is said to admit *cancellation* and is called a *cancellation semigroup* if for all x, y, $z \in S$

$$xz = yz \quad \text{or} \quad zx = zy \quad \text{implies} \quad x = y. \tag{1}$$

Clearly every free semigroup admits cancellation. We can now characterize free semigroups as follows:

THEOREM 6.1. *Let S be a cancellation semigroup and X its set of atoms. Then S is free if and only if the following conditions hold*:

(i) *S has no unit other than 1,*

(ii) *S is generated by X,*

(iii) *given a, a', b, $b' \in S$, if $ab' = ba'$, then either $a = bc$ or $b = ac$ for some $c \in S$ i.e. S is rigid.*

Moreover, when these conditions hold, X is a free generating set of S (in fact the only one).

Proof. Let S be free, then we know that its free generating set is uniquely determined as the set of its atoms. Now (i), (ii) are evident and (iii) follows easily from the normal form described above.

Conversely, assume that (i)–(iii) hold; by cancellation, every one-sided unit is two-sided, hence by (i), S is conical. It will be enough to show that each element of S can be written in just one way as a product of elements of X. Let $a \in S$, then by (ii) there is at least one way of so expressing a; suppose that

$$a = x_1 x_2 \ldots x_r = y_1 y_2 \ldots y_s \qquad x_i, y_j \in X. \tag{2}$$

By (iii), $x_1 = y_1 b$ or $y_1 = x_1 b$ for some $b \in S$, say the former holds. Since x_1, y_1 are atoms, b must be a unit, and by (i) $b = 1$, i.e. $x_1 = y_1$. Cancelling the factor x_1 and applying induction on $\max(r, s)$ we find that the two expressions for a in (2) agree, so S is indeed free on X. ∎

If we introduce the *length* $l(a)$ of an element a, as the number of atomic factors in a complete decomposition of a, we see that when $l(a) \geqslant l(b)$, then $ab' = ba'$ implies $a = bc$.

Theorem 6.1 may also be used to give criteria for a subsemigroup of a free semigroup to be free. Consider for example the free semigroup on one free generator x (the free cyclic semigroup); the subsemigroup generated by x^2 and x^3 is commutative, and if free must be cyclic, which is clearly not the case. The conditions of the next theorem make it easy to find other examples of non-free subsemigroups of free semigroups.

THEOREM 6.2. *A subsemigroup T of a free semigroup S is free if and only if one of the following three equivalent conditions holds*:

(a) *Given $a \in T$, $s \in S$, if $as \in T$ and $sa \in T$ then $s \in T$,*

(b) *given a, b, a', $b' \in T$, if $ab' = ba'$, then $a = bc$ or $b = ac$ for some $c \in T$, i.e. T is rigid,*

(c) *given a, b, $b' \in T$, if $ab' = ba$, then $a = bc$ or $b = ac$ for some $c \in T$.*

Proof. If (b) holds, we can apply Theorem 6.1 to show that T is free: clearly T can have no units apart from 1, and to express a non-unit of T as a product of atoms of T, we simply write the element as a product of members of T with as many factors as possible; since S is free, there is a bound on the number of factors. Conversely it is clear that a free subsemigroup T satisfies (b). Further, (c) is a special case of (b), so assume that (c) holds and let $ab' = ba'$, then $b'a . b'b = b'b . a'b$. By (c) we have either $b'a = b'bc$ or $b'b = b'ac$ for some $c \in T$, and by cancelling b' we obtain (b).

We complete the proof by showing that (a) $\Leftrightarrow$ (c). If (c) holds and $au = b$, $ua = b'$, where a, b, $b' \in T$, then $aua = ba = ab'$, hence by (c), $a = bc$ or $b = ac$ for some $c \in T$. If $a = bc$, then $a = auc$ and hence $u = c = 1$; if $b = ac$, then $ac = au$ and so $u = c$. In either case $u \in T$, which proves (a). Conversely, assume (a) and let $ab' = ba$ (a, b, $b' \in T$); clearly it is enough to consider the case $b \neq 1$. Let us put $a = b^r u$ ($u \in S$) with r chosen as large as possible; this is permissible since r is bounded by the length of a (in S). Then $b^r ub' = b^{r+1} u$, hence

$$bu = ub'. \tag{3}$$

By (c) applied to S we have $b = uv$ or $u = bv$ for some $v \in S$; here the second alternative conflicts with the definition of r, hence

$$b = uv, \qquad b' = vu. \tag{4}$$

We conclude that $av = b^{r+1} \in T$ and $va = b'^{r+1} \in T$; applying (a) we find that $v \in T$ and hence applying (a) again to (4) we conclude that $u \in T$. Thus if $r > 0$ we have $a = b . b^{r-1} u$, while for $r = 0$, $a = u$ and $b = av$. This shows that (c) holds. ∎

We note that whereas (a) refers to S, conditions (b) and (c) are intrinsic in T. In a free semigroup, let us associate with every element $u \neq 1$ the shortest element of which it is a power (the "least repeating segment") and call this the *root* of *u*. E.g. *xyxyxy* has the root *xy*. Every element $u \neq 1$ has a unique root, which may also be characterized as an element of shortest positive length commuting with *u*. From the criteria of Theorem 6.2 we easily obtain the

COROLLARY. *In a free semigroup two elements different from* 1 *commute if and only if they have the same root.* ∎

This result shows that commutativity is an equivalence relation on the set of elements different from 1, and the equivalence classes, with 1 adjoined, are free cyclic semigroups.

In **0.7** we saw that the free associative algebra $k\langle X \rangle$ on X over a field k may be defined as the semigroup algebra over k of the free semigroup S_X; the weak algorithm, holding in the free algebra, may be regarded as the analogue of condition (iii) of Theorem 6.1. Using the criterion of Theorem 6.1 we can show that the homogeneous elements of $k\langle X \rangle$ form a free semi-group. Let us take X totally ordered, and order the monomials in X by increasing length, while monomials of the same length are ordered lexicographically. Using this ordering we can speak of the highest (= last) term of any non-zero element of $k\langle X \rangle$.

PROPOSITION 6.3. *In any free associative algebra* $R = k\langle X \rangle$, *the homogeneous elements in which the coefficient of the highest term is* 1 *form a free semigroup.*

Proof. Denote the set of homogeneous elements with coefficient of the highest term 1 by S; it is clear that S is a semigroup. Since R is an integral domain, S satisfies cancellation and from the definition it is clear that S has no units other than 1. In any factorization of an element $f \in S$ the number of factors $\neq 1$ cannot exceed the degree of f, so by taking the maximum number of factors we can express f as a product of atoms in S. This shows that (i), (ii) of Theorem 6.1 hold, and it remains to verify (iii). Let $ab' = ba'$ in S and $d(a) \geqslant d(b)$ say. Then by the weak algorithm, there exists $c \in R$ such that

$$d(a - bc) < d(a). \qquad (5)$$

If in c we only keep terms of highest degree, then (5) is unaffected and a, bc are now homogeneous; by (5) they have the same degree. Thus $a - bc$ is homogeneous of degree $d(a)$ and (5) shows that $a - bc = 0$. Clearly $c \in S$ and so (iii) also holds and we conclude that S is free. ∎

If we do not want to use the total ordering on the free generators, we can express the result by saying that the set of all homogeneous elements of R is the direct product of k^* (*qua* multiplicative group) and a free semigroup S.

Exercises 6.6

1. Let S be a free semigroup. Show that two elements of S form a free generating set if and only if they do not commute.

2. Prove the following generalization of Theorem 6.1: Let G be a group, F a free semigroup and denote their free product $G*F$ by S. Then (i') given units u, v, if $ua = av$ for some non-unit a, then $u = v = 1$, (ii') given any element c of S, the number of non-unit factors in a factorization of c is bounded, (iii) as in Theorem 6.1.
   Conversely, show that any cancellation semigroup satisfying (i'), (ii'), (iii) is of the form $G*F$, where G is a group and F is a free semigroup. Here G is uniquely determined (as the group of units of S) while F is determined up to isomorphism.

3. Let S be a semigroup in which the relation "$a \leqslant b$ if and only if $ax = b$ for some $x \in S$" is a well-ordering. If S also has right cancellation ($xz = yz$ implies $x = y$), show that S either consists of one element or is isomorphic to $\mathbf{N}$, the additive semigroup of non-negative integers.

4*. Let S be a semigroup that is well-ordered under the relation in Exercise 3 and has left cancellation. Show that S has a presentation with generators x_α $(\alpha < \tau)$ and defining relations $x_\alpha x_\beta = x_\beta$ if $\alpha < \beta$, where τ is some ordinal.

5. Use Theorem 6.2 to find a procedure for reducing a finite subset X of a free semigroup to a set X' such that (i) X and X' generate the same subsemigroup T, and (ii) if T is free, X' is a free generating set.

6^0. Let F be the free group on a finite set X and S the subsemigroup generated by X. Classify the subsemigroups between S and F. Does this set of subsemigroups satisfy ACC?

6.7 Subalgebras of free associative algebras

A subalgebra of a free associative algebra is not necessarily free. This is already clear by considering the subalgebra of the polynomial ring $k[x]$ which is generated by x^2 and x^3. We therefore look for criteria for a subalgebra to be free. In the case of a polynomial ring a simple criterion is given by

PROPOSITION 7.1. *A subalgebra R of the polynomial ring $k[x]$ is free if and only if it is integrally closed.*

Proof. Assume that R is a free subalgebra; then either $R = k$ or $R = k[y]$ for some y transcendental over k. Clearly R is integrally closed in each case.
   Conversely, let R be integrally closed, and denote its field of fractions by K. Since $k \subseteq R$, we have $k \subseteq R \subseteq k(x)$. Now if $K = k$, then $R = k$ and the result follows. Otherwise, by Lüroth's theorem, $K = k(y)$, where y is

transcendental over k. Since R is integrally closed, it is an intersection of valuation rings of $k(y)$ (*cf.* **6.2**). Now any valuation ring of $k(y)$ is of the form $\mathfrak{o}_p \cap k(y)$, where $\mathfrak{o}_p$ is a valuation ring of $k(x)$ (*cf.* e.g. Lang [58]). If x is finite at p, then $\mathfrak{o}_p \supseteq k[x] \supseteq R$, whence $R \subseteq \mathfrak{o}_p \cap k(y)$. Thus R is contained in all valuation rings of $k(y)$ over k, except at most one, namely the one obtained from the pole of x. Since $R \neq k$, R is not contained in all valuation rings; if $\mathfrak{o}_q$ is the exceptional one, we can by a suitable choice of the generator y in K ensure that q is a pole of y, and since R is the intersection of all the other valuation rings of $k(y)$ over k, it follows that $R = k[y]$. ∎

For free algebras on more than one free generator, no such convenient criterion is available. We can of course use the characterization in terms of the weak algorithm given in **2**.4. Applied to subalgebras, this yields

PROPOSITION 7.2. *Let $A = k\langle X \rangle$ be a free associative algebra; then a sub-algebra B of A is free if and only if there is a filtration on B (over k) for which B satisfies the weak algorithm.* ∎

The difficulty in applying this criterion lies in the fact that the different degree-functions in a free associative algebra are not related in any obvious way. We obtain sufficient conditions by observing that the usual degree-function in A (defined in terms of X) defines a degree-function in B:

COROLLARY. *Let $A = k\langle X \rangle$; then any subalgebra of A which satisfies the weak algorithm relative to the X-degree in A is free.* ∎

The converse is false (*cf.* Exercise 1); however we do obtain necessary and sufficient conditions if we impose homogeneity:

PROPOSITION 7.3. *Let $A = k\langle X \rangle$ be free on X over k. Then any subalgebra B generated by homogeneous elements is free if and only if it satisfies the weak algorithm relative to the X-degree.*

For the condition is clearly sufficient to ensure that B is free. Conversely, if B is free on Y, we may clearly take Y to consist of homogeneous elements. If with each $y_i \in Y$ we associate a positive integer n_i, then B satisfies the weak algorithm relative to the degree-function defined by

$$d\left(\Sigma y_{i_1} \ldots y_{i_r} \alpha_{i_1 \ldots i_r} \right) = \max \{ n_{i_1} + \ldots + n_{i_r} | \alpha_{i_1 \ldots i_r} \neq 0 \}. \tag{1}$$

Now take n_i to be the X-degree of y_i; then (1) just gives the X-degree of the general element of B and the result follows. ∎

Generally if A is a free associative algebra, any free generating set X defines a degree-function d_X for which the weak algorithm holds. Conversely, every degree-function (over k) for which the weak algorithm holds, leads to a free generating set of A, and two free generating sets of A define the same degree function if and only if they are related by a linear transformation. Further, the free generating sets of A are permuted transitively and regularly by $\mathscr{A} = \text{Aut}\,(A)$, the group of automorphisms of A, i.e. taking the free generating sets to be families indexed by the same set I, if (x_i), (y_i) are any two families freely generating A, then there is one and only one automorphism $\alpha \in \mathscr{A}$ such that $x_i^\alpha = y_i$. It follows that the degree-functions for which the weak algorithm holds are permuted transitively by $\mathscr{A}$ and the stabilizer of any one of them, d_X say, is the subgroup L_X of $\mathscr{A}$ which is realized by linear transformations of the corresponding free generating set X. Thus the introduction of a degree-function reduces the problem of studying the group $\mathscr{A}$ to a study of the coset space $\mathscr{A}/L_X$. However, very little is known about this set.

If B is a free subalgebra of a free algebra A, there are always degree-functions on B for which the weak algorithm holds (Proposition 7.2), but the question remains whether a degree-function exists on A for which both A and B satisfy the weak algorithm (in that case B might be called "regularly embedded" in A). Clearly the existence of such a function would be of practical interest since it would lead to free generating sets of A and B related in a specially simple way.

There is one case in which more information is available, namely that in which $A = k[x]$ is free on one generator. We already know a simple criterion for subalgebras to be free (Proposition 7.1), and as we shall see, there is also a simple criterion in terms of the algorithm. For in this case (and only here) Aut (A) consists entirely of linear transformations $x \mapsto ax + b$ $(a, b \in k, a \neq 0)$, hence there is a unique degree-function in A. We thus obtain the following criterion for subalgebras of $k[x]$ to be free:

PROPOSITION 7.4. *A subalgebra B of $k[x]$ is free if and only if it has the division algorithm (relative to the degree in x).* ∎

We now return to free algebras on any generating sets and apply Proposition 7.2, Corollary to obtain a sufficient condition for a subalgebra to be free. If A is any filtered ring (filtered by v) and B is a subring, then it is clear that B is again filtered, using the filtration v of A. Moreover, in defining v-dependence in A we may restrict the coefficients to lie in B, and it makes sense to speak of a family of elements of A being right v-dependent over B, or an element of A being v-dependent over B on a given family of elements of A. This situation is really a special case of a filtered module over a filtered ring; we shall not spell out the definitions here as we shall have no occasion to use this general concept.

THEOREM 7.5. *Let $A = k\langle X \rangle$ be free on X over k. If B is a subalgebra of A such that A is free as right B-module, with a right d_X-independent basis (over B), then B is a free associative algebra over k.*

Proof. We shall verify that B satisfies the weak algorithm relative to the degree-function $d = d_X$ and apply Proposition 7.2, Corollary.

Let $U = (u_v)$ be a right d-independent basis of A_B; this must include an element of degree 0 which we may take to be $u_0 = 1$. Given any left d-independent set $b_1, \ldots, b_k \in B$, with $d(b_1) \geqslant \ldots \geqslant d(b_k)$ say, then by the weak algorithm, in A, some b_i is left d-dependent on $b_{i+1}, \ldots, b_k$, say,

$$b_i = \Sigma c_j b_j + c' \qquad (c_j, c' \in A), \tag{2}$$

where

$$d(c_j b_j) \leqslant d(b_i), \qquad d(c') < d(b_i). \tag{3}$$

Putting $c_j = \Sigma u_v a_{vj}$, $c' = \Sigma u_v a_v'$, where a_{vj}, $a_v' \in B$, and inserting these values in (2), we obtain

$$u_0 b_i = \Sigma\Sigma u_v a_{vj} b_j + \Sigma u_v a_v',$$

whence

$$b_i = \Sigma a_{0j} b_j + a_0'. \tag{4}$$

Moreover, by (3) and the right d-independence of U over B,

$$d(a_{0j} b_j) \leqslant d(c_j b_j) \leqslant d(b_i), \qquad d(a_0') \leqslant d(c') < d(b_i),$$

this shows b_i to be left d-dependent on $b_{i+1}, \ldots, b_k$ over B; hence B satisfies the weak algorithm, as asserted. ∎

It is clear how to define the weak algorithm in a filtered module over a filtered ring; when it is satisfied we know that the module has a d-independent basis. Therefore we obtain

COROLLARY 1. *If $A = k\langle X \rangle$ is free on X over k and B is a subalgebra such that the module A_B satisfies the weak algorithm (relative to d_X), then B is a free subalgebra.* ∎

For homogeneously generated subalgebras we have a somewhat stronger result:

COROLLARY 2. *If $A = k\langle X \rangle$ is free on X over k and B is a subalgebra generated by homogeneous elements (relative to X), such that A_B is free, with a d-independent basis over B, then B has a free generating set consisting of homogeneous elements.*

For by Theorem 7.5, B is free; moreover the homogeneous components of any element of B again lie in B. Now any right d-independent generating set (over B) of the augmentation ideal of B is a free generating set of B as algebra (by Proposition 2.4.2) and it is a simple matter to choose such a generating set to consist of homogeneous elements. ∎

Examples of non-free subalgebras of free algebras are obtained from almost any two-sided ideal, by the following result:

PROPOSITION 7.6. *Let $A = k\langle X \rangle$ be a free associative algebra. If $\mathfrak{a}$ is any non-zero ideal of A such that the subalgebra B generated by $\mathfrak{a}$ is different from A, then B is not free.*

Proof. Suppose that B is free on Y. Take $x \in X$, $y \in Y$ and let $y = y_0 + \alpha$, where $y_0 \in \mathfrak{a}$ and $\alpha \in k$. Put $a = xy_0$, $b = y_0 x$, then $a, b \in \mathfrak{a}$ and

$$y_0 a = b y_0, \tag{5}$$

i.e. y_0 and b are right d_Y-dependent. By the weak algorithm in B, and the fact that $d_Y(y_0) = 1$, we find that

$$b = y_0 c + \beta, \qquad c \in B, \ \beta \in k.$$

Thus $y_0(x - c) = \beta$; it follows that $\beta = 0$ and $x = c \in B$. Since x was any element of X, we conclude that $B = A$. ∎

For one-sided ideals this no longer holds; see Exercises 0.7.6–7 for examples of this fact, and other examples of non-free subalgebras.

Exercises 6.7

1. Let $R = k\langle x, y \rangle$; show that the subalgebra B generated by $x + y^2$ and y^3 is free on these generators but does not satisfy the weak algorithm relative to the filtration by (x, y)-degree.
   By a suitable change of variables show that B is regularly embedded in R.

2^0. Is every free subalgebra of $R = k\langle X \rangle$ regularly embedded in R?

3. Let R be a finitely generated graded k-algebra (k a field), S a subalgebra generated by homogeneous elements and $f:S \to R$ a degree-preserving surjective homomorphism. Show that $S = R$ and f is an automorphism.
   Deduce that any augmentation preserving surjective endomorphism of a free algebra of finite rank is an automorphism.

4. Let $R = k\langle x_1, ..., x_n \rangle$, $S = k\langle y_1, ..., y_m \rangle$. If there is a surjective homomorphism $f:R \to S$, show that $m \leqslant n$, with equality if and only if f is an isomorphism.

5^0. (Bergman). Let R, S be as in Exercise 4, and $f:R \rightarrow S$ a surjective homomorphism. Does there exist an automorphism g of R such that $(\ker f)^g$ is the ideal of R generated by $x_{m+1}, ..., x_n$?

6^0. (Clark). Is every retract of a free algebra of rank n free? (For $n = 1, 2$ see **6.8**.)

7^0. Does the set of subalgebras of a free algebra satisfy ACC_n?

8^0. Let R be a free algebra and B a subalgebra generated by m elements $y_1, ..., y_m$. Is it true that either B is free on $y_1, ..., y_m$ or there is a strictly larger subalgebra on m generators? (A positive answer to Exercises 7 and 8 would provide an answer to Exercise 6).

6.8 Centralizers in power series rings and in free algebras

In a free algebra, few elements commute with each other, and one would expect the centralizer of a non-scalar element to be small. This expectation is borne out, as we shall show in Theorem 8.5 below; a similar question can be raised for the free power series ring, and since this is rather easier to answer, we begin with it.

THEOREM 8.1. *Let R be a negatively filtered ring which is complete and satisfies $\mu_v(R) > 2$. Then the centralizer C of any $a \in R$, not zero or a unit, is a complete discrete valuation ring.*

If moreover, R is a connected k-algebra, then the centralizer of any non-scalar element of R is a formal power series ring in one variable over k and hence is commutative.

Proof. Let x, $y \in R$ be right v-dependent and $v(x) \leqslant v(y)$, then $x - yz$ and y are right v-dependent for any $z \in R$, hence by Lemma **2.8.3**, $x = yz$ for some $z \in R$, and so $xR \subseteq yR$.

Now let x, $y \in C$, then for any $n \geqslant 0$, $xa^n = a^n x$, hence for sufficiently large n, $a^n R \subseteq xR$ and $a^n R \subseteq yR$. Thus $xR \cap yR \neq 0$, and so if $v(x) \leqslant v(y)$, then $x = yz$. Clearly $z \in C$, so $xC \subseteq yC$, i.e. the right ideals of C are totally ordered by inclusion. Since R is atomic (by Proposition **2.8.6**), and any non-unit of C remains a non-unit in R, it follows that C is atomic and therefore a rigid UFD (Theorem **3.4.6**, Corollary). It contains a large non-unit, namely a; hence it is a right (and by symmetry left) discrete valuation ring (Theorem **2.4.7**), and it is complete, since R is.

Now assume that R is connected, and let x be an element of maximal (negative) degree in C, so that every ideal of C has the form $Cx^n = x^n C$. Then every element of C has the form $x^n v$, where v is a unit. The additional hypothesis allows us to write $v = \alpha + v_1 x^r$ $(\alpha \in k)$ and an induction argument (using the completeness) shows that $C = k[[x]]$. ∎

In particular, the free power series ring satisfies all the hypotheses, and we obtain the

COROLLARY. *The centralizer of a non-scalar element in the free power series ring $k\langle\!\langle X \rangle\!\rangle$ is of the form $k[[c]]$.* ∎

We know from Proposition 2.8.11 that any non-scalar element in a free power series ring generates a free power series ring in one variable. Two elements need not generate a free power series algebra; e.g. the complete algebra generated by x^2 and x^3 is not free. But we have

PROPOSITION 8.2. *Let R be a negatively filtered ring, which is complete and satisfies $\mu_v(R) > 2$, and is a connected k-algebra. If x, y are any two non-commuting elements of R, then the complete algebra generated by x, y is the free power series ring $k\langle\!\langle x - \alpha, y - \beta \rangle\!\rangle$, where α, β are the constant terms of x, y respectively.*

Proof. By hypothesis $xy - yx \neq 0$, hence $-v(xy - yx)$ is a positive integer n, say; we shall use induction on n. Note that n remains unchanged if we replace x, y by $x - \alpha$, $y - \beta$ respectively (α, $\beta \in k$), so we may assume that $v(x)$, $v(y) < 0$.

If x, y are not free, we have a relation

$$xf + yg + \lambda = 0, \tag{1}$$

where f, g are expressions in x and y and $\lambda \in k$. Equating constant terms in (1) we find $\lambda = 0$, and so x, y are right v-dependent. If $v(x) \leqslant v(y)$ say, then

$$x = yz. \tag{2}$$

If we can show that y, z are free, then so are x, y by (2), for the elements y, yz are clearly free in $k\langle\!\langle y, z \rangle\!\rangle$ (since they are left v-independent). Now

$$0 \neq xy - yx = yzy - yyz = y(zy - yz),$$

hence $yz - zy \neq 0$ and $v(yz - zy) > v(xy - yx)$. The result follows by induction, since we must have $v(xy - yx) < 0$ for any x, $y \in R$. ∎

The result can be expressed more strikingly by saying: x and y are free if and only if they do not commute. In particular, since any free algebra can be embedded in a power series ring, we have the

COROLLARY. *Any two non-commuting elements of a free associative algebra form a free set.* ∎

We now go on to consider centralizers in a free algebra $R = k\langle X \rangle$. Let $a \in R$, $a \notin k$, and let C be the centralizer of a in R. The embedding of R in

$k\langle\!\langle X\rangle\!\rangle$ shows that C is commutative (by Theorem 8.1, Corollary); more-over, C is finitely generated, as module over $k[a]$ or as algebra. For if $d(a) = n$ say, we choose for each integer $v = 0, ..., n - 1$ such that an element of degree $\equiv v \pmod n$ occurs in C, an element of least degree $\equiv v \pmod n$ in C. Calling these elements $c_0, ..., c_r$ $(r \leqslant n - 1)$, we see that the elements $c_i a^j$ $(i = 0, ..., r, \ j = 0, 1, ...)$ form a k-basis of C, hence $c_0, ..., c_r$ generate C as $k[a]$-module, and together with a they generate C as k-algebra by Theorem 6.2, Corollary and Proposition 6.3.

Our aim is to show that C is a polynomial ring over k. In order to establish this fact we shall study homomorphisms of C into polynomial rings. If we look at leading terms of elements of C we note that all are powers of a given one and it turns out that in order to achieve a homo-morphism of C into a polynomial ring we need a preordering of the free semigroup which lists each word together with all its powers. This is done by introducing 'infinite' words.

Let W be the free semigroup on a totally ordered set X and let $\overline{W}$ be the set of all right infinite words in X, i.e. infinite sequences of letters from X. Given $u \in W$, $u \neq 1$, we denote by u^∞ the word obtained by repeating u indefinitely: $u^\infty = uuu \ldots \in \overline{W}$. Clearly the mapping $u \mapsto u^\infty$ is a mapping from W to $\overline{W}$ which identifies two words if and only if they are powers of the same root.† We shall take W to be ordered lexicographically.

LEMMA 8.3. *Given* $u, v \in W \setminus \{1\}$, *if* $u^\infty > v^\infty$, *then*

$$u^\infty > (uv)^\infty > (vu)^\infty > v^\infty, \tag{3}$$

and similarly with $>$ *replaced by* $<$ *or* $=$ *throughout.*

Proof. Suppose that $(uv)^\infty > (vu)^\infty$; then

$$(vu)^\infty = v(uv)^\infty > v(vu)^\infty = v^2(uv)^\infty > v^2(vu)^\infty = \ldots \to v^\infty,$$

since lexicographic order is 'continuous'. Similarly we find that $(uv)^\infty < u^\infty$, therefore (3) follows whenever $(uv)^\infty > (vu)^\infty$. Likewise $(uv)^\infty < (vu)^\infty$ implies

$$u^\infty < (uv)^\infty < (vu)^\infty < v^\infty, \tag{4}$$

while $(uv)^\infty = (vu)^\infty$ implies

$$u^\infty = (uv)^\infty = (vu)^\infty = v^\infty. \tag{5}$$

Now for any u, v exactly one of (3), (4), (5) holds and the assertion follows. ∎

Let R be the semigroup algebra on W over k. This is just the free associative algebra $k\langle X\rangle$. Given any periodic word z in $\overline{W}$, i.e. an

† E.g. can and cancan.

infinite power of a word in W, let us define A_z as the k-subspace of R spanned by words u satisfying $u = 1$ or $u^\infty \leqslant z$, and let I_z be the k-subspace spanned by words u such that $u \neq 1$ and $u^\infty < z$. By Lemma 8.3, A_z is a subalgebra of R in which I_z is a two-sided ideal.

The set of words u in W with $u^\infty = z$, together with 1, will form the set of non-negative powers of an element v which we shall call again the *root* of z. It follows that $A_z/I_z \cong k[v]$.

PROPOSITION 8.4. *Let C be a finitely generated subalgebra of a free associative algebra R over k. If $C \neq k$, then there is a homomorphism f of C into the polynomial ring over k in one variable, such that $f(C) \neq k$.*

Proof. Let $R = k\langle X\rangle$ and let us totally order X. Take a finite generating set Y for C and let z be the maximum of u^∞ as u ranges over all monomials $\neq 1$ occurring with non-zero coefficient in members of Y. Then $Y \subseteq A_z$, hence $C \subseteq A_z$ and the quotient mapping $f: A_z \to A_z/I_z \cong k[v]$ is non-trivial on C. ∎

When C is not finitely generated, the result need no longer hold (*cf.* Exercise 4).

Consider now a free associative algebra $R = k\langle X\rangle$. The centralizer C of a non-scalar element is a commutative finitely generated k-algebra, as we have seen. Therefore it can be mapped non-trivially into a polynomial algebra; now C as finite extension of $k[a]$ has transcendence degree 1 over k and so it must be embedded in the polynomial algebra. Being integrally closed (by Proposition 1.5, Corollary) it must be free (Proposition 7.1). This proves Bergman's centralizer theorem:

THEOREM 8.5. *Let $R = k\langle X\rangle$ be free on X over k; then the centralizer of any non-scalar element of R is a polynomial ring in one variable over k.* ∎

Exercises 6.8

1. (Bergman [67]) Let $R = k\langle X\rangle$ and let C be the centralizer of an element $a \in R^*$. Show that the valuation on $k[a]$ given by the degree in a is totally ramified on C.

2. (Schur [04]). Let R be the differential ring $k[x; 1, ']$, where $k = F(t)$ is a rational function field and $'$ denotes differentiation with respect to t. Show that the centralizer of any element outside F is commutative. (Hint: Use Theorem 8.1 to form the completion of R by negative powers of x.)

3. (Bergman [67]). Let W be the free semigroup on $x_1, ..., x_{n-1}$ and $\overline{W}$ the set of infinite sequences. With each $u = a_1a_2 ...$ in W or $\overline{W}$ associate the "decimal

expansion" $\lambda(u) = \Sigma a_i n^{-i}$, and obtain a formula for $\lambda(u^\infty)$ in terms of $\lambda(u)$ and the length of u. Hence express $\lambda((uv)^\infty)$ as a convex linear combination of $\lambda(u^\infty)$ and $\lambda(v^\infty)$.

4. Let R be a free algebra of rank > 1. Show that the subalgebra of R generated by the commutator ideal cannot be mapped into a polynomial ring. Use this fact to show that the hypothesis that C is finitely generated in Proposition 8.4 cannot be omitted.

5. In a free algebra over a field of characteristic 0, if $ab + ba = 0$, show without using Theorem 8.5 that either a or b is 0.

6. Let $R = k\langle X \rangle$ be a free algebra and B the subalgebra on two elements $y, z \in R$. Show that either B is free on y and z or B is contained in a polynomial subring of R of the form $k[c]$.

7^0. (Bergman [67]). Let R be a free algebra and $\hat{R}$ its power series completion. Given $a \in R$, denote by C, C' its centralizers in R and $\hat{R}$ respectively. Is C' the closure of C in $\hat{R}$?

8*. (Koševoi [71]). Let R be the algebra generated over a field k by a, b, c, d with the defining relation $ad = bc$. Show that the mapping $a \mapsto x$, $b \mapsto xy$, $c \mapsto z$, $d \mapsto yz$ extends to a homomorphism of R into $A = k\langle x, y, z \rangle$ which is injective.

For any $u \in A$, show that if a non-scalar polynomial in u belongs to R, then u itself does. Deduce that the conclusion of Bergman's centralizer theorem extends to R.

Notes and comments on Chapter 6

The main result of Bergman's thesis was his centralizer theorem (Theorem 8.5; *cf.* Bergman [67, 69]). This had been conjectured by the author in the early '60's (*cf.* Cohn [62] for Proposition 8.1 and Cohn [63'] for the fact that the centralizer is commutative). Problem 7.5 is the non-commutative analogue of a long-standing conjecture in algebraic geometry which has recently been proved by Abhyankar and Moh when $m = 1$, $n = 2$, and k has characteristic 0.

Condition (a) of Theorem 6.2 was obtained by Schützenberger [59] in connexion with comma-free codes. The treatment in sections 6.6 and 6.7 follows Cohn [62'], [64] respectively. The notions of invariant and bounded element are treated for principal ideal domains in Jacobson [43]; our account in 6.2, 6.5 follows Bowtell–Cohn [71]; for Theorem 5.8 see Carcanague [71]. Section 6.1 is taken from Bergman [67], and 6.3–4 and Theorem 2.4 from Bergman–Cohn [71], though with some modification of proofs. For Cohen's Theorem see Kaplansky [70].

The structure of automorphisms of free algebras has not been mentioned so far, mainly because very little was known at the time of writing (*cf.* Cohn [69]). But recently one of the main conjectures has been proved by A. Czerniakiewicz for free algebras of rank 2: Every automorphism is tame (an automorphism is *tame* if it can be expressed as a product of linear automorphisms and transformations of the form $x_1 \mapsto x_1 + f(x_2, ..., x_n)$, $x_i \mapsto x_i$ ($i \neq 1$)). The proof depends on the theorem of Jung [42] that every automorphism of the polynomial ring $k[x, y]$ is tame.

7. Skew fields of fractions

This chapter studies ways of embedding rings in fields, or more generally, the homomorphisms of rings into fields. For a commutative ring such homomorphisms can be completely described in terms of the set of its prime ideals, and in the course of this chapter we shall see that the same description applies to quite general rings.

After some generalities on the rings of fractions obtained by inverting matrices (7.1) and on R-fields and their specializations (7.2), we introduce in 7.3 the notion of a matrix ideal. This corresponds to the concept of an ideal in a commutative ring, but has no direct interpretation. The analogue of a prime ideal, the prime matrix ideal defined in 7.4, has properties corresponding closely to those of prime ideals; what is more, the prime matrix ideals can be used to describe the homomorphisms of general rings into fields, just as prime ideals are used in the commutative case. This follows from Theorem 5.3, which characterizes the prime matrix ideals as the sets of matrices which become singular under a homomorphism into some field.

This characterization is applied in 7.6 to derive criteria for a general ring to be embeddable in a field, or to have a universal field of fractions. Finally these results are used to show that every semifir has a universal field of fractions.

7.1 The Σ-rational closure of a homomorphism

Let R be a ring; our basic problem will be to study the possible ways of embedding R in a field. Of course there may be no such embedding, and it is more natural to treat the wider problem of finding homomorphisms of R into a field. Even this problem may have no solution, e.g. if $R = K_n$ is a total matrix ring over a field, where $n > 1$, then R is simple, so any homomorphism f of R into a field must be injective (f cannot be zero, because $1^f = 1 \neq 0$), and this is impossible because R has zero-divisors.

As a step towards the solution we may take a subset M of R and consider M-inverting homomorphisms. In the commutative case, once we have an R^*-inverting homomorphism, we have achieved the embedding in a field, but in general, if we have an R^*-inverting homomorphism, we do not necessarily get an embedding in a field. Thus in the general case the M-inverting homomorphisms are not very good approximations to homomorphisms into a field. We shall remedy this defect by inverting, instead of a set of elements, a set of matrices.

Let R be a ring and Σ a set of square matrices over R. A homomorphism $f: R \to S$ is said to be Σ-*inverting* if every matrix in Σ is mapped by f to an invertible matrix in S. Given any Σ-inverting homomorphism $f: R \to S$, we define the Σ-*rational closure of R in S* as the set $R_\Sigma(S)$ of all components of inverses of matrices in Σ^f, the image of Σ under f; the elements of $R_\Sigma(S)$ are also said to be Σ-*rational* over R. When Σ is the set of all matrices whose images under f have an inverse in S, we omit reference to it and speak simply of the *rational closure* and *rational* elements, and denote the rational closure by $R^f(S)$.

As we shall see, the rational closure of a ring R under a homomorphism f is always a subring containing im f. For general sets Σ the Σ-rational closure need not be a subring, as we know from the commutative case. If M is a multiplicative subset of a commutative ring R, then as we have seen in **0.5**, the localization R_M is a ring. Let us call a set Σ of square matrices *multiplicative* if $1 \in \Sigma$, and whenever $A, B \in \Sigma$, then $\begin{pmatrix} A & C \\ 0 & B \end{pmatrix} \in \Sigma$ for any matrix C of the appropriate size.

We first check that the set of all matrices inverted in a homomorphism is multiplicative.

PROPOSITION 1.1. *Given any homomorphism $f: R \to S$, where S is a ring with IBN, let Σ be the set of all matrices of R whose image under f is invertible in S. Then Σ is multiplicative.*

Proof. Clearly 1^f is invertible, and if A, B are invertible matrices, then for any matrix C of suitable size,

$$\begin{pmatrix} A & C \\ 0 & B \end{pmatrix} \quad \text{has the inverse} \quad \begin{pmatrix} A^{-1} & -A^{-1}\,CB^{-1} \\ 0 & B^{-1} \end{pmatrix}.$$

Further, the IBN in S ensures that all invertible matrices in S are square. ∎

We next characterize the Σ-rational closure in various ways. In stating the result we shall use the notation e_i for the column vector with 1 in the ith place and 0's elsewhere.

THEOREM 1.2. *Let R be a ring and Σ a multiplicative set of matrices over R. Given any Σ-inverting homomorphism $f: R \to S$, the Σ-rational closure $R_\Sigma(S)$*

is a subring of S containing im f, *and for any* $x \in S$ *the following conditions are equivalent*:

(a) $x \in R_\Sigma(S)$,

(b) x *is a component of the solution* u *of a matrix equation*

$$Au + a = 0, \tag{1}$$

where $A \in \Sigma^f$ *and* a *is a column with components in* im f,

(c) x *is a component of the solution* u *of a matrix equation*

$$Au - e_j = 0, \tag{2}$$

where $A \in \Sigma^f$.

Proof. We first prove the equivalence of the three conditions.

(a) $\Rightarrow$ (b). By definition $R_\Sigma(S)$ consists of the components of the inverses of matrices in Σ^f. If x occurs as (i, j)-element in A^{-1}, then it is the i-th component of the solution of (2), which is a special case of (1). Thus (b) holds.

(b) $\Rightarrow$ (c). The equation (1) is equivalent to the equation

$$\begin{pmatrix} A & a \\ 0 & 1 \end{pmatrix} \begin{pmatrix} u \\ 1 \end{pmatrix} - \begin{pmatrix} 0 \\ 1 \end{pmatrix} = 0, \tag{3}$$

which is of the form (2).

(c) $\Rightarrow$ (a). Clearly the ith component in the solution of (2) is the (i, j)-element of A^{-1}.

To prove that the Σ-rational closure $R_\Sigma(S)$ is a ring containing im f we shall use property (b). Let $c \in$ im f, then c satisfies the equation $1 \cdot u - c = 0$, which is of the form (1), hence $R_\Sigma(S) \supseteq$ im f. Next if u_1 is the first component of the solution of (1) and v_1 the first component of the solution of

$$Bv + b = 0,$$

then $u_1 - v_1$ is the first component of the solution of

$$\left(\begin{array}{c|cc} A & a_1 & 0 \\ \hline 0 & & B \end{array} \right) w + \begin{pmatrix} a \\ b \end{pmatrix} = 0,$$

where a_1 is the first column of A, while $u_1 v_1$ is the first component of the solution of

$$\left(\frac{A \mid a \quad 0}{0 \quad\quad B}\right) z + \binom{0}{b} = 0.$$

Corresponding equations hold for components other than the first. This shows that $R_\Sigma(S)$ is closed under subtraction and multiplication, and we have already seen that it contains 1, therefore it is a subring, as claimed. ∎

Any property of the solution u of (1) (or its components) can generally be expressed in terms of A and a; we illustrate this by considering zero-divisors and invertibility in S. Since we are only concerned with the image under f, we lose nothing by confining ourselves to the case of an embedding; thus we may take R to be a subring of S. The result to be stated is essentially *Cramer's rule*:

PROPOSITION 1.3. *Given a ring S and a subring R of S, let u_1 be the first component of the solution of*

$$Au \mid a = 0, \tag{1}$$

where A is invertible over S, and write $A = (a_1, ..., a_n)$, $A_1 = (a, a_2, ..., a_n)$. Then u_1 is (i) *a left zero-divisor,* (ii) *a right zero-divisor,* (iii) *left invertible,* (iv) *right invertible in S if and only if A_1 has the corresponding property in S_n.*
  Further, $u_1 = 0$ if and only if a is a right linear combination of the last $n - 1$ columns of A over S.

Proof. We firstly note that if an element c in a ring has one of the properties (i)–(iv), then any associate has the same property. Secondly, we note that A_1 is associated to $\begin{pmatrix} u_1 & 0 \\ 0 & I \end{pmatrix}$ in S_n (where I is of order $n - 1$); for on writing $u = \binom{u_1}{u'}$, we have

$$A_1 = A(A^{-1}A_1) = A \begin{pmatrix} u_1 & 0 \\ u' & I \end{pmatrix} = A \begin{pmatrix} u_1 & 0 \\ 0 & I \end{pmatrix} \begin{pmatrix} 1 & 0 \\ u' & I \end{pmatrix},$$

and here the outer factors on the right are units in S_n. Finally it is clear that u_1 has any one of the properties stated precisely when $\begin{pmatrix} u_1 & 0 \\ 0 & I \end{pmatrix}$ does.

To prove the last sentence we rewrite (1) as

$$a_1 u_1 + a_2 u_2 + ... + a_n u_n + a = 0. \tag{4}$$

If $u_1 = 0$, (4) is a dependence relation of the required kind. Conversely, let $v_2, ..., v_n \in S$ be such that

$$a_2 v_2 + ... + a_n v_n + a = 0,$$

then (1) has the solutions $(u_1, ..., u_n)^T$ and $(0, v_2, ..., v_n)^T$, hence by uniqueness, $u_1 = 0$. ∎

Exercises 7.1

1. Let R be a commutative ring and Σ a subset of R. Find conditions on Σ for the set $\{a^f(s^f)^{-1} | a \in R, s \in \Sigma\}$, under any Σ-inverting homomorphism f, to be a subring.

2. Show that if in a homomorphism $f : R \to S$, S is the rational closure of f, then f is epimorphic. Give an example to show that the converse is false.

3. Let K be any field. If the transpose of every invertible matrix is invertible, show that K is commutative. (Hint: Try a 2×2 matrix with 1 in the $(1, 1)$-position.)

4. A set of matrices satisfying the conditions for a multiplicative set but including rectangular matrices will be called *pseudo-multiplicative*. Show that in any homomorphism $f : R \to S$ the set of all matrices of R whose image under f is invertible (in the sense of having a two-sided inverse) form a multiplicative or pseudo-multiplicative set. State and prove an analogue of Theorem 1.2 for pseudo-multiplicative sets.

5*. Given a homomorphism $f : R \to S$, define the *division closure* of R in S as the least subring of S containing im f and closed under forming inverses (when they exist in S). If every matrix over R has an image in S which is either a left zero-divisor or invertible, show that the division closure is contained in the rational closure. Give examples to show that in general neither of these two closures is contained in the other.

6. Given a homomorphism $f : R \to S$, if every non-zero square matrix from R maps either to a left and right zero-divisor (or 0) or to an invertible matrix over S, show that the rational closure of R in S is such that every non-zero element is a right and left zero-divisor or invertible in S.

If further, S is an integral domain, deduce that the rational closure of R is a field. Show that under a homomorphism of R into a field, the rational closure is a subfield.

7. Let u_1 be the first component of the solution u of $Au = e_1$ and denote by A_{11} the matrix obtained by omitting the first row and column from A. Show that when A is invertible, then u_1 is a (left or right) zero-divisor (respectively invertible) if and only if A_{11} is.

8. Given a homomorphism $f : R \to S$, where S is weakly finite, let Σ be the set of matrices inverted by f. Show that for any square matrices A, B over R, if $\begin{pmatrix} A & C \\ O & B \end{pmatrix} \in \Sigma$ for some C, then $A, B \in \Sigma$.

9*. Let R and S be algebras over an infinite field. Given a homomorphism $f : R \to S$, show that for any finite set of elements of the rational closure there exists a matrix A and columns $c_1, ..., c_r$ such that the given elements are the first components of the solutions of the equations $Ax - c_i = 0$ $(i = 1, ..., r)$.

7.2 The category of R-fields and specializations

Given a ring R, by an *R-ring* we understand a ring L with a homo-morphism $R \to L$. The R-rings (for fixed R) form a category in which the maps are the ring-homomorphisms $L \to L'$ such that the triangle

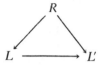

commutes.

By an *R-field* we shall mean an R-ring K which is a field, and such that K is the least field containing the image of R. If, moreover, the canonical mapping $R \to K$ is injective, we call K a *field of fractions* for R. Of course, as we saw earlier (in 7.1), for some rings R there may be no R-fields at all.

The only R-ring homomorphism possible between R-fields is an isomorphism. For any homomorphism between fields must be injective, and in this case the image will be a field containing the image of R, hence we have a surjection, and therefore an isomorphism. This shows the need to consider more general maps. Let us define a *specialization* between R-fields K, L as an R-ring homomorphism $f: K_0 \to L$ from an R-subring K_0 of K to L such that any element of K_0 not in the kernel of f has an inverse in K_0. The definition shows that K_0 is a local ring with maximal ideal $\ker f$, hence $K_0/\ker f$ is a field, isomorphic to a subfield of L, namely $\operatorname{im} f$. The latter is a subfield of L containing the image of R in L, and hence $\operatorname{im} f = L$, by the definition of L as R-field. Thus any specialization of R-fields is surjective. Two specializations from K to L are considered equal if they agree on a subring K_0 of K and the common restriction to K_0 is again a specialization.

The R-fields and specializations again form a category $\mathscr{F}_R$ say. Here it is only necessary to check that the composition of maps is defined. Given specializations $f: K \to L$, $g: L \to M$, let K_0, L_0 be the domains of f and g respectively, and put $K_1 = \{x \in K_0 | xf \in L_0\}$, $f_1 = f|K_1$. We assert that $f_1 g: K_1 \to M$ defines a specialization. Let us denote the canonical mapping

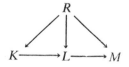

$R \to K$ by μ_K, then we have $\mu_K f = \mu_L$, hence $R\mu_K \subseteq K_1$, so that K_1 is an R-ring. Moreover, if $x \in K_1$ and $xf_1 g \neq 0$, then $xf = xf_1 \neq 0$, so $x^{-1} \in K_0$

and $(x^{-1})f_1 = (xf_1)^{-1} \in L_0$, hence $x^{-1} \in K_1$. This shows that $f_1 g$ in fact defines a specialization.

An initial object in the category $\mathscr{F}_R$ is called a *universal R-field*. Explicitly a universal R-field is an R-field U such that for any R-field K there is a unique specialization $U \to K$. Clearly a universal R-field, if it exists at all, is unique up to isomorphism.

In general a ring R need not have a universal R-field (e.g. a commutative ring has a universal R-field if and only if its nil radical is prime and we shall obtain an analogous condition for general rings later on, *cf.* Theorem 6.4). Suppose that R has a universal R-field U. Then R has a field of fractions if and only if U is a field of fractions, as a glance at the above triangle shows; in that case we call it the *universal field of fractions*.

Let us give some examples to illustrate these definitions.

1. Let R be a commutative ring. The R-fields correspond precisely to the prime ideals of R. Thus, given any R-field K, the kernel of the canonical mapping $\mu_K : R \to K$ is a prime ideal, and conversely, if $\mathfrak{p}$ is a prime ideal of R, then the mapping $R \to F(R/\mathfrak{p})$ (where $F(A)$ is the field of fractions of A) gives us an R-field. The category $\mathscr{F}_R$ in this case is isomorphic to the set of prime ideals of R, with inclusion mappings as the maps. There is a universal R-field if and only if there is a least prime ideal (i.e. the nil radical is prime), and when this is 0 (i.e. when R is an integral domain) we have a universal field of fractions. A similar correspondence exists in the general case, and will be described in 7.5 below, once we have identified the objects to be used in place of prime ideals.

2. Let R be a (left or right) Noetherian ring. Any R-field is obtained as a field of fractions of $R/\mathfrak{a}$, for a suitable ideal $\mathfrak{a}$. But $R/\mathfrak{a}$ is again Noetherian and an integral domain, as subring of a field, and hence (by 0.5.4) it is a (left or right) Ore domain and the field of fractions of $R/\mathfrak{a}$ is unique up to isomorphism. Let us call an ideal $\mathfrak{a}$ *strongly prime* if $R/\mathfrak{a}$ is an integral domain; what has been said shows that the category $\mathscr{F}_R$, for a left or right Noetherian ring, corresponds to the set of all strongly prime ideals of R, with inclusions as maps.

In order to study R-fields, we relate them to certain multiplicative sets of matrices, but before we can do so, we need the obvious but important remark that for any set Σ of matrices there always exists a *universal Σ-inverting homomorphism*: by this term we understand a homomorphism

$\lambda: R \rightarrow R_\Sigma$ which is Σ-inverting and such that any Σ-inverting homomorphism f can be factored uniquely by λ, i.e. given a Σ-inverting homomorphism $f: R \rightarrow S$, there is a unique homomorphism $\bar{f}: R_\Sigma \rightarrow S$ such that the accompanying triangle commutes. The ring R_Σ is clearly determined up to isomorphism by these conditions; it is called the *universal Σ-inverting ring*.

Such a ring always exists (for any choice of R and Σ) and it may be constructed as follows. For each $n \times n$ matrix $A = (a_{ij})$ in Σ we take a set of n^2 symbols $A' = (a_{ij}')$ and take a ring presentation of R_Σ consisting of all the elements of R, as well as all the a_{ij}' as generators, and as defining relations take all the relations holding in R, together with the relations, in matrix form,

$$AA' = A'A = I \qquad \text{for each } A \in \Sigma. \tag{1}$$

The mapping taking each element of R to the corresponding element of R_Σ is clearly a homomorphism $\lambda: R \rightarrow R_\Sigma$, which is Σ-inverting, by construction. If $f: R \rightarrow S$ is any Σ-inverting homomorphism, we define a homomorphism $\bar{f}: R_\Sigma \rightarrow S$ by putting $x\bar{f} = xf$ for any $x \in R$ and for any matrix $A \in \Sigma$ defining $\bar{f}$ on Af^{-1} by putting $(A\bar{f})^{-1} = (Af)^{-1}$. This gives a well-defined homomorphism $\bar{f}$, because any relation in R_Σ is a consequence of the defining relations in R and the relations (1). But all these relations also hold in S.

Of course the canonical homomorphism $\lambda: R \rightarrow R_\Sigma$ need not be injective and in fact it may be zero; this is so if and only if $R_\Sigma = 0$, and it will happen e.g. if $0 \in \Sigma$. But if there is a Σ-inverting homomorphism f which is injective, then λ must be injective, as we see from the commutative triangle shown above. We sum up these results in

THEOREM 2.1. *Let R be any ring and Σ any set of square matrices over R; then there is a universal Σ-inverting homomorphism*

$$\lambda: R \rightarrow R_\Sigma,$$

where the ring R_Σ is unique up to isomorphism. Moreover, λ is injective if and only if R can be embedded in a ring in which all the elements of Σ have inverses. ∎

With the help of the universal Σ-inverting rings the R-fields can be described as follows.

THEOREM 2.2. *Let R be any ring. Then*

(i) *if Σ is a set of matrices such that the universal Σ-inverting ring R_Σ is a local ring, then the residue-class field of R_Σ is an R-field,*

(ii) *if K is an R-field and Σ the set of all matrices over R whose images in K are invertible, then Σ is multiplicative and R_Σ is a local ring with residue-class field isomorphic to K.*

Proof. Let Σ be a set of matrices such that R_Σ is a local ring, and denote the residue-class field by K. By composing the natural mappings we get a homomorphism $R \to R_\Sigma \to K$, and K is generated by the inverses of matrices in Σ, hence it is an R-field.

Conversely, let K be any R-field and Σ the set of all matrices over R whose images in K are invertible. Then Σ is multiplicative, by Proposition 1.1. Further, by the definition of Σ and by Theorem 2.1, we have a commutative triangle as shown, and it will be enough to prove that any element of R_Σ not in ker α is invertible.

Let $u_1 \in R_\Sigma$ be the first component of the solution of

$$Au + a = 0,$$

where $A \in \Sigma$, and write $A = (a_1, ..., a_n)$, $A_1 = (a, a_2, ..., a_n)$. If $u_1{}^\alpha \neq 0$, then $u_1{}^\alpha$ is invertible, hence $A_1{}^\alpha$ is not a zero-divisor, by Proposition 1.3, and so is invertible. Therefore $A_1 \in \Sigma$, and by applying Cramer's rule (Proposition 1.3) we find that u_1 is invertible over R_Σ; thus every element of R_Σ not in ker α has an inverse. It follows that R_Σ is a local ring with maximal ideal ker α, and its residue-class field is therefore isomorphic to K, as claimed. ∎

This theorem shows that any R-field is completely determined by the multiplicative set of matrices which becomes invertible. Let us call a set Σ of matrices in R *localizing* if

(i) any matrix A which becomes invertible under the canonical homomorphism $R \to R_\Sigma$, already lies in Σ, and

(ii) R_Σ is a local ring.

Then Theorem 2.2 may be expressed by saying that R-fields correspond to localizing sets of matrices. In this correspondence inclusion relations correspond to specializations:

THEOREM 2.3. *In any ring R, let Σ_i $(i = 1, 2)$ be localizing sets of matrices and K_Σ the corresponding R-fields. Then there is a specialization $K_{\Sigma_1} \to K_{\Sigma_2}$ if and only if $\Sigma_1 \supseteq \Sigma_2$.*

Proof. Let R_i $(i = 1, 2)$ be the local ring corresponding to Σ_i, with maximal ideal $\mathfrak{m}_i$ and residue-class field K_i. First assume that $\Sigma_1 \supseteq \Sigma_2$ and let R_{12} be the subring of R_1 generated by the inverses of matrices in Σ_2. Then R_{12} is a homomorphic image of R_2, the universal Σ_2-inverting ring, and the natural homomorphism $R_1 \to K_1$ maps R_{12} to $R'_{12} = R_{12}/(R_{12} \cap \mathfrak{m}_1)$. Now R_{12} is a local ring and $R_{12} \cap \mathfrak{m}_1$ is a proper ideal, therefore the natural homomorphism $R_2 \to K_2$ can be taken via R'_{12}, giving a homomorphism from a local subring (namely R'_{12}) of K_1 onto K_2; this is the required specialization.

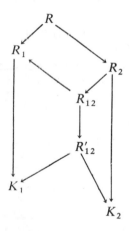

Conversely, let $\alpha: K_1 \to K_2$ be a specialization of R-fields, with corresponding localizing sets Σ_1, Σ_2 and let $\mu_i: R \to K_i$ be the canonical homomorphism. Take $A \in \Sigma_2$, then A^{μ_2} has an inverse which is the image of a matrix B in $K_1: (A^{\mu_2})(B^\alpha) = I$, hence $A^{\mu_1} B = I + C$, where $C^\alpha = 0$. Hence $I + C$ has an inverse, and therefore so does A^{μ_1}, i.e. $A \in \Sigma_1$, and this shows that $\Sigma_2 \subseteq \Sigma_1$. ∎

In order to pursue the analogy with the commutative case we need to consider the complements of localizing sets. To this end we shall develop the notion of a matrix ideal in the next section.

Exercises 7.2

1. Show that every commutative integral domain has a universal R-field. (By contrast there are noncommutative integral domains with no R-field at all, e.g. an integral domain without IBN may be embedded in a simple integral domain R, cf. Cohn [66, 59] and this cannot have an R-field.)

2. Show that a commutative ring R has a universal R-field if and only if R has a least prime ideal (i.e. the nil radical of R is prime). What is the corresponding condition for right Noetherian rings?

3. Let R be a commutative ring and Σ a set of square matrices over R. Show that R_Σ is isomorphic to R_S, where S is the set consisting of the determinants of elements of Σ.

4. If R is a right Bezout domain, show that for any multiplicative set Σ of matrices, $R_\Sigma = R_S$ for a suitable subset S of R. (Hint: Use Exercise 0.5.6)

5*. Let R be a ring, Φ the set of all full matrices over R and suppose there exists a Φ-inverting homomorphism of R into a weakly finite ring. Show that Φ is a localizing set.

6. Show that a homomorphism $R \to K$ of a ring into a field is epimorphic if and only if K is an R-field.

7. For which values of n does the ring R of integers mod n have (i) a field of fractions, (ii) a universal R-field?

8*. A set Σ of square matrices over a ring R is said to be *saturated* if any square matrix A over R and invertible in R_Σ lies in Σ. Show that any saturated set is multiplicative; it is moreover localizing if and only if it satisfies the following condition: For any matrix $A \in \Sigma$, denote by A_{11} the matrix obtained by omitting the first row and column, then either $A_{11} \in \Sigma$ or $A + e_{11} \in \Sigma$.

9. Given any set Σ of square matrices over a ring R, let Σ' be the set of all square matrices inverted in R_Σ. Show that Σ' is saturated, and that $R_\Sigma \cong R_{\Sigma'}$. ($\Sigma'$ is called the *saturation* of Σ).

10*. (J. Fisher [71]). Let k be a commutative field and $P = k[t]$ the polynomial ring with the endomorphism α_n induced by $t \mapsto t^n$. Show that the k-subalgebra of $P[x; \alpha_n]$ generated by x and $y = xt$ is free on x and y. Hence obtain an embedding of the free algebra $R = k\langle x, y \rangle$ in a field, viz. $K(x; \alpha_n)$ where $K = k(t)$. Show that the fields so obtained are final objects in the category of fields of fractions of R and are non-isomorphic, as R-fields, for different n.

11*. (Bergman). Let R be a ring with two non-isomorphic fields of fractions K, L; say L is not a specialization of K. Denote by $\bar{R}$ the subring of $K[t]$ consisting of polynomials with constant term in R. Show that $\bar{R}$ has a field of fractions, viz. $K(t)$, but no universal $\bar{R}$-field.

12⁰. For which specializations of R-fields is the domain a valuation ring?

13⁰. Let Σ be a localizing set of matrices, so that R_Σ is a local ring. Under what conditions on Σ is R_Σ actually a field? (For a case where this happens, see Theorem 6.5.)

14^0. If R is a semifir and Σ a multiplicative set of matrices, when is R_Σ again a semifir?

7.3 Matrix ideals

It is well known that any square matrix A over a field (even skew) is either invertible or is a two-sided zero-divisor (or 0); accordingly we call A *non-singular* or *singular*, respectively. In the last section we examined the set of matrices which become invertible under a homomorphism into a field; we now want to look at the complementary set, i.e. the set of all matrices which become singular under a homomorphism into a field. These sets show a close analogy to ideals in commutative rings, but instead of sum and product we have two operations on matrices, the determinantal sum and the diagonal sum, which we must now define.

Given two matrices A, B over a ring R, we define their *diagonal sum* as

$$A \dotplus B = \begin{pmatrix} A & 0 \\ 0 & B \end{pmatrix}.$$

This sum is always defined, and for square matrices of orders r, s the diagonal sum is square of order $r + s$. We shall use notations like $A_1 \dotplus A_2 \dotplus \dots \dotplus A_r$ for repeated diagonal sums; if in the last sum all the A_i are equal, we also write $\overset{r}{\dotplus} A$.

In Chapter 5 we defined a square matrix to be full if it is of inner rank equal to its order. Thus A is non-full if it can be written in the form PQ, where P is $n \times r$ and Q is $r \times n$, and $r < n$. Hence a non-full matrix maps to a singular matrix under any homomorphism into a field. If A is non-full, then its diagonal sum with any square matrix is again non-full. For if $A = PQ$ as above, then

$$A \dotplus B = \begin{pmatrix} P & 0 \\ 0 & B \end{pmatrix} \begin{pmatrix} Q & 0 \\ 0 & I \end{pmatrix}.$$

On the other hand, if A is full, its diagonal sum with another full matrix need not be full. The example of rings without IBN shows that for each integer n there is a ring in which the diagonal sum of $n - 1$ copies of 1 is full, but not the diagonal sum of n copies (*cf.* e.g. Cohn [66]). However, we note that if A is a matrix (full or not) such that $\overset{r}{\dotplus} A$ is not full, for some r, then $\overset{r}{\dotplus} A$ and hence A itself map to a singular matrix under any homomorphism into a field. The same is true for any diagonal sum of I and copies of A.

The second operation to be described is defined only for certain pairs of matrices. Let A, B be two $n \times n$ matrices which differ at most in the first row. Then we shall say that the *determinantal sum* of A and B with respect to the first row exists; it is the matrix C whose first row is the sum of the first rows of A and B and whose other rows agree with those of A (and B). The determinantal sum with respect to another row (or column) is defined similarly, when it exists. We write

$$C = A \bigtriangledown B, \tag{1}$$

indicating (if necessary) in words the row or column whose elements are being added.

When R is commutative, so that determinants are defined, we have $\det C = \det A + \det B$, whenever A, B, C are related as in (1). On the other hand, over a field, even skew, if A, B are singular, then so is C, as is easily seen. Further, any ring homomorphism preserves determinantal sums, therefore if A and B both map to singular matrices under a homomorphism into a field, then so does C.

Repeated determinantal sums need to be used with care, since the operation is not everywhere defined and a fortiori not associative. Thus to say that C is a determinantal sum of matrices $A_1, ..., A_r$ means that we can replace two of $A_1, ..., A_r$ by their determinantal sum (with respect to some row or column) and repeat this process on two matrices in the resulting set, and so on until we are only left with one matrix, namely C.

Let R be any ring. We define a *matrix pre-ideal* in R as a set $\mathcal{P}$ of square matrices over R satisfying the following three conditions:

1. *$\mathcal{P}$ includes all non-full matrices,*

2. *If A, $B \in \mathcal{P}$ and their determinantal sum C (with respect to some row or column) exists, then $C \in \mathcal{P}$,*

3. *If $A \in \mathcal{P}$, then $A \dotplus B \in \mathcal{P}$ for all square matrices B,*

If further, we have

4. *$A \dotplus 1 \in \mathcal{P}$ implies $A \in \mathcal{P}$,*

we call $\mathcal{P}$ a *matrix ideal*. A matrix pre-ideal is said to be *proper* if it does not contain the unit matrix of any size; clearly a matrix ideal is proper precisely when it does not contain the element 1.

We note the following simple consequences of the definitions. Let $\mathcal{P}$ be any matrix pre-ideal. Then

(a) Any square matrix with a zero row or column lies in $\mathcal{P}$. For if $A = (0 \, A')$, where 0 is a zero column and A' is an $n \times (n-1)$ matrix, then $A = A'(0 \, I)$, where I is of order $n - 1$, and this shows that A is not full. Similarly for other columns or for rows.

(b) Let $A \in \mathscr{P}$, then the result of adding any right multiple of one column of A (or any left multiple of a row) to another again lies in $\mathscr{P}$.

Write $A = (a_1, a_2, ..., a_n)$, then

$$(a_1 + a_2 c, a_2, ..., a_n) = A \nabla (a_2, ..., a_n) \begin{pmatrix} c \\ 0 & I_{n-1} \end{pmatrix};$$

on the right we have the determinantal sum with respect to the first column) of A and a non-full matrix, hence the result lies in $\mathscr{P}$ by 1 and 2.

(c) Let $A \in \mathscr{P}$, then the result of interchanging any two columns (or rows) of A and changing the sign of one of them again lies in $\mathscr{P}$.

This follows in familiar fashion from (b). Writing only the two columns in question, we have, by repeated application of (b),

$$(a_1, a_2) \to (a_1 + a_2, a_2) \to (a_1 + a_2, -a_1) \to (a_2, -a_1).$$

(d) Let A, B be any square matrices, say of orders m, n respectively. Then for any $n \times m$ matrix C,

$$\begin{pmatrix} A & 0 \\ C & B \end{pmatrix} \in \mathscr{P} \Leftrightarrow \begin{pmatrix} A & 0 \\ 0 & B \end{pmatrix} \in \mathscr{P}. \tag{2}$$

For given A, B, C, let a_1, c_1 be the first columns of A, C respectively, and write $A = (a_1 \ A')$, $C = (c_1 \ C')$, then

$$\begin{pmatrix} A & 0 \\ C & B \end{pmatrix} = \begin{pmatrix} a_1 & A' & 0 \\ 0 & C' & B \end{pmatrix} \nabla \begin{pmatrix} 0 & A' & 0 \\ c_1 & C' & B \end{pmatrix}, \tag{3}$$

where the determinantal sum is with respect to the first column. We have

$$\begin{pmatrix} 0 & A' & 0 \\ c_1 & C' & B \end{pmatrix} = \begin{pmatrix} A' & 0 \\ 0 & I \end{pmatrix} \begin{pmatrix} 0 & I & 0 \\ c_1 & C' & B \end{pmatrix};$$

thus the second matrix on the right of (3) is not full and so lies in $\mathscr{P}$. Now (3) may be rewritten

$$\begin{pmatrix} a_1 & A' & 0 \\ 0 & C' & B \end{pmatrix} = \begin{pmatrix} A & 0 \\ C & B \end{pmatrix} \nabla \begin{pmatrix} 0 & A' & 0 \\ -c_1 & C' & B \end{pmatrix},$$

where the second matrix on the right again is not full. Hence

$$\begin{pmatrix} A & 0 \\ C & B \end{pmatrix} \in \mathscr{P} \Leftrightarrow \begin{pmatrix} a_1 & A' & 0 \\ 0 & C' & B \end{pmatrix} \in \mathscr{P}.$$

In a similar way we can vary the other columns of C, and so prove the assertion. An entirely analogous argument, using rows, shows that for any $m \times n$ matrix C,

$$\begin{pmatrix} A & C \\ 0 & B \end{pmatrix} \in \mathscr{P} \Leftrightarrow \begin{pmatrix} A & 0 \\ 0 & B \end{pmatrix} \in \mathscr{P}.$$

(e) If $\mathscr{P}$ is actually a matrix ideal, then for any two square matrices A, B of the same size, $AB \in \mathscr{P}$ whenever either A or B belongs to $\mathscr{P}$.

Assume that $A \in \mathscr{P}$, say. Then $A \dotplus B \in \mathscr{P}$ by (3); using (d) and (b) several times, we obtain in turn

$$\begin{pmatrix} A & 0 \\ 0 & B \end{pmatrix} \to \begin{pmatrix} A & 0 \\ I & B \end{pmatrix} \to \begin{pmatrix} 0 & -AB \\ I & B \end{pmatrix} \to \begin{pmatrix} AB & 0 \\ -B & I \end{pmatrix} \to \begin{pmatrix} AB & 0 \\ 0 & I \end{pmatrix}.$$

Now an application of 4 shows that $AB \in \mathscr{P}$.

(f) If A belongs to a matrix ideal $\mathscr{P}$, then the result of permuting the rows or columns of A in any way again belongs to $\mathscr{P}$.

For we can permute the rows (or columns) by (c) and use (e) to get rid of the minus signs.

(g) A matrix ideal $\mathscr{P}$ is proper if and only if some square matrix does not belong to $\mathscr{P}$.

If $\mathscr{P}$ is proper, then $1 \notin \mathscr{P}$, by definition. Conversely, if $\mathscr{P}$ is improper, then $I \in \mathscr{P}$ and hence by (e), $A = AI \in \mathscr{P}$, for any square matrix A.

Let $(\mathscr{P}_\lambda)$ by any family of matrix ideals, then it is clear that $\mathscr{P} = \bigcap \mathscr{P}_\lambda$ is again a matrix ideal. We can therefore speak of the 'least' matrix ideal containing a given set $\mathscr{X}$ of square matrices, namely the intersection of all matrix ideals containing $\mathscr{X}$. This least matrix ideal containing $\mathscr{X}$ is also called the matrix ideal generated by $\mathscr{X}$. Similarly we can define the matrix pre-ideal generated by $\mathscr{X}$. Explicitly, this is obtained by forming the determinantal sum of matrices $X \dotplus A$ ($X \in \mathscr{X}$, A any square matrix) and of non-full matrices. For this set is contained in any matrix pre-ideal containing $\mathscr{X}$, and it satisfies 1 and 2; let us show that it also satisfies 3. If the set contains C, then (for suitable bracketing),

$$C = B_1 \nabla \ldots \nabla B_r \qquad (B_i = X \dotplus A \text{ or non-full}), \qquad (4)$$

hence for any square matrix P,

$$C \dotplus P = (B_1 \dotplus P) \nabla \ldots \nabla (B_r \dotplus P),$$

with the same bracketing.

Thus the matrix pre-ideal generated by a set $\mathscr{X}$ consists precisely of all determinantal sums C of the form (4). Now the connexion between matrix pre-ideals and matrix ideals is given by

PROPOSITION 3.1. *Let $\mathscr{P}$ be any matrix pre-ideal, then the least matrix ideal containing $\mathscr{P}$ is given by*

$$\overline{\mathscr{P}} = \{A | A \dotplus I \in \mathscr{P} \text{ for some unit-matrix } I\}. \tag{5}$$

Proof. If $\overline{\mathscr{P}}$ is defined by (5), then $\overline{\mathscr{P}} \supseteq \mathscr{P}$ and any matrix ideal containing $\mathscr{P}$ must also contain $\overline{\mathscr{P}}$, so it only remains to show that $\overline{\mathscr{P}}$ is itself a matrix ideal. Properties 1–3 are clear. If $A \dotplus 1 \in \overline{\mathscr{P}}$ then $A \dotplus 1 \dotplus I \in \mathscr{P}$, i.e. $A \dotplus I \in \mathscr{P}$ (for a unit-matrix of larger order than before), and hence $A \in \overline{\mathscr{P}}$. Thus 4 also holds, and the result follows. ■

COROLLARY. *A matrix pre-ideal generates a proper matrix ideal if and only if it is itself proper. More specifically, the matrix ideal generated by a set $\mathscr{X}$ is proper if and only if the unit matrix I (of any size) cannot be expressed as a determinantal sum of non-full matrices and matrices of the form $X \dotplus A$, where $X \in \mathscr{X}$ and A is any square matrix.* ■

Let $\mathscr{Z}$ be the matrix pre-ideal generated by the empty set. Clearly this is the least matrix pre-ideal, and it consists precisely of all determinantal sums of non-full matrices. By the last corollary we find

PROPOSITION 3.2. *Let R be any ring, then R has proper matrix ideals if and only if no unit-matrix can be written as a determinantal sum of non-full matrices.* ■

For example, any ring without IBN has no proper matrix ideals, because the unit matrix (of a certain size) is actually non-full.

Exercises 7.3

1. Give an example to show that property (e) need not hold for matrix pre-ideals.

2. Let $\mathfrak{a}$ be an ideal in a commutative ring R. Show that the set $\mathfrak{a}^*$ of matrices A such that $\det A \in \mathfrak{a}$ is a matrix ideal. If $\mathscr{P}$ is a matrix ideal, show that the set $\mathscr{P}^*$ of elements $\det A$, where $A \in \mathscr{P}$, is an ideal. Verify that $\mathfrak{a}^{**} = \mathfrak{a}$, $\mathscr{P}^{**} \supseteq \mathscr{P}$, and that equality holds here, when R is Euclidean.
What is the relation between $\mathscr{P}^*$ and the 1×1 matrices in $\mathscr{P}$?

3. Show that a field has precisely one proper matrix ideal.

4^0. Show that over a commutative ring, a matrix which is a non-zerodivisor is full; for which rings does the converse hold?

5. A homomorphism of rings is called *honest* if it keeps every full matrix full. Show that every honest homomorphism is injective. Give an example to show that an injective homomorphism, even between commutative rings, need not be honest.

6. Verify that determinantal sums are preserved under homomorphisms.

7. If $C = A \triangledown B$, where the determinantal sum is with respect to a column, show that $PC = PA \triangledown PB$, for any P of the appropriate size, but that in general $CP \neq AP \triangledown BP$. However, show that $CP = AP \triangledown BP$ when P is diagonal.

8^0. Investigate the notion of a 'sum' or 'join' of matrix ideals; is the lattice of matrix ideals modular? For what rings is the lattice of matrix ideals distributive?

9^0. Develop a theory of one-sided matrix ideals, admitting determinantal sums only with respect to rows (or only columns), and find an application.

7.4 Prime matrix ideals

In the study of homomorphisms of commutative rings into fields prime ideals play an important role, and we shall find that in the general case there is an analogue in the prime matrix ideals, now to be defined. It will be convenient to begin by defining a multiplication of matrix ideals; this is an associative and commutative operation, like the multiplication of ideals in commutative rings.

Definition. Given two matrix ideals $\mathscr{P}_1$, $\mathscr{P}_2$ in a ring R, their *product*, denoted by $\mathscr{P}_1 \mathscr{P}_2$, is defined as the matrix ideal generated by all $A_1 \dotplus A_2$ with $A_i \in \mathscr{P}_i$ $(i = 1, 2)$.

The product so defined is easily seen to be associative. We write $\mathscr{P}_1 \mathscr{P}_2 \mathscr{P}_3$ etc. for repeated products, and abbreviate $\mathscr{P}\mathscr{P}$, $\mathscr{P}\mathscr{P}\mathscr{P}$, ... as $\mathscr{P}^2$, $\mathscr{P}^3$, ... From property (f) it follows that the product is commutative, and by 3 it is contained in the intersection of the factors:

$$\mathscr{P}_1 \mathscr{P}_2 = \mathscr{P}_2 \mathscr{P}_1 \subseteq \mathscr{P}_1 \cap \mathscr{P}_2.$$

The following lemma is often useful in constructing products:

LEMMA 4.1. *Let* $\mathscr{X}_1$, $\mathscr{X}_2$ *be any sets of square matrices,* $\mathscr{X}$ *the set of matrices* $A_1 \dotplus A_2$ $(A_i \in \mathscr{X}_i)$, *and* $\mathscr{P}_1$, $\mathscr{P}_2$, $\mathscr{P}$ *the matrix ideals generated by* $\mathscr{X}_1$, $\mathscr{X}_2$, $\mathscr{X}$ *respectively, then* $\mathscr{P} = \mathscr{P}_1 \mathscr{P}_2$.

Proof. Clearly $\mathscr{X} \subseteq \mathscr{P}_1 \mathscr{P}_2$, hence $\mathscr{P} \subseteq \mathscr{P}_1 \mathscr{P}_2$; to establish equality it is enough to show that $A_1 \dotplus A_2 \in \mathscr{P}$ for any $A_i \in \mathscr{P}_i$. Take $A_i \in \mathscr{P}_i$, then $A_1 \dotplus I$ is a determinantal sum of matrices B_μ which are either non-full or diagonal sums with a term in $\mathscr{X}_1$; similarly $A_2 \dotplus I$ is a determinantal sum of matrices C_ν which are either non-full or diagonal sums with a term in $\mathscr{X}_2$.

Therefore $A_1 \dotplus A_2 \dotplus I$ is a determinantal sum of the matrices $B_\mu \dotplus C_\nu$ which are either non-full or a diagonal sum with a term in $\mathscr{X}$. Hence $A_1 \dotplus A_2 \in \mathscr{P}$, and it follows that $\mathscr{P} = \mathscr{P}_1 \mathscr{P}_2$ as asserted. ■

A matrix ideal $\mathscr{P}$ is said to be *prime* if it is proper and

$$A \dotplus B \in \mathscr{P} \Rightarrow A \in \mathscr{P} \quad \text{or} \quad B \in \mathscr{P}.$$

A matrix ideal $\mathscr{P}$ is said to be *semiprime* if

$$\overset{2}{\dotplus} A \in \mathscr{P} \Rightarrow A \in \mathscr{P}.$$

An alternative description of these notions is given in

PROPOSITION 4.2. *Let $\mathscr{P}$ be a matrix ideal in a ring R; then*

(i) *$\mathscr{P}$ is prime if and only if $\mathscr{P}$ is proper and for any matrix ideals $\mathscr{P}_1, \mathscr{P}_2$ we have $\mathscr{P}_1 \mathscr{P}_2 \subseteq \mathscr{P} \Rightarrow \mathscr{P}_1 \subseteq \mathscr{P}$ or $\mathscr{P}_2 \subseteq \mathscr{P}$,*

(ii) *$\mathscr{P}$ is semiprime if and only if for any matrix ideal $\mathscr{Q}$, $\mathscr{Q}^2 \subseteq \mathscr{P} \Rightarrow \mathscr{Q} \subseteq \mathscr{P}$.*

Proof. (i) Let $\mathscr{P}$ be prime and $\mathscr{P}_1 \mathscr{P}_2 \subseteq \mathscr{P}$ but $\mathscr{P}_i \nsubseteq \mathscr{P}$ ($i = 1, 2$). Then there exists $A_i \in \mathscr{P}_i$ but $A_i \notin \mathscr{P}$. Since $\mathscr{P}$ is prime, $A_1 \dotplus A_2 \notin \mathscr{P}$, but $A_1 \dotplus A_2 \in \mathscr{P}_1 \mathscr{P}_2 \subseteq \mathscr{P}$, a contradiction.

Conversely, assume that $\mathscr{P}$ satisfies the given conditions, and let $A_1 \dotplus A_2 \in \mathscr{P}$. Write (A_i) for the matrix ideal generated by A_i, then by Lemma 4.1, the product $(A_1)(A_2)$ is generated by $A_1 \dotplus A_2$, and hence $(A_1)(A_2) \subseteq \mathscr{P}$. Therefore $(A_i) \subseteq \mathscr{P}$ for $i = 1$ or 2, say $i = 1$, and so $A_1 \in \mathscr{P}$, showing that $\mathscr{P}$ is prime.

(ii) Similarly if $\mathscr{P}$ is semiprime and $\mathscr{Q}^2 \subseteq \mathscr{P}$, let $A \in \mathscr{Q}$, then $\overset{2}{\dotplus} A \in \mathscr{Q}^2 \subseteq \mathscr{P}$, hence $A \in \mathscr{P}$, and so $\mathscr{Q} \subseteq \mathscr{P}$, as claimed. Conversely, if $\mathscr{P}$ satisfies the given condition, and $\overset{2}{\dotplus} A \in \mathscr{P}$, then $(A)^2$ is generated by $\overset{2}{\dotplus} A$ and so $(A)^2 \subseteq \mathscr{P}$, hence $(A) \subseteq \mathscr{P}$, whence $A \in \mathscr{P}$. This shows that $\mathscr{P}$ is semiprime. ■

Let $\mathscr{P}$ be any matrix ideal and define its *radical* as the set

$$\sqrt{\mathscr{P}} = \{A | \overset{r}{\dotplus} A \in \mathscr{P} \quad \text{for some} \quad r \geqslant 1\}.$$

Clearly $\sqrt{\mathscr{P}} \supseteq \mathscr{P}$; we assert that $\sqrt{\mathscr{P}}$ is the least semiprime matrix ideal containing $\mathscr{P}$. In the first place, any semiprime matrix ideal containing $\mathscr{P}$ must contain $\sqrt{\mathscr{P}}$; for if $A \in \sqrt{\mathscr{P}}$, then $\overset{r}{\dotplus} A \in \mathscr{P}$ for some r; hence this holds for any $r' \geqslant r$, and taking $r' = 2^k$, we have $\overset{2^k}{\dotplus} A \in \mathscr{P}$. Now it follows (by induction on k) that any semiprime matrix ideal containing $\mathscr{P}$ also contains A, and hence contains $\sqrt{\mathscr{P}}$. Thus if we can show that $\sqrt{\mathscr{P}}$ is a semi-prime matrix ideal we will have shown that it is the least semiprime containing $\mathscr{P}$. Properties 1, 2, 4 clearly hold; to prove 3 we note that $\overset{n}{+} (A \nabla B)$ is a determinantal sum of terms $C_1 \overset{}{+} \dots \overset{}{+} C_n$ where each C_i is A or B. Hence if $\overset{r}{+} A$ and $\overset{s}{+} B$ lie in $\mathscr{P}$, then $\overset{r+s-1}{+} (A \nabla B) \in \mathscr{P}$; therefore $\sqrt{\mathscr{P}}$ also satisfies 3 and so is a matrix ideal. If $\overset{2}{+} A \in \sqrt{\mathscr{P}}$, say $\overset{2r}{+} A \in \mathscr{P}$, then $A \in \sqrt{\mathscr{P}}$, so $\sqrt{\mathscr{P}}$ is semiprime, as claimed.

To relate semiprime and prime matrix ideals we first show that the familiar method of constructing prime ideals as maximal ideals still works for matrix ideals.

THEOREM 4.3. *In any ring R, let Σ be a non-empty set of square matrices closed under diagonal sums; then any matrix ideal $\mathscr{P}$ which is maximal disjoint from Σ is prime.*

Proof. Let $\mathscr{P}_1 \mathscr{P}_2 \subseteq \mathscr{P}$ but $\mathscr{P}_i \nsubseteq \mathscr{P}$, then $\mathscr{P}_i \cap \Sigma \neq \varnothing$. Take $A_i \in \mathscr{P}_i \cap \Sigma$, then $A_1 \overset{}{+} A_2 \in \Sigma$, and $A_1 \overset{}{+} A_2 \in \mathscr{P}_1 \mathscr{P}_2 \subseteq \mathscr{P}$, which is a contradiction. So we have $\mathscr{P}_i \subseteq \mathscr{P}$ for $i = 1$ or 2. Moreover, $\mathscr{P}$ is proper because $\Sigma \neq \varnothing$; this shows that $\mathscr{P}$ is prime. ∎

Whether maximal matrix ideals as in Theorem 4.3 exist or not, depends on whether there are any proper matrix ideals in R. Thus let Σ be any non-empty set of square matrices closed under diagonal sums, then any matrix ideal disjoint from Σ must be proper. If there is one, i.e. if Σ is disjoint from the least matrix ideal, then the collection $\mathscr{C}$ of all matrix ideals disjoint from Σ is non-empty. Clearly $\mathscr{C}$ is inductive, and hence there are maximal matrix ideals disjoint from Σ, to which the theorem can be applied.

The connexion between semiprime and prime matrix ideals is given in

THROREM 4.4. *In any ring R, a matrix ideal is semiprime if and only if it is an intersection of prime matrix ideals.*

Proof. Let $\mathscr{P} = \bigcap \mathscr{P}_\lambda$, where $\mathscr{P}_\lambda$ is prime. If $\overset{2}{+} A \in \mathscr{P}$, then $\overset{2}{+} A \in \mathscr{P}_\lambda$ for all λ, hence $A \in \mathscr{P}_\lambda$ for all λ, so $A \in \mathscr{P}$ and $\mathscr{P}$ is semiprime.

Conversely, let $\mathscr{P}$ be semiprime. It will be enough to find, for each $A \notin \mathscr{P}$, a prime matrix ideal $\mathscr{P}_A$ containing $\mathscr{P}$ but not A, for then

$$\mathscr{P} = \bigcap \{\mathscr{P}_A | A \notin \mathscr{P}\}.$$

Let $A \notin \mathscr{P}$ be given and consider the set Σ_A of all diagonal sums of copies of A. Since $\mathscr{P}$ is semiprime, $\Sigma_A \cap \mathscr{P} = \varnothing$, and clearly Σ_A is non-empty and closed under diagonal sums. Hence there is a maximal matrix ideal $\mathscr{P}_A$

containing $\mathscr{P}$ and disjoint from Σ_A. By Theorem 4.3, $\mathscr{P}_A$ is prime and $A \notin \mathscr{P}_A$. Hence $\bigcap \mathscr{P}_A = \mathscr{P}$, and the result follows. ∎

It is clear how under suitable maximum conditions, every matrix ideal will have a primary decomposition with the usual uniqueness properties. E.g. it would be enough to assume that the ring is right Noetherian; but in that case there is no need to consider prime matrix ideals, since the homomorphisms into fields can then be described by strongly prime ideals. In any case, we shall mainly be interested in non-Noetherian rings, in which the necessary chain conditions are unlikely to hold.

At this point we expect to find a relation between prime matrix ideals and multiplicative sets. If Σ is the complement of a prime matrix ideal, then clearly Σ is multiplicative and it is not hard to see that the universal Σ-inverting ring R_Σ is a local ring provided that it is non-zero. To show that $R_\Sigma \neq 0$ it will be enough to construct an R-field with Σ as the precise set of non-singular matrices. This will be done in the next section, while the following one will explore the consequences of this result.

Exercises 7.4

1. Let $\mathscr{P}$ be a matrix ideal and $\mathscr{X}$ any set of square matrices. Show that the set of square matrices $\mathscr{P} : \mathscr{X} = \{A | A \dotplus X \in \mathscr{P} \text{ for all } X \in \mathscr{X}\}$ is a matrix ideal, the *quotient* of $\mathscr{P}$ by $\mathscr{X}$, and may be characterized as the largest matrix ideal satisfying $\mathscr{X}(\mathscr{P} : \mathscr{X}) \subseteq \mathscr{P}$. If the matrix ideal generated by $\mathscr{X}$ is denoted by $(\mathscr{X})$, show that $\mathscr{P} : \mathscr{X} = \mathscr{P} : (\mathscr{X})$.

2. If $\mathscr{P}, \mathscr{P}_\lambda$ are matrix ideals and $\mathscr{X}$ is any set of square matrices, show that $\mathscr{P} : \mathscr{X} = \bigcap \{\mathscr{P} : \{X\} | X \in \mathscr{X}\}$, $(\mathscr{P}_1 : \mathscr{P}_2) : \mathscr{P}_3 = \mathscr{P}_1 : \mathscr{P}_2 \mathscr{P}_3$, $(\bigcap \mathscr{P}_\lambda) : \mathscr{X} = \bigcap (\mathscr{P}_\lambda : \mathscr{X})$.

3. A matrix ideal $\mathscr{Q}$ is said to be *primary* if $A \dotplus B \in \mathscr{Q}$, $A \notin \mathscr{Q}$ implies that $\overset{r}{\dotplus} B \in \mathscr{Q}$ for some r. Show that for any primary matrix ideal $\mathscr{Q}$, $\sqrt{\mathscr{Q}}$ is a prime (in this case $\mathscr{Q}$ is also called $\sqrt{\mathscr{Q}}$-primary). Give an example of a matrix ideal $\mathscr{Q}$ such that $\sqrt{\mathscr{Q}}$ is prime but $\mathscr{Q}$ is not primary.

Show that for any prime matrix ideal $\mathscr{P}$, the intersection of $\mathscr{P}$-primary matrix ideals is again $\mathscr{P}$-primary.

If $\mathscr{Q}$ is $\mathscr{P}$-primary, show that for any set $\mathscr{X}$, $\mathscr{Q} : \mathscr{X}$ is $\mathscr{P}$-primary, and is proper if and only if $\mathscr{X} \not\subseteq \mathscr{Q}$.

4. Let $\mathscr{A}$ be a matrix ideal which has a primary decomposition: $\mathscr{A} = \mathscr{Q}_1 \cap \ldots \cap \mathscr{Q}_r$, where $\mathscr{Q}_i$ is $\mathscr{P}_i$-primary. Show that $\sqrt{\mathscr{A}} = \mathscr{P}_1 \cap \ldots \cap \mathscr{P}_r$, and show that $\mathscr{A}$ also has a primary decomposition in which all the $\mathscr{P}_i$ are different and no $\mathscr{Q}_i$ contains the intersection of all the others (such a primary decomposition is called *irredundant*). Show that any two irredundant primary decompositions of $\mathscr{A}$ have the same number of terms, and these can be numbered so that corresponding primes are equal. If $\mathscr{P}_{i_1}, \ldots, \mathscr{P}_{i_t}$ are the primes disjoint from a given set $\mathscr{X}$, show that the corresponding intersection $\mathscr{Q}_{i_1} \cap \ldots \cap \mathscr{Q}_{i_t}$ is independent of the particular decomposition.

5. Show that the union and intersection of any chain of prime matrix ideals are again prime. Given prime matrix ideals $\mathscr{P} \subset \mathscr{P}'$, show that there exist prime matrix ideals $\mathscr{P}_1, \mathscr{P}_2$ such that $\mathscr{P} \subseteq \mathscr{P}_1 \subset \mathscr{P}_2 \subseteq \mathscr{P}'$ but there is no prime matrix ideal between $\mathscr{P}_1$ and $\mathscr{P}_2$.

6*. Let $\mathscr{P}$ be a minimal prime matrix ideal in a ring R. Show that for each $A \in \mathscr{P}$ there exists a matrix $C \notin \mathscr{P}$ and a positive integer r such that $C \overset{r}{+} A$ is a determinantal sum of non-full matrices.

7^0. Let $\mathscr{A}$ be a matrix ideal and $\mathscr{P}_1, \mathscr{P}_2$ prime matrix ideals of a ring R. If $\mathscr{A} \subseteq \mathscr{P}_1 \cup \mathscr{P}_2$, does it follow that $\mathscr{A} \subseteq \mathscr{P}_1$ or $\mathscr{A} \subseteq \mathscr{P}_2$?

8^0. Investigate rings in which every finitely generated matrix ideal can be generated by a single matrix.

7.5 Localization at a prime matrix ideal

Let R be any ring with a prime matrix ideal $\mathscr{P}$, and denote by Σ the complement of $\mathscr{P}$ (in the set of all square matrices). We would like to show that R_Σ is a local ring; this follows fairly easily once we know that $R_\Sigma \neq 0$. In fact we shall prove that there is an R-field K in which the set of singular matrices is precisely $\mathscr{P}$. By Theorem 2.2, it then follows that R_Σ is a local ring with residue-class field K.

Some preparations are necessary. In the first place, we need a lemma which expresses the field operations in terms of division and the operation $x\theta = 1 - x$.

LEMMA 5.1. *Let G be a group with a mapping θ of the set $G_1 = \{x \in G | x \neq 1\}$ into itself such that*

(i) $(yxy^{-1})\,\theta = y(x\theta)y^{-1}$ $(x \in G_1, \, y \in G)$,

(ii) $x\theta^2 = x$,

(iii) $(xy^{-1})\,\theta = [(x\theta)(y\theta)^{-1}]\,\theta \, . \, (y^{-1})\theta$ $(x, y \in G_1, \, x \neq y)$,

(iv) $e = (x^{-1})\,\theta \, . \, x \, . \, (x\theta)^{-1}$ *is independent of x.*

Then there exists a unique field K, whose multiplicative group of non-zero elements coincides with G, and on G, $x\theta = 1 - x$, $e = -1$.

Proof. If G has only one element, the result is trivially true, so we may take G non-trivial, i.e. G_1 is non-empty.

Let K be a field with the required properties; then the multiplicative group of K must coincide with G as a group. Thus K is obtained by adjoining 0 to G and extending the multiplication on G by the rule

$$0x = x0 = 0 \qquad (x \in K). \tag{1}$$

If we extend the mapping θ to K by putting $1\theta = 0$, $0\theta = 1$, then for K to satisfy the conditions of the lemma we must have $1 - x = x\theta$ for all $x \in K$, and it follows that the subtraction on K, which together with the multiplication determines K completely, must be given by

$$
x - y = \begin{cases} ey & \text{if} \quad x = 0, \\ (yx^{-1})\,\theta \,.\, x & \text{if} \quad x \neq 0. \end{cases} \tag{2}
$$

Thus the field K, it it exists, is uniquely determined. We now show that (1) and (2) always define a field structure of $K = G \cup \{0\}$ when the hypotheses of the lemma are satisfied.

To define a field structure on K it is enough to define the additive group structure and verify the distributive laws. Now an abelian group may be defined in terms of subtraction as a non-empty set with a binary operation $a - b$ satisfying

$$
a - (a - b) = b, \tag{3}
$$

$$
(a - b) - c = (a - c) - b. \tag{4}
$$

In terms of subtraction, the distributive laws become

$$
(a - b)c = ac - bc, \tag{5}
$$

$$
a(b - c) = ab - ac. \tag{6}
$$

Thus we have to verify (3)–(6), using the definitions (1) and (2). We begin with (5) and (6). If a or c vanishes, the two sides of (5) both reduce to ebc, so we may assume that $ac \neq 0$. Then we have

$$
ac - bc = (bcc^{-1}a^{-1})\,\theta \,.\, ac = (ba^{-1})\,\theta \,.\, ac = (a - b)c,
$$

by the definition (2), hence (5) holds.

Before proving (6) we note that e lies in the centre of G. For by (i), x commutes with $x\theta$ and $(x^{-1})\theta$, for any $x \in G_1$, and hence by (iv), $ex = (x^{-1})\,\theta x (x\theta)^{-1} x = xe$. If in (6) a or b vanishes, the two sides reduce to eac ($= aec$), so we may assume that $ab \neq 0$. Then (6) asserts that $a(cb^{-1})\,\theta \,.\, b = (acb^{-1}a^{-1})\,\theta \,.\, ab$, i.e.

$$
a(cb^{-1})\,\theta \,.\, a^{-1} = (acb^{-1}a^{-1})\theta. \tag{7}
$$

But this is just (i), provided that $c \neq 0$, b. In these exceptional cases both sides of (7) reduce to 1 or 0 respectively, thus (7) and with it (6) holds in all cases.

The condition (ii) may be restated as

$$1 - (1 - x) = x,$$

and in this form it still holds for $x = 1$ and $x = 0$. Multiplying on the left by any element a of K, we obtain by (6),

$$a - (a - ax) = ax.$$

As a varies over K, so does ax, provided that $a \neq 0$. Hence putting $ax = b$ we obtain (3) in case $a \neq 0$. When $a = 0$, (3) reduces to the form $-(-b) = b$, i.e. $e^2 b = b$, so that it is enough to show that

$$e^2 = 1, \tag{8}$$

to complete the proof of (3). Now by (iv) we have $e = (x^{-1}) \theta \cdot x(x\theta)^{-1}$ for any $x \in G_1$, and such elements exist, because $G_1 \neq \emptyset$. Therefore $e^2 = (x^{-1}) \theta \cdot x(x\theta)^{-1} \cdot (x\theta) x^{-1}(x^{-1} \theta)^{-1} = 1$. This proves (8) and hence (3) for all $a, b \in K$.

Now (iv) with x replaced by x^{-1}, may be written as

$$1 - x = - (x - 1);$$

this still holds for $x = 1$, and by (8) it also holds for $x = 0$. Multiplying both sides by a, we find

$$a - ax = - (ax - a).$$

As x varies over K, so does ax if $a \neq 0$; hence we obtain

$$a - b = - (b - a), \tag{9}$$

provided that $a \neq 0$; but for $a = 0$ (9) holds trivially, so it holds for all $a, b \in K$.

To prove (4) we write out (iii) with y replaced by $y\theta = 1 - y$:

$$1 - x(1 - y)^{-1} = [1 - (1 - x)y^{-1}][1 - (1 - y)^{-1}].$$

Now multiply by $1 - y$ on the right, using (5), (9) and (3) in turn:

$$(1 - y) - x = [1 - (1 - x)y^{-1}][(1 - y) - 1]$$
$$= - [1 - (1 - x)y^{-1}][1 - (1 - y)]$$
$$= - [1 - (1 - x)y^{-1}] y = - [y - (1 - x)];$$

hence

$$(1 - y) - x = (1 - x) - y. \tag{10}$$

Clearly this still holds for $x = 1 - y$ and $x = 0$, while for $x = 1$ it follows by applying (9) and (3). Thus (10) holds for all x and similarly for all y in K. Putting $x = a^{-1}b$ and $y = a^{-1}c$ and multiplying by a on the left we thus obtain (4) when $a \neq 0$. But when $a = 0$, we have by (9) and (8),

$$eb - c = e(c - eb) = ec - b,$$

which establishes (4) in this case. ∎

Secondly we shall need a method of constructing a group as a quotient of a semigroup. This is described in Lemma 5.2; contrary to our usual convention, the semigroup occurring in that lemma is *not* assumed to have a unit-element.

LEMMA 5.2. *Let M be a non-empty semigroup and M' a subsemigroup of M such that*

(i) *$xy \in M' \Rightarrow yx \in M'$, and*

(ii) *for each $x \in M$ there exists $\bar{x} \in M$ such that $\bar{\bar{x}} = x$, $x\bar{x} \in M'$.*

Then the relation defined on M by

$$x \sim y \quad whenever \quad xz\bar{y} \in M' \quad for some \quad z \in M' \tag{11}$$

is a congruence on M whose quotient is a group.

Proof. Clearly M' is non-empty. If $x\bar{y} \in M'$ and $u \in M'$, then by (i), $\bar{y}x \in M'$, hence $\bar{y}xu \in M'$ and so $xu\bar{y} \in M'$, therefore $x \sim y$ whenever $x\bar{y} \in M'$.

To prove that (11) is an equivalence, we have for any $x \in M$, $x\bar{x} \in M'$, hence $x \sim x$ by the remark just made. Next if $x \sim y$, $y \sim z$, say u, $xu\bar{y}$, v, $yv\bar{z} \in M'$, then by (ii) $\bar{y}y \in M'$ and so $u\bar{y}yv \in M'$. Now $x . u\bar{y}yv . \bar{z} \in M'$, therefore $x \sim z$, i.e. '$\sim$' is transitive. To establish symmetry, assume that $x \sim y$, say u, $xu\bar{y} \in M'$. Then $\bar{y}xu$, $u\bar{y}x \in M'$ and hence $\bar{y}xu^2 \bar{y}x \in M'$. Further, $\bar{y}xu . u \in M'$ hence $xu^2 \bar{y} \in M'$ and so $y\bar{y} . xu^2 \bar{y} . x\bar{x} \in M'$. But $\bar{y}xu^2 \bar{y}x \in M'$, therefore $y \sim x$ and the symmetry of '$\sim$' is proved.

Next we observe that

$$x \sim y \Rightarrow \bar{x} \sim \bar{y}. \tag{12}$$

For if u, $xu\bar{y} \in M'$, then $\bar{x}xu\bar{y}y \in M'$, hence $\bar{x} \sim \bar{y}$.

Given any x, $y \in M$, we have $y\bar{y}\bar{x}x \in M'$, hence $\bar{y}\bar{x}xy \in M'$ and so, for any $u \in M'$, $xy\bar{y}\bar{x}u \in M'$, and therefore $\bar{y}\bar{x}uxy \in M'$. This proves

$$\bar{y}\bar{x} \sim \overline{xy}. \tag{13}$$

Now let $x \sim x_1$, $y \sim y_1$, say u, $xu\bar{x}_1$, v, $yv\bar{y}_1 \in M'$. Then $\bar{x}x \cdot yv\bar{y}_1 \in M'$ hence $xyv\bar{y}_1\,\bar{x} \in M'$ and $xu\bar{x}_1 \in M'$, so

$$xy \cdot v\bar{y}_1\,\bar{x} \cdot xu\bar{x}_1 \cdot x_1 y_1 \cdot \overline{x_1 y_1} \in M'. \tag{14}$$

On the other hand, $w = \bar{x}xu\bar{x}_1\,x_1 \in M'$ and $v\bar{y}_1\,y_1 \in M'$, hence $y_1\,v\bar{y}_1 \in M'$, therefore $y_1 v\bar{y}_1\,w \in M'$, and $v\bar{y}_1 wy_1 \in M'$. But this is just the central factor in (14), so $xy \sim x_1 y_1$ and we have established that '$\sim$' is a congruence.

Let $G = M/\!\sim$ be the quotient semigroup and denote the class corresponding to M' by e. Then in G, $ex = xe = x$, $\bar{x}x = x\bar{x} = e$ for all $x \in G$ (where $\bar{x}$ is the class of all $\bar{a}$, for a in x; this makes sense by (12)). This shows G to be a group. ∎

We can now state the main result of this section.

THEOREM 5.3. *Let R be any ring and $\mathscr{P}$ a prime matrix ideal in R. Then there exists an R-field K such that $\mathscr{P}$ is the precise class of matrices mapped to singular matrices under the canonical homomorphism $R \to K$.*

The idea of the proof is to construct the elements of K as classes of solutions of matrix equations of the form $Au + a = 0$. We make the collection of all such equations into a semigroup, form a group by the procedure of Lemma 5.2 and then use Lemma 5.1 to obtain the field K from it.

Let us call a square matrix over R *singular* if it lies in $\mathscr{P}$ and *non-singular* otherwise. We know that if a field K of the required form exists, then each element $x \in K$ can be obtained as the first component u_1 of the solution of a system

$$Au + a = 0, \tag{15}$$

where A is non-singular. If x is associated with the system (15) in this way, then for $x \neq 0$, x^{-1} is associated with the system

$$A_1 v + a_1 = 0,$$

where $A = (a_1, ..., a_n)$, $A_1 = (a, a_2, ..., a_n)$. We shall generally use these abbreviations for any systems of equations, thus given a system $Bv + b = 0$, the first column of B will be written b_1 and B_1 is the matrix obtained by exchanging b_1 against b. Further, let us write, for the system (15),

$$A^+ = (a_1 + a, a_2, ..., a_n), \tag{16}$$

$$A^* = (-a_1, a_2, ..., a_n), \qquad a^* = a_1 + a. \tag{17}$$

Then $A^* v + a^* = 0$ has as its first component $1 - x$ and $A^+ v + a = 0$ has as its first component $x(1 - x)^{-1}$. These facts are not actually used in the proof, but they will help to motivate the steps.

Let M be the set of all matrices $(A|a) = (a_1, ..., a_n, a)$ such that both A and A_1 are non-singular. Here n, the *order* of the system, can have any value $\geqslant 1$. On M we define a multiplication by the formula

$$(A|a)(B|b) = \begin{pmatrix} A & a\ 0 & 0 \\ 0 & B & b \end{pmatrix}. \tag{18}$$

If $(A|a)$ and $(B|b)$ have orders n, m respectively, then the right-hand side of (18) has order $m + n$. To verify that we get an element of M we note that $\begin{pmatrix} A & a\ 0 \\ 0 & B \end{pmatrix}$ is non-singular; further, on writing $A' = (a_2, ..., a_n)$, we have

$$\begin{pmatrix} 0 & A' & a\ 0 \\ b\ 0 & B \end{pmatrix} \rightarrow \begin{pmatrix} A_1 & 0 \\ b_1 0 & B_1 \end{pmatrix}$$

by column interchange, and the latter is non-singular.

The multiplication defined by (18) is associative; we have

$$(A|a)(B|b)(C|c) = \begin{pmatrix} A & a\ 0 & 0 & 0 \\ 0 & B & b\ 0 & 0 \\ 0 & 0 & C & c \end{pmatrix}$$

for either bracketing.

Let us define M' as the subset of M consisting of all $(A|a)$ for which A^+ (as defined in (16)) is singular. This set is a subsemigroup of M: Given $(A|a)$ and $(B|b)$ in M', if we form the operation (16) on their product, we get the matrix

$$\begin{pmatrix} A & a\ 0 \\ b\ 0 & B \end{pmatrix} \rightarrow \begin{pmatrix} A^+ & a\ 0 \\ b{*}0 & B \end{pmatrix} = \begin{pmatrix} A^+ & 0 \\ b{*}0 & B \end{pmatrix} \triangledown \begin{pmatrix} A^+ & a\ 0 \\ b{*}0 & 0\ B' \end{pmatrix}.$$

Here the arrow indicates a column operation, and on the right we have a determinantal sum. The first term is singular because A^+ is, and the second term, after a column interchange, becomes

$$\begin{pmatrix} A_1 & a{*}0 \\ 0 & B^+ \end{pmatrix}$$

which is also singular, because B^+ is.

Thus M' is indeed a subsemigroup of M. We claim that M' satisfies the conditions of Lemma 5.2. Given $(A|a)$, $(B|b)$, if their product lies in M', then $\begin{pmatrix} A & a\ 0 \\ b\ 0 & B \end{pmatrix}$ is singular. Hence by column interchanges and row interchanges,

$$\left(\begin{array}{c|c} A & a\ 0 \\ \hline b\ 0 & B \end{array} \right) \rightarrow \left(\begin{array}{c|c} a\ 0 & A \\ \hline B & b\ 0 \end{array} \right) \rightarrow \left(\begin{array}{c|c} B & b\ 0 \\ \hline a\ 0 & A \end{array} \right),$$

so the matrix on the right is singular, i.e. $xy \in M' \Rightarrow yx \in M'$.

We now define

$$\overline{(A|a)} = (A_1|a_1). \tag{19}$$

Clearly $\bar{\bar{x}} = x$, and if $x = (A|a)$, then $x\bar{x} = \left(\begin{array}{cc|c} A & a\ 0 & 0 \\ 0 & A_1 & a_1 \end{array} \right)$. By row and column operations we have

$$\left(\begin{array}{c|c} A & a\ 0 \\ \hline a_1 0 & A_1 \end{array} \right) \rightarrow \left(\begin{array}{c|c} A^+ & a\ 0 \\ \hline a^*0 & A_1 \end{array} \right) \rightarrow \left(\begin{array}{c|c} 0\ A' & 0\ -A' \\ \hline a^*0 & A_1 \end{array} \right) \rightarrow \left(\begin{array}{c|c} 0\ A' & 0 \\ \hline a^*0 & A_1 \end{array} \right),$$

and the matrix on the right is clearly singular.

Thus all the conditions of Lemma 5.2 are satisfied, and we obtain a group G. To complete the proof of the theorem, we shall define an operation θ as in Lemma 5.1. Let us write $G_1 = \{x \in G | x \neq 1\}$, $M_1 = \{x \in M | x \notin M'\}$, and for any $(A|a)$ in M_1 write

$$(A|a)f = (A^*|a^*),$$

where the right-hand side is defined as in (17). The resulting system again lies in M, for $A^* = (-a_1, a_2, ..., a_n)$ and $(a^*, a_2, ..., a_n)$ are both non-singular, and the system is not in M' because $-a_1 + a^* = a$, and $(a, a_2, ..., a_n)$ is non-singular.

Let $x \sim y$ (in the notation of Lemma 5.2); we must show that $xf \sim yf$. Write $x = (A|a)$, $y = (B|b)$, then we are given that for some $(C|c)$,

both C^+ and $\begin{pmatrix} A & a\ 0 & 0 \\ 0 & C & c\ 0 \\ b_1 0 & 0 & B_1 \end{pmatrix}$ are singular, and we shall show that

$\begin{pmatrix} A^* & a^*0 & 0 \\ 0 & C & c\ 0 \\ -b_1 0 & 0 & B^+ \end{pmatrix}$ is singular, which will prove that $xf \sim yf$. We have

$$\begin{pmatrix} A^* & a^*0 & 0 \\ 0 & C & c\ 0 \\ -b_1 0 & 0 & B^+ \end{pmatrix} \rightarrow \begin{pmatrix} A & a^*0 & 0 \\ 0 & C & c\ 0 \\ b_1 0 & 0 & B^+ \end{pmatrix} \rightarrow \begin{pmatrix} A & a^*0 & 0 \\ 0 & C^+ & c\ 0 \\ b_1 0 & b^*0 & B^+ \end{pmatrix} \rightarrow$$

$$\begin{pmatrix} A & a\ 0 & 0 \\ 0 & C^+ & c\ 0 \\ b_1 0 & b\ 0 & B^+ \end{pmatrix} = \begin{pmatrix} A & a\ 0 & 0 \\ 0 & C^+ & 0 \\ b_1 0 & b\ 0 & B \end{pmatrix} \triangledown \begin{pmatrix} A & a\ 0 & 0 \\ 0 & C^+ & c\ 0 \\ b_1 0 & b\ 0 & B_1 \end{pmatrix}.$$

Here the first term on the right is singular because C^+ is, and the second term is of the form

$$\begin{pmatrix} A & a\,0 & 0 \\ 0 & C & c\,0 \\ b_1 0 & 0 & B_1 \end{pmatrix} \triangledown \begin{pmatrix} A & 0 & 0 \\ 0 & C_1 & c\,0 \\ b_1 0 & b\,0 & B_1 \end{pmatrix}.$$

here the first term is singular by hypothesis, while the second has two columns equal.

This shows that $x \sim y$ implies $xf \sim yf$, and it follows that f induces a mapping θ of G_1 into itself. Clearly $f^2 = 1$, hence $\theta^2 = 1$ and it remains to verify the other conditions of Lemma 5.1. In order to do this we note that we can perform the following operations on a system $(A|a)$ in M without affecting its congruence class mod M': (a) any row operation, (b) any column operation of the form $c \rightarrow c + d\lambda$, where $\lambda \in R$, c is any column of $(A|a)$ and d is any column other than c and the first and last columns in $(A|a)$, (c) changing the sign of the first and last column in $(A|a)$, or changing the sign of any column other than the first or last column.

(i) $(yxy^{-1})\theta = y(x\theta)y^{-1}$. Let $x = (A|a)$, $y = (B|b)$, then $x\theta = (A^*|a^*)$,

$$yx\bar{y} = \begin{pmatrix} B & b\,0 & 0 & | & 0 \\ 0 & A & a\,0 & | & 0 \\ 0 & 0 & B_1 & | & b_1 \end{pmatrix}, \qquad (yx\bar{y})\theta = \begin{pmatrix} B^* & b\,0 & 0 & | & b_1 \\ 0 & A & a\,0 & | & 0 \\ 0 & 0 & B_1 & | & b_1 \end{pmatrix}.$$

$$y(x\theta)\bar{y} = \begin{pmatrix} B & b\,0 & 0 & | & 0 \\ 0 & A^* & a^*0 & | & 0 \\ 0 & 0 & B_1 & | & b_1 \end{pmatrix}.$$

Using operations (a)–(c), we have

$$(yx\bar{y})\theta = \begin{pmatrix} B^* & b\,0 & 0 & | & b_1 \\ 0 & A & a\,0 & | & 0 \\ 0 & 0 & B_1 & | & b_1 \end{pmatrix} \rightarrow \begin{pmatrix} B^* & b\,0 & -B_1 & | & 0 \\ 0 & A & a\,0 & | & 0 \\ 0 & 0 & B_1 & | & b_1 \end{pmatrix}$$

$$\rightarrow \begin{pmatrix} -B & -b\,0 & -B_1 & | & 0 \\ 0 & A^* & a\,0 & | & 0 \\ 0 & 0 & B_1 & | & b_1 \end{pmatrix} \rightarrow \begin{pmatrix} B & b\,0 & B_1 & | & 0 \\ 0 & A^* & a\,0 & | & 0 \\ 0 & 0 & B_1 & | & b_1 \end{pmatrix}$$

$$\rightarrow \begin{pmatrix} B & b\,0 & b\,0 & | & 0 \\ 0 & A^* & a\,0 & | & 0 \\ 0 & 0 & B_1 & | & b_1 \end{pmatrix} \rightarrow \begin{pmatrix} B & b\,0 & 0 & | & 0 \\ 0 & A^* & a^*0 & | & 0 \\ 0 & 0 & B_1 & | & b_1 \end{pmatrix} = y(x\theta)\bar{y}.$$

We have already proved (ii); to establish (iii) we must show that

$$(xy^{-1})\,\theta = [x\theta(y\theta)^{-1}]\theta \cdot (y^{-1})\theta \qquad \text{for} \qquad x, y \neq 1, \quad x \neq y. \tag{20}$$

Let $x = (A|a)$, $y = (B|b)$, then

$$(x\bar{y})\,\theta = \begin{pmatrix} A^* & a & 0 & a_1 \\ 0 & & B_1 & b_1 \end{pmatrix}, \qquad (x\theta)\overline{(y\theta)} = \begin{pmatrix} A^* & a^*0 & 0 \\ 0 & B^+ & -b_1 \end{pmatrix},$$

$$\bar{y}\theta = (B_1{}^*|b^*),$$

where $B_1{}^* = (-b, b_2, \ldots, b_m)$. Now

$$[(x\theta)\overline{(y\theta)}]\,\theta = \begin{pmatrix} A & a^*0 & -a_1 \\ 0 & B^+ & -b_1 \end{pmatrix},$$

hence for the right-hand side of (20) we have

$$[(x\theta)\overline{(y\theta)}]\,\theta \cdot \bar{y}\theta = \begin{pmatrix} A & a^*0 & -a_1 0 & 0 \\ 0 & B^+ & -b_1 0 & 0 \\ 0 & 0 & B_1{}^* & b^* \end{pmatrix} \rightarrow \begin{pmatrix} A & a^*0 & a_1 0 & 0 \\ 0 & B^+ & b_1 0 & 0 \\ 0 & 0 & B_1 & b^* \end{pmatrix}$$

$$\rightarrow \begin{pmatrix} A & a^*0 & a_1 0 & 0 \\ 0 & B^+ & B^+ & b^* \\ 0 & 0 & B_1 & b^* \end{pmatrix} \rightarrow \begin{pmatrix} A & a^*0 & -a\,0 & 0 \\ 0 & B^+ & 0 & b^* \\ 0 & 0 & B_1 & b^* \end{pmatrix}$$

$$\rightarrow \begin{pmatrix} A & a^*0 & -a\,0 & -a_1 \\ 0 & B^+ & 0 & 0 \\ 0 & 0 & B_1 & b_1 \end{pmatrix} \rightarrow \begin{pmatrix} A & -a\,0 & a^*0 & -a_1 \\ 0 & 0 & B^+ & 0 \\ 0 & B_1 & 0 & b_1 \end{pmatrix}$$

$$\rightarrow \begin{pmatrix} A & -a\,0 & a^*0 & -a_1 \\ 0 & B_1 & 0 & b_1 \\ 0 & 0 & B^+ & 0 \end{pmatrix} \rightarrow \begin{pmatrix} A^* & -a\,0 & a^*0 & a_1 \\ 0 & B_1 & 0 & -b_1 \\ 0 & 0 & B^+ & 0 \end{pmatrix}$$

$$\rightarrow \begin{pmatrix} A^* & -a\,0 & a^*0 & a_1 \\ 0 & -B_1 & 0 & b_1 \\ 0 & 0 & B^+ & 0 \end{pmatrix} \rightarrow \begin{pmatrix} A^* & a\,0 & a^*0 & a_1 \\ 0 & B_1 & 0 & b_1 \\ 0 & 0 & B^+ & 0 \end{pmatrix}.$$

To show that this is in the same class (mod M') as $(x\bar{y})\theta$, we consider the matrix for $(x\bar{y})\,\theta\,.\,\bar{z}$, where z is the system just found. This is

$$
\begin{pmatrix}
A^* & a\ 0 & a_1 0 & 0 & 0 \\
0 & B_1 & b_1 0 & 0 & 0 \\
-a_1 0 & 0 & A & a\ 0 & a^*0 \\
0 & 0 & b_1 0 & B_1 & 0 \\
0 & 0 & 0 & 0 & B^+
\end{pmatrix}
$$

We must show that this matrix is singular. Since $y \neq 1$, B^+ is non-singular and we may omit the last row and column block. This leaves the matrix

$$
\begin{pmatrix}
A^* & a\ 0 & a_1 0 & 0 \\
0 & B_1 & b_1 0 & 0 \\
-a_1 0 & 0 & A & a\ 0 \\
0 & 0 & b_1 0 & B_1
\end{pmatrix}
\rightarrow
\begin{pmatrix}
0\ A' & a\ 0 & a_1 0 & 0 \\
b_1 0 & B_1 & b_1 0 & 0 \\
0 & 0 & A & a\ 0 \\
b_1 0 & 0 & b_1 0 & B_1
\end{pmatrix}
$$

$$
\rightarrow
\begin{pmatrix}
0\ A' & a\ 0 & a_1 0 & 0 \\
b_1 0 & B_1 & b_1 0 & 0 \\
0 & 0 & A & a\ 0 \\
0 & -B_1 & 0 & B_1
\end{pmatrix}
\rightarrow
\begin{pmatrix}
0\ A' & a\ 0 & a_1 0 & 0 \\
b_1 0 & B_1 & b_1 0 & 0 \\
0 & a\ 0 & A & a\ 0 \\
0 & 0 & 0 & B_1
\end{pmatrix}.
$$

Again B_1 is non-singular, so there only remains

$$
\begin{pmatrix}
0\ A' & a\ 0 & a_1 0 \\
b_1 0 & B_1 & b_1 0 \\
0 & a\ 0 & A
\end{pmatrix}
\rightarrow
\begin{pmatrix}
0\ A' & a\ 0 & a_1 0 \\
b_1 0 & B_1 & b_1 0 \\
0-A' & 0 & 0\ A'
\end{pmatrix}
\rightarrow
\begin{pmatrix}
A^* & a\ 0 & a_1 0 \\
0 & B_1 & b_1 0 \\
0 & 0 & 0\ A'
\end{pmatrix}
$$

and this is clearly singular.

To prove (iv) we show that $e = x(\bar{x}\theta)\overline{(x\theta)}$ is represented by the system $(1|1)$ of order 1. Let $x = (A|a)$, then $\bar{x}\theta = (A_1^*|a^*)$, $\overline{x\theta} = (A^+|-a_1)$, and the required result will follow if we prove that the matrix

$$
\begin{pmatrix}
A & a\ 0 & 0 & 0 \\
0 & A_1^* & a^*0 & 0 \\
0 & 0 & A^+ & -a_1 \\
1\ 0 & 0 & 0 & 1
\end{pmatrix}
$$

is singular. By subtracting the last column from the first, we reduce the last row to 0 except in the last place, and so obtain

$$\begin{pmatrix} A & a\,0 & 0 \\ 0 & A_1{}^* & a*0 \\ a_1 0 & 0 & A^+ \end{pmatrix} \rightarrow \begin{pmatrix} A & 0\,A' & a*0 \\ 0 & A_1{}^* & a*0 \\ a_1 0 & 0 & A^+ \end{pmatrix}$$

$$\rightarrow \begin{pmatrix} 0\,A' & 0\,A' & 0-A' \\ 0 & A_1{}^* & a*0 \\ a_1 0 & 0 & A^+ \end{pmatrix} \rightarrow \begin{pmatrix} 0\,A' & 0 & 0 \\ 0 & A_1{}^* & a*0 \\ a_1 0 & 0 & A^+ \end{pmatrix}$$

and the latter matrix is clearly singular.

We have now verified all the conditions of Lemma 5.1 and so obtain a field K, say. We now define a mapping $\lambda: R \to K$ as follows. If $\alpha \in R$ is singular (as 1×1 matrix), then $\alpha\lambda = 0$; otherwise we map $\alpha \mapsto (1|-\alpha)$ and define $\alpha\lambda$ as the congruence class (mod M') of $(1|-\alpha)$. To show that this is a homomorphism, we first check that it preserves multiplication: $(xy)\lambda = x\lambda . y\lambda$. If x or y is singular, this is clear; otherwise x, y are both non-singular, and we must show that

$$\left(\begin{matrix} 1 & -x \\ 0 & 1 \end{matrix} \middle| \begin{matrix} 0 \\ -y \end{matrix} \right) \sim (1|-xy),$$

i.e. that the matrix

$$\begin{pmatrix} 1 & -x & 0 \\ 0 & 1 & -y \\ 1 & 0 & -xy \end{pmatrix}$$

is singular. Now we have

$$\begin{pmatrix} 1 & -x & 0 \\ 0 & 1 & -y \\ 1 & 0 & -xy \end{pmatrix} \rightarrow \begin{pmatrix} 1 & 0 & -xy \\ 0 & 1 & -y \\ 1 & 0 & -xy \end{pmatrix} \rightarrow \begin{pmatrix} 1 & 0 & -xy \\ 0 & 1 & -y \\ 0 & 0 & 0 \end{pmatrix}$$

and the latter matrix is singular, which proves our claim. Moreover, if $x = (1|-a)$ and a, $1 - a$ are non-singular, then $x\theta = (-1|1 - a)$, which coincides with $(1 - a)\lambda$. Thus $a\lambda\theta = (1 - a)\lambda$, i.e. θ corresponds to the operation $a \mapsto 1 - a$, hence λ is a homomorphism of R into K.

Finally we have to show that a matrix A over R maps to a singular matrix over K precisely when $A \in \mathscr{P}$, i.e. when A is 'singular'. Denote the set of square matrices mapping to singular matrices over K by $\mathscr{P}'$, then if $A \in \mathscr{P}$, the system $Au + a = 0$ has a solution in K for all a, by the construction of K. Taking $a = -e_i$ $(i = 1, ..., n)$ in turn, we see that A has a right inverse over K, and so is invertible. Hence $A \notin \mathscr{P}'$ and we have shown that $\mathscr{P}' \subseteq \mathscr{P}$. To establish equality we use induction on n, the order of A.

The elements of order 1 in $\mathscr{P}$ and $\mathscr{P}'$ are the same, by the definition of the mapping λ. Now let $n > 1$ and assume that the matrices of order less than n in $\mathscr{P}$ and $\mathscr{P}'$ agree. Take $A \in \mathscr{P}$ of order n, say $A = (a_1, ..., a_n)$, and suppose first that there exists a column a such that $A_1 = (a, a_2, ..., a_n) \notin \mathscr{P}$. Then the equation

$$A_1 u + a_1 = 0 \tag{21}$$

has a solution in K, and since $A \in \mathscr{P}$, it follows that $u_1 = 0$, so that (21) reads $a_2 u_2 + ... + a_n u_n + a_1 = 0$. Thus the columns of A are linearly dependent over K, i.e. $A \in \mathscr{P}'$. Next if $(a, a_2, ..., a_n) \in \mathscr{P}$ for all choices of a, let us take $a = e_i$ $(i = 1, ..., n)$. Each of the resulting n matrices is in $\mathscr{P}$, hence the matrix obtained from A by omitting any one row and the first column is in $\mathscr{P}$, and by the induction hypothesis, also in $\mathscr{P}'$. On expanding A by its first column (as a determinantal sum) we find that $A \in \mathscr{P}'$, and this shows that $\mathscr{P}' = \mathscr{P}$. Thus all the assertions of Theorem 5.3 are now established. ∎

Exercises 7.5

1. If G is as in Lemma 5.1, verify directly that the least subset of G containing a given element x and admitting the operations θ and $x \mapsto x^{-1}$ has at most six elements.

2*. (Rabinow [37]). Let A be a non-empty set with a binary operation $x - y$ satisfying

$$x - (x - y) = y, \qquad (x - y) - z = (x - z) - y.$$

Show that $(x - y) - (x - y) = y - y$ for all $x, y \in A$. Given $a, b \in A$, solve the equation $x - a = b$. (Hint: try $x = u - [(u - a) - b]$, where u is arbitrary.) Deduce that the element $0 = y - y$ is independent of the choice of y, and complete the proof that A is an abelian group with addition $x - (0 - y)$ and zero element 0.

3°. Prove Theorem 5.3 by constructing the field directly (i.e. without going via Lemmas 5.1–2).

4°. If Σ is a multiplicative set in a ring R, construct the localization R_Σ in a form which is sufficiently explicit to allow one to decide when $R_\Sigma \neq 0$.

5°. If Σ is a localizing set in a ring R, find conditions for the canonical homomorphism $R \to R_\Sigma$ to be injective. Give an example of a pair R, Σ where this homomorphism is not injective.

7.6 The embedding of rings in skew fields

The theorem proved in the last section can be used to answer various questions about the embeddability of rings in fields. In essence the result showed that for every prime matrix ideal $\mathscr{P}$ in a ring R there is an R-field K in which $\mathscr{P}$ is the precise set of singular matrices. Writing Σ for the complement of $\mathscr{P}$ (in the set of all square matrices over R) we see that R_Σ is a local ring with residue class field K; thus Σ is a localizing set. Conversely, if Σ is any localizing set and the residue class field of the local ring R_Σ is denoted by K_Σ, then Σ is the precise class of matrices inverted in K_Σ. This establishes

THEOREM 6.1. *In any ring R the localizing sets are precisely the complements of the prime matrix ideals.* ■

Taken in conjunction with Theorem 2.3, this result shows that the category of R-fields and specializations is equivalent to the category of prime matrix ideals and inclusions, in precise analogy with the prime ideals in a commutative ring. For example, R has a universal R-field if and only if there is a unique least prime matrix ideal.

Of course, a ring need not have any prime matrix ideals at all; to find if it has any we go back to the method of generating matrix ideals described in 7.3. In any ring R, denote by $\mathscr{Z}$ the set of all determinantal sums of non-full matrices. Thus $A \in \mathscr{Z}$ if and only if

$$A = C_1 \nabla \ldots \nabla C_t,$$

where each C_i is non-full and the right-hand side is suitably bracketed. Clearly $\mathscr{Z}$ is the least matrix pre-ideal in R (*cf.* 7.3). Let $\overline{\mathscr{Z}}$ be the matrix ideal generated by $\mathscr{Z}$ and put $\mathscr{N} = \sqrt{\overline{\mathscr{Z}}}$, then $\mathscr{N}$ is proper if and only if $\mathscr{Z}$ is proper. By Theorem 4.4, $\mathscr{N}$ is the intersection of all prime matrix ideals, so $\mathscr{N}$ is proper if and only if R has prime matrix ideals. But the latter correspond to field homomorphisms, so we find that $\mathscr{N}$ is proper if and only if R has a homomorphism into a field. We therefore obtain the following criterion for the existence of field homomorphisms:

THEOREM 6.2. *Let R be any ring. Then there exists a homomorphism of R into a field if and only if no unit matrix in R can be written as a determinantal sum of non-full matrices.* ■

This includes the well known necessary condition: If a ring is embeddable in a field, then the unit matrix (of any size) is full (*cf.* Cohn [66], where this is called condition II).

Secondly we obtain a criterion for the invertibility of a matrix:

THEOREM 6.3. *Let R be a ring and A any square matrix over R. Then*

(i) *there is a homomorphism of R into a field mapping A to an invertible matrix if and only if no diagonal sum $I \overset{r}{+} A$ can be written as a determinantal sum of non-full matrices,*

(ii) *there is a homomorphism of R into a field mapping A to a singular matrix if and only if no unit matrix I can be written as a determinantal sum of non-full matrices and matrices of the form $A \dotplus B$ (where B is any square matrix).*

Proof. Both parts will follow if we prove that for any matrices P, Q there is a homomorphism to a field mapping P to an invertible matrix and Q to a singular matrix if and only if no diagonal sum $I \overset{r}{+} P$ can be written as a determinantal sum of non-full matrices and matrices $Q \dotplus B$ (where B is any square matrix). For (i) we take $P = A$, $Q = 0$, and for (ii) we take $P = I$, $Q = A$.

The condition for a homomorphism of the required sort to exist is that there should be a prime matrix ideal containing Q but not P. Let (Q) be the matrix ideal generated by Q, then there is a prime matrix ideal containing Q but not P if and only if $P \notin \sqrt{(Q)}$. So the required condition is that $\overset{r}{+} P \notin (Q)$ for all r, i.e. there is no equation

$$I \overset{r}{+} P = C_1 \triangledown \ldots \triangledown C_t \qquad (C_i \text{ non-full or of form } Q \dotplus B_i). \ \blacksquare \qquad (1)$$

From this theorem it is easy to obtain a criterion for the embeddability of a ring in a field. We recall that an integral domain R is embeddable in a field if and only if for each $a \in R^*$ there is an a-inverting homomorphism into a field. By Theorem 6.3 (i) this holds if and only if there is no equation

$$I \overset{r}{+} a = C_1 \triangledown \ldots \triangledown C_t \qquad (C_i \text{ non-full}). \qquad (2)$$

In particular, $\overset{r}{+} a = aI$ cannot be expressed as in (2). If there is an expression (2) with I of order s say, we multiply both sides by $\overset{s}{+} a \dotplus I$ and observe that the determinantal sum is distributive with respect to multiplication by diagonal matrices. Thus we obtain aI as a determinantal sum of non-full matrices. This proves

COROLLARY 1. *A ring R can be embedded in a field if and only if it is an integral domain and no non-zero scalar matrix aI can be written as a determinantal sum of non-full matrices.* $\blacksquare$

An alternative formulation is given in

COROLLARY 2. *A ring R is embeddable in a field if and only if no diagonal matrix with non-zero elements on the main diagonal can be written as a determinantal sum of non-full matrices.*

For if $ab = 0$, then

$$\begin{pmatrix} a & 0 \\ 0 & b \end{pmatrix} = \begin{pmatrix} a & 0 \\ 1 & b \end{pmatrix} \triangledown \begin{pmatrix} 0 & 0 \\ -1 & b \end{pmatrix} = \begin{pmatrix} a \\ 1 \end{pmatrix}^{(1, b)} \triangledown \begin{pmatrix} 0 \\ 1 \end{pmatrix}^{(-1, b)},$$

and here both matrices on the right are non-full. Thus the condition of Corollary 2 is sufficient to exclude zero-divisors, and by Corollary 1 is therefore sufficient for embeddability in a field. The converse is clear. ∎

It may be of interest to note that the condition in Corollary 1, apart from the absence of zero-divisors, is in the form of quasi-identities, as required by general theory.

We have already found the condition for the existence of a universal R-field: it was that the radical of the least matrix ideal, the set $\mathcal{N}$ constructed before Theorem 6.2, should be prime. Moreover, there will be a universal field of fractions if and only if, further, $\mathcal{N}$ contains no non-zero elements of R. But frequently we are interested in the special case when there is a universal field of fractions in which every full matrix is invertible.

THEOREM 6.4. *Let R be any ring. Then there is an R-field in which every full matrix of R can be inverted (and which is therefore a universal field of fractions) if and only if*

(i) *the diagonal sum of any full matrices is full, and*

(ii) *the determinantal sum of any non-full matrices (where defined) is non-full.*

For the conclusion requires that the set of non-full matrices should be the unique least prime matrix ideal. If this is to be the case, (i) and (ii) must hold. Conversely, when they are satisfied, then the non-full matrices form a matrix ideal by (ii), necessarily the least, and it is prime by (i). ∎

As an application of Theorem 6.4 let us show that every semifir has a universal field of fractions, in which every full matrix of the ring can be inverted.

Let R be a semifir. Then any $n \times n$ matrix A corresponds to an n-generator submodule H of ^nR, where H is generated by the columns of A, and A is full if and only if H is not contained in any $(n-1)$-generator submodule of ^nR. Let A, B be two non-full matrices whose determinantal sum with respect to the first column exists, and write

$$C = A \nabla B.$$

Then A, B correspond to submodules H, K respectively of ^nR; by hypothesis the vectors spanning H and K may be taken to be $a_1, \ldots, a_n$ and $a_1', a_2, \ldots, a_n$ respectively. Moreover, since A and B are not full, we can find submodules H' and K' of ^nR, each generated by $n-1$ elements, such that $H \subseteq H'$ and $K \subseteq K'$. Clearly $H' + K'$ is finitely generated and hence free. If $rk(H' + K') < n$, then the submodule corresponding to the matrix C, which is spanned by $a_1 + a_1', a_2, \ldots, a_n$ is contained in $H' + K'$ of rank less than n, and it follows that C is not full. In what follows we may therefore assume

$$rk(H' + K') \geqslant n. \tag{3}$$

If $rk(H' \cap K') < n - 1$, we observe that $a_2, \ldots, a_n \in H' \cap K'$; hence $a_2, \ldots, a_n$ lie in a submodule of rank less than $n-1$, and therefore $a_1 + a_1'$, $a_2, \ldots, a_n$ lie in a submodule of rank less than n, so we conclude again that C is not full. So we may also assume that

$$rk(H' \cap K') \geqslant n - 1. \tag{4}$$

But we have

$$rk(H' + K') + rk(H' \cap K') = rk\, H' + rk\, K'.$$

The left-hand side is at least $2n - 1$, by (3) and (4), while the right-hand side is at most $2n - 2$, because each of H', K' can be generated by $n - 1$ elements. This contradiction shows that C is not full, so (ii) of Theorem 6.4 holds. Clearly (i) also holds (cf. Theorem 5.6.2).

Thus the conditions of Theorem 6.4 hold and we conclude that every semifir has a universal field of fractions in which every full matrix can be inverted.

Despite the application just made, the criterion of Theorem 6.4 is not easy to verify in other cases, especially condition (ii), and it is of some interest to have an alternative condition for the existence of a universal field of fractions. In what follows we shall denote the set of all full matrices (over the given ring R) by Φ; a ring (or homomorphism) will be called *fully inverting* if it is Φ-inverting. Any fully inverting homomorphism $f : R \to S$ into a non-zero ring must be injective, for every non-zero element of R is full, as a 1×1 matrix, and so maps to an invertible element of S. We shall want to know under what conditions the rational closure, or even the Φ-rational closure, is a field. The next result gives two conditions for this to happen.

THEOREM 6.5. *Let $f: R \to S$ be a fully inverting homomorphism and assume that either*

(i) *$S \neq 0$ and the set Φ of full matrices over R is multiplicative, or*

(ii) *every unit matrix in S is full,*

then f is injective and the Φ-rational closure is a field of fractions of R.

Proof. In either case $S \neq 0$, so by the remark just made, f must be injective. Now assume that (i) holds; we must show that $\bar{R}$, the Φ-rational closure, is a field. Let $x \in \bar{R}$, say $x = u_1$ is the first component of the solution of the equation

$$Au + a = 0,$$

where A is a full matrix over R. Denote the columns of A by $a_1, \ldots, a_n$ and write $A_1 = (a, a_2, \ldots, a_n)$, $v = (1, u_2, \ldots, u_n)^T$, then

$$A_1 v + a_1 u_1 = 0. \tag{5}$$

We assume that $u_1 \neq 0$ and assert that in that case A_1 is full. For if not, then we write $A_1 = BC$, where B is $n \times r$ and C is $r \times n$, for some $r < n$. Let C_1 be the $r \times (n-1)$ matrix consisting of the last $n - 1$ columns of C, then

$$A = (a_1, a_2, \ldots, a_n) = (a_1, BC_1) = (a_1, B) \begin{pmatrix} 1 & 0 \\ 0 & C_1 \end{pmatrix}.$$

Here (a_1, B) is $n \times (r+1)$ and $r + 1 \leqslant n$. Since A is full, we see that $r + 1 = n$ and (a_1, B) is full. Now (5) can be written

$$(a_1, B) \begin{pmatrix} u_1 \\ Cv \end{pmatrix} = 0,$$

and since every full matrix over R is invertible over S, we find that $u_1 = 0$, a contradiction. This shows that A_1 in (5) is full, hence invertible over S, and so u_1 has an inverse in $\bar{R}$. Thus every non-zero element of $\bar{R}$ has an inverse, hence $\bar{R}$ is a field, as asserted.

In case (ii), let Φ' be the set of all square matrices over R mapping to invertible matrices under f. By hypothesis Φ' includes all full matrices, and since the unit matrix in S is always full, it includes no others, thus $\Phi' = \Phi$ consists precisely of the full matrices over R. By Proposition 1.1, Φ is multiplicative, so the hypotheses of (i) are satisfied and we can apply the first part of the proof. ∎

The way in which Theorem 6.5 is usually applied is to take a ring R in which Φ is multiplicative. If there is a fully inverting homomorphism into a non-zero ring, it must be injective, and hence so is the universal fully inverting homomorphism. This yields the

COROLLARY. *Let R be any ring in which the set Φ of all full matrices is multiplicative. Then the universal fully inverting homomorphism*

$$\lambda : R \to R_\Phi \tag{6}$$

is either 0 or an embedding of R into the universal field of fractions.

For by Theorem 6.5, if $f \neq 0$, it is injective and R_Φ must be a field. Now Φ is the greatest set of matrices to be inverted under homomorphism into a field, hence R_Φ is the universal field of fractions. ∎

It is possible to prove directly that for a fir the mapping (6) is injective, and this provides another proof that a fir has a universal field of fractions (*cf.* Cohn [71]) without having to use Theorem 5.3.

Let R be a ring with a universal R-field K; even when the natural mapping $R \to K$ is an embedding, it is not necessarily true that every full matrix over R is invertible over K (*cf.* the commutative case). The case where K is *fully* inverting is a special and rather favourable one; we list one important consequence. We recall that a homomorphism is said to be *honest* if it maps every full matrix to a full matrix.

THEOREM 6.6. *Let R be a ring with a fully inverting R-field K, so that K is necessarily the universal field of fractions. Then every honest endomorphism of R extends to a unique endomorphism of K. In particular, every automorphism of R extends to a unique automorphism of K.*

Proof. By Theorem 6.5, Corollary, K is just the universal fully inverting ring R_Φ. Now any honest endomorphism α of R maps the set Φ of all full matrices into itself, and so α can be extended to R_Φ by putting $(A^{-1})^\alpha = (A^\alpha)^{-1}$; clearly this is the only way of extending α to an endomorphism of R_Φ, and it yields an automorphism when α is an automorphism on R. ∎

Exercises 7.6

1. Let R be a ring with a fully inverting R-field K. Show that every derivation on R extends to one on K.

2. Show that a ring R has a universal R-field if and only if the radical of the least matrix ideal is prime.

3. Show that R is embeddable in a direct product of fields if and only if no non-zero scalar matrix can be written as a determinantal sum of non-full matrices.

4. Show that any ring with an inverse weak algorithm has a universal field of fractions. If K is the universal field of fractions of the power series ring $k\langle\!\langle X\rangle\!\rangle$, show that the subfield of K generated by $k\langle X\rangle$ is the universal field of fractions of the latter. (Hint: Use the inertia Theorem 2.8.12.)

5. Let $R = k\langle x, y, z, t\rangle$ and define an endomorphism α of R by the rules: $x \mapsto xz$, $y \mapsto xt$, $z \mapsto yz$, $t \mapsto yt$. Show that α is injective, but not honest.

6*. Let $R = k\langle x, y, z\rangle$, denote its universal field of fractions by K and the k-subalgebra generated by x, y, z and $t = zx^{-1}y$ by S. Show that the matrix $\begin{pmatrix} x & y \\ z & t \end{pmatrix}$ is full in S but not invertible in K.

7. Let R be a ring with a 2×2 matrix C satisfying $C^n = 0$ (for some $n \geqslant 1$). If c is any coefficient of C^2, show that for some unit matrix I, cI is a determinantal sum of non-full matrices.

8^0. Find an explicit expression for cI in Exercise 7 when R is generated (over some field k) by the coefficients of C, and $C^3 = 0$.

9^0. Given $n > 1$, find a criterion for a ring to be embeddable in an $n \times n$ matrix ring over a field.

10^0. Let K be the universal field of fractions of $k\langle X\rangle$, and K^{ab} the group K^* made abelian. Determine K^{ab}, or at least show that the canonical mapping $K \to K^{ab}$, restricted to X, is injective.

11^0. Let A be a matrix over a free algebra. If the matrix ideal generated by A is improper, is A necessarily invertible? (*cf.* Exercise 4.3.10.)

Notes and comments on Chapter 7

Necessary and sufficient conditions for the embeddability of a semigroup in a group were found by Malcev [39], but they gave no hint for the problem of embedding rings in fields; in fact it was only recently that examples were found of rings whose non-zero elements are embeddable in groups but which are not embeddable in fields (Bokut [69], Bowtell [67'], Klein [67]). Until 1970 the only purely algebraic methods of embedding rings in fields were based on Ore's method (for a survey which includes the topological methods, see Cohn [71']). However, it followed from general principles in universal algebra that a criterion for embedding integral domains in fields could be written down in terms of 'quasi-identities' (= universal Horn sentences), *cf.* Cohn [65], p. 235.

The basic idea underlying the method of this chapter is to invert matrices rather than elements. This was inspired by the rationality criteria of Schützenberger [62] and Nivat [69] for non-commutative power series rings. The observation that these criteria had quite general validity was exploited in Cohn [71] to embed firs in fields. These criteria survive in Theorem 1.2 (the forms given by Schützenberger and Nivat correspond to (c) and (b) respectively). The application to firs is based on the following theorem (Cohn [71]): Let R be a ring and M a subset. Then the mapping $R \to R_M$ (into the universal M-inverting ring) is injective provided that the elements of M are non-zerodivisors and for any $p, q \in M$, $\text{Hom}_R(R/pR, R/qR)$ is 0 or a field according as $p \neq q$ or $p = q$. As a further consequence, for any atomic 2-fir R, the semigroup R^* is embeddable in a group. Other interesting consequences depend on the fact that the free product of fields is a fir and so is embeddable in a field.

The notion of universal field of fractions was defined by Amitsur [66] as a result of studying generalized rational identities (cf. also Bergman [70]). The direct description in 7.2, in terms of matrices, is based on Cohn [72]; in fact it can be used to give an alternative proof of Amitsur's result (cf. Cohn [72″]). For applications to radical rings see Cohn [71‴].

The development of 7.3–5 follows Cohn [72‴] (cf. Cohn [61′] for Lemma 5.1); that reference includes a description of the 'affine scheme' $(X, \tilde{R})$ associated with a general ring R. For any ring R one has a homomorphism

$$R \to \Gamma(X, \tilde{R}) \tag{1}$$

into the ring of global sections, but of course one cannot expect an isomorphism here (as in the commutative case, cf. Macdonald [68]). Thus if R has no homomorphisms into fields, the scheme will be trivial. Besides the sections arising from R one also has the following 'rational' sections: Let A be a matrix over R; if A maps to an invertible matrix under every localization, then the entries of A^{-1} define *rational sections* and one might ask: (i) for which rings is every global section rational, and (ii) for a free algebra, is every global section integral?

An affirmative answer to (ii) would also solve Exercise 6.11 and Exercise 4.3.10.

Let us call a ring *tidy* if every global section is integral, i.e. if (1) is surjective. Then it is not hard to show that a principal ideal domain is tidy if and only if it is invariant and one might conjecture that a semifir is tidy if and only if it has a distributive factor lattice.

8. Principal ideal domains

There is one important property of principal ideal domains which does not carry over to firs: the diagonal reduction of matrices. This is discussed in **8**.1, and its consequences for modules are described in **8**.2. In particular, these results apply to skew polynomial rings (**8**.3–4) and they allow us to analyse certain simple types of skew field extensions. Many of the earlier results in the book can be used here, and this has been done wherever it simplified matters. The concluding section **8**.6 describes Jategaonkar's iterated skew polynomial rings and some of their properties.

8.1 The diagonal reduction for matrices over a principal ideal domain

Let R be a fir and A an $m \times n$ matrix over R. Then we can always find $P \in \mathbf{GL}_m(R)$ and $Q \in \mathbf{GL}_n(R)$ such that $PAQ = \begin{pmatrix} A_1 & 0 \\ 0 & 0 \end{pmatrix}$, where A_1 is a non-zerodivisor, but in general no further reduction is possible. However, when R is a principal ideal domain, we can reduce A to diagonal form. In order to state the result, we need a definition. In an integral domain, an element a is said to be a *total divisor* of b, written $a\|b$, if there exists an invariant element c such that $a|c|b$. We observe that an element is not generally a total divisor of itself; in fact $a\|a$ if and only if a is invariant. If the ring R is simple, it has no non-unit invariant elements and $a\|b$ implies that a is a unit or $b = 0$.

We write diag $(a_1, ..., a_r)$ for a matrix with $a_1, ..., a_r$ on the main diagonal and 0's elsewhere. This notation will be used even for matrices that are not square; the exact size of the matrix will be indicated explicitly, unless it is clear from the context, as in (1) below.

THEOREM 1.1. *Let R be a principal ideal domain and let A be an $m \times n$ matrix over R. Then the row rank and column rank of A are the same; denoting the common value by r, we can find $P \in \mathbf{GL}_m(R)$ and $Q \in \mathbf{GL}_n(R)$ such that*

$$P^{-1} AQ = \text{diag} (e_1, ..., e_r, 0, ..., 0), \qquad e_i\|e_{i+1}, \ e_r \neq 0. \tag{1}$$

Proof. We have the following four types of operations on the columns of A, of which the first three are the well known elementary operations:

1. *Interchange two columns,*

2. *Multiply a column on the right by a unit factor,*

3. *Add a right multiple of one column to another,*

4. *Replace the first element in each of two given columns by their highest common left factor and 0 respectively.*

As is well known, each of 1–3. corresponds to right multiplication by an elementary matrix; only 4. requires further justification. Thus let the first element in each of two columns be a, b respectively, and assume that a, b are not both zero. We have $aR + bR = kR$, say $a = ka_1$, $b = kb_1$. This means that $a_1 R + b_1 R = R$, and since $a_1 R \cap b_1 R$ is principal, it follows by Theorem 3.3.2, Corollary 2, that there is an invertible 2×2 matrix C with (a_1, b_1) as first row, hence $(k, 0) C = (a, b)$, and $(a, b) C^{-1} = (k, 0)$, as claimed. Corresponding operations can of course be carried out on the rows, acting on the left.

We can now proceed with the reduction. If $A = 0$, there is nothing to prove; otherwise we bring a non-zero element to the $(1, 1)$-position in A, by permuting rows and permuting columns (op.1). Using op. 4, we replace a_{11} successively by the HCLF of a_{11} and a_{12}, then by the HCLF of the new a_{11} and a_{13}, etc. After $n - 1$ steps we have transformed A to a form where $a_{12} = \ldots = a_{1n} = 0$. By symmetry the same process can be applied to the first column of A; in the course of the reduction the first row of A may again become non-zero, but this can happen only if the length (i.e. the number of factors) of a_{11} is reduced, therefore by an induction on the length of a_{11} we reach the form

$$A = \begin{pmatrix} a_{11} & 0 & 0 \ldots 0 \\ 0 & & \\ \vdots & & A_1 \\ 0 & & \end{pmatrix}.$$

By another induction, on $\max (m, n)$, we reach the form

$$A = \operatorname{diag} (a_1, a_2, \ldots, a_k, 0, \ldots, 0).$$

Consider a_1 and a_2; for any $d \in R$ we have

$$\begin{pmatrix} 1 & d \\ 0 & 1 \end{pmatrix} \begin{pmatrix} a_1 & 0 \\ 0 & a_2 \end{pmatrix} = \begin{pmatrix} a_1 & da_2 \\ 0 & a_2 \end{pmatrix},$$

and now we can again diminish the length of a_1 unless a_1 is a left factor of da_2 for all $d \in R$, i.e. unless $a_1 R \supseteq Ra_2$. But in that case $a_1 R \supseteq Ra_2 R \supseteq$

Ra_2; thus $a_1|c|a_2$, where c is the invariant generator of the ideal $Ra_2 R$. Hence $a_1\|a_2$, and by repeating the argument we obtain the form diag $(e_1, ..., e_k, 0, ..., 0)$, where $e_i\|e_{i+1}$. If $e_k \neq 0$, this has row rank and column rank k; but clearly A and $P^{-1}AQ$ have the same column rank. Similarly the row rank of A is k and the assertion follows. ∎

We shall return to this theorem in the next section, to see how far the e's are unique. For the moment we note the following

COROLLARY. *Let R be a simple principal ideal domain. Then any matrix A over R is associated to* diag $(1, 1, ..., 1, a, 0, ..., 0)$, *where $a \in R$.*

For if $a\|b$, then either $b = 0$ or a is a unit. Now any unit may be reduced to 1, and we see that there can be at most one diagonal element not zero or a unit. ∎

In the case of a field, every non-zero element is a unit and we see that every matrix is associated to diag $(1, 1, ..., 1, 0, ..., 0)$, where the number of 1's equals the rank of A.

Exercises 8.1

1. What simplifications are possible in the proof of Theorem 1.1 if R is either (i) Euclidean or (ii) commutative? In particular, prove the uniqueness in this case (Theorem 2.3).
2. (Kaplansky [49]). By an *elementary divisor ring* is meant a ring over which every matrix admits a diagonal reduction as in Theorem 1.1. Show that a ring over which every $m \times n$ matrix, with $m, n \leqslant 2$, admits a diagonal reduction is an elementary divisor ring.
3. (Kaplansky [49]). Show that an elementary divisor ring which is an integral domain is weakly finite.
4°. Is every commutative Bezout domain an elementary divisor ring?
5. Show that an elementary divisor ring which is a fir is necessarily principal.

8.2 Finitely generated modules over principal ideal domains

Let R be a principal ideal domain. Then R is in particular a right Ore ring, and so every right R-module M has a torsion submodule tM with torsion-free quotient M/tM (*cf.* 0.6). Suppose now that M is finitely generated torsion free over R, then M can be embedded in a free R-module (Proposition 0.6.4) and hence is free (Theorem 1.1.1). Thus we have proved

PROPOSITION 2.1. *Every finitely generated torsion free R-module over a principal ideal domain is free.* ∎

Let us return to an arbitrary finitely generated R-module M. Then M/tM is finitely generated torsion free and hence free; therefore it can be lifted to a free submodule of M complementing tM:

COROLLARY. *Let R be a principal ideal domain. If M is any finitely generated R-module, there exists a free submodule F of M such that*

$$M = tM \oplus F.$$

Here tM is uniquely determined as the torsion submodule of M, while F is unique up to isomorphism. ∎

This result shows that in classifying finitely generated modules over principal ideal domains we can restrict ourselves to torsion modules, but there would be no gain in making this simplification. However, there is another important (and well known) fact which will be useful to us.

PROPOSITION 2.2. *Let R be a principal ideal domain, and M an n-generator R-module. Then each submodule of M can be generated by n elements.*

Proof. Let $M = F/G$, where $F = {}^nR$, then any submodule of M has the form H/G, where H is a submodule of F. Therefore it is enough to prove the result for free modules. Thus let H be any submodule of $F = {}^nR$ and denote the projection on the last factor by π. Then $\pi(H)$ is a right ideal of R, i.e. free of rank 0 or 1. Thus $H = \ker \pi \oplus F_1$ where F_1 is free of rank at most 1. Now $\ker \pi \subseteq {}^{n-1}R$ and the result follows by induction on n. ∎

In terms of the characteristic defined in Chapter 5 we can express the result by saying that over a principal ideal domain, every module has non-negative characteristic.

Proposition 2.2 shows in particular that every finitely generated module is finitely presented. Let M be a finitely generated module with presentation

$$^nR \xrightarrow{\alpha} {}^mR \to M \to 0, \tag{1}$$

and let A be the $m \times n$ matrix over R which represents the homomorphism $\alpha: {}^nR \to {}^mR$, relative to the standard bases in ${}^nR$, ${}^mR$. Then M is completely specified by A, and if we change the bases in ${}^nR$ and ${}^mR$, this amounts to replacing A by $P^{-1}AQ$, where $P \in \mathbf{GL}_m(R)$, $Q \in \mathbf{GL}_n(R)$. Using Theorem 1.1 we now obtain the following generalization of the fundamental theorem for finitely generated abelian groups:

THEOREM 2.3. *Let R be a principal ideal domain. Then any finitely generated R-module M is a direct sum of cyclic modules*:

$$M \cong R/e_1 R \oplus ... \oplus R\, e_r R \oplus {}^{m-r}R, \tag{2}$$

where $e_i \| e_{i+1}$, and this condition determines the e_i up to similarity.

Proof. Let M be defined by a presentation (1) with matrix A. By Theorem 1.1, A is associated to $\mathrm{diag}\,(e_1, ..., e_r, 0, ..., 0)$ with $e_i \| e_{i+1}$, and since this does not affect the module, we obtain (2). It only remains to prove the uniqueness.

We begin with the remark that finitely generated modules can be cancelled. If

$$M \oplus N = M \oplus N', \tag{3}$$

where M, N, N' are finitely generated R-modules, then $N \cong N'$. For N, N' clearly have the same rank, and by applying Proposition 2.1, Corollary we may assume that M, N, N' all are torsion modules. Now $N \cong N'$ follows on taking complete decompositions on both sides of (3) and applying the Krull–Schmidt theorem 3.6.3. Let us write $R/a_1 R \oplus ... \oplus R/a_k R$ as $[a_1, ..., a_k]$ for short and use $\sim$ to indicate isomorphism, then $[a] \sim [b]$ if and only if $a \sim b$, and what we have shown is that

$$[a, b_1, ..., b_r] \sim [a, b_1', ..., b_s'] \quad \text{implies} \quad [b_1, ..., b_r] \sim [b_1', ..., b_s'].$$

Let us take two representations of M as direct sums of cyclic modules:

$$M \sim [d_1, ..., d_r] \sim [e_1, ..., e_r], \qquad d_i \| d_{i+1}, \; e_i \| e_{i+1}. \tag{4}$$

There is no loss of generality in assuming the same number of summands on both sides, since we can always add zero summands, represented by unit factors: $R/R = 0$. In addition we may suppose that the torsion free part of M has been split off, so that the d_i, e_i are all different from 0. If $r = 1$, the result follows since then $d_1 \sim e_1$, so let $r > 1$ and use induction on r. We shall write $l(a)$ for the length of a and assume that $l(d_1) \geqslant l(e_1)$; further, let $d_1 | c | d_2$, where c is invariant. If N is any module, Nc is a submodule; more specifically, if $N = R/aR$, and c is invariant, then $N/Nc \cong R/(aR + cR)$. Now consider M/Mc; writing $e_i R + cR = e_i' R$, we have by (4),

$$M/Mc \sim [d_1, c, ..., c] \sim [e_1', e_2', ..., e_r'], \tag{5}$$

and $l(e_i') \leqslant l(c)$ $(i = 1, ..., r)$, $l(e_1') \leqslant l(e_1) \leqslant l(d_1)$. Comparing lengths in (5) (which must be equal, as the length of a composition series of M/Mc), we find $l(d_1) + (r - 1) l(c) = \Sigma l(e_i')$, i.e.

$$l(d_1) - l(e_1') + \Sigma_2^r (l(c) - l(e_i')) = 0.$$

Since each term is non-negative, all are zero and $l(e_1') = l(e_1) = l(d_1)$, $l(e_i') = l(c)$. It follows that $e_1' = e_1$, $e_i' = c$ $(i > 1)$, and (5) now reads

$$[d_1, c, ..., c] \sim [e_1, c, ..., c].$$

By cancellation, $e_1 \sim d_1$; so we may cancel the first term on both sides of (4) and obtain $[d_2, ..., d_r] \sim [e_2, ..., e_r]$. Now an induction on r gives the result. ∎

This result shows that the e_i in Theorem 1.1 are determined up to similarity. They are called the *invariant factors* of the matrix A or also of the module M. The condition imposed on the e_i (that each e_i be a total divisor of the next) ensures that (2) is a decomposition of M into cyclic modules with as *few* factors as possible, and it may be shown that this proviso ensures that conversely we get a decomposition of the form (2). At the other extreme we have a decomposition into as *many* factors as possible, i.e. a complete direct decomposition, into indecomposable modules. The indecomposable parts must then be cyclic, by Theorem 2.3, and they are unique up to order, by the Krull–Schmidt theorem. The factors in this case are called the *elementary divisors* of the module (or the matrix). E.g., over the integers, $M = \mathbf{Z}/30\mathbf{Z} \oplus \mathbf{Z}/150\mathbf{Z} \oplus \mathbf{Z}/7500\mathbf{Z}$ has the invariant factors 30, 150, 7500 and the elementary divisors 2, 3, 5; 2, 3, 5^2; 2^2, 3, 5^4.

As an application of Theorem 2.3 let us take R to be commutative and consider the decomposition of a torsion module. Given a torsion module M over a commutative principal ideal domain R and an atom $p \in R$, write $M(p)$ for the set of $x \in M$ which are annihilated by some power of p, and call $M(p)$ the *p-primary component* of M.

PROPOSITION 2.4. *Let R be a commutative principal ideal domain and M a torsion module over R, then there is a direct sum decomposition*

$$M = \Sigma M(p), \tag{6}$$

where the summation is over a set of representatives of associated classes of atoms of R.

Proof. Let P be a set containing one representative for each class of associated atoms, choose $a \in M$ and write the annihilator of a in the form

$p_1^{r_1} \ldots p_k^{r_k} R$, where the p_i are distinct elements of P. If we put $q_i = \prod_{j \neq i} p_j^{r_j}$,

then $\Sigma q_i R = R$, for the common factor of the q_i divides a and is divisible by no p_i. Hence we have an equation $\Sigma q_i u_i = 1$. Put $a_i = a q_i$, then $a_i p_i^{r_i} = 0$ and $a = \Sigma a_i u_i$, hence $M = \Sigma M(p)$, and it only remains to show that the sum is direct. Let $a_1 + \ldots + a_k = 0$, where $a_i p_i^{r_i} = 0$, say, and $a_1 \neq 0$. Then $-a_1 = \Sigma_2^k a_i$, therefore $a_1 p_2^{r_2} \ldots p_k^{r_k} = 0$, but $(p_1^{r_1}, p_2^{r_2} \ldots p_k^{r_k}) = 1$, hence $a_1 = 0$, a contradiction, and this shows the sum (6) to be direct. ∎

This result shows that for a module over a commutative principal ideal domain, the elementary divisors are necessarily prime powers. The same is true more generally for invariant principal ideal domains, but in no other cases.

Let M be a right R-module, where R is arbitrary, and denote the endomorphism ring $\text{End}_R(M)$ by S, then M is a left S-module in a natural way. If $\text{End}_S(M) = R'$, then again M is a right R'-module, and moreover there is a natural mapping $R \to R'$, since the action on M by an element of R is an S-endomorphism. If this mapping is surjective, R is said to act *bicentrally* on M. E.g. R always acts bicentrally on itself, and more generally, R acts bicentrally on any finitely generated free module, by a double application of Lemma **0**.1.1.

PROPOSITION 2.5. *Let R be a principal ideal domain and M a finitely generated R-module, then R acts bicentrally on M.*

Proof. By Theorem 2.3, $M = u_1 R \oplus \ldots \oplus u_r R$, where u_i has annihilator $e_i R$ say, and $e_i \| e_{i+1}$. It follows that for $k = 1, \ldots, r$ there is an endomorphism α_k of M such that $\alpha_k u_i = \delta_{ir} u_k$. Now take any f in the bicentralizer of R, and suppose $u_r f = u_r a + u_{r-1} b + \ldots$. Then $u_r f = (\alpha_r u_r) f = \alpha_r (u_r f) = \alpha_r (u_r a + \ldots) = u_r a$. Further, $u_k f = (\alpha_k u_r) f = \alpha_k (u_r a) = u_k a$. Thus f is just the right multiplication by a. ∎

Exercises 8.2

1. Let R be a principal ideal domain. If the elementary divisors of any module are powers of atoms, show that R is invariant. (Hint: Use Theorem 4.2.7.)

2. Obtain an analogue of Proposition 2.4 for bounded modules over non-commutative principal ideal domains. More generally, consider *locally bounded* modules, i.e. modules in which every finitely generated submodule is bounded.

3*. Let M be a finitely generated bounded module over a principle ideal domain. Prove directly that the bounds of the elementary divisors of M are independent of the decomposition of M.

4⁰. Let R be a 2-fir; if every finitely generated torsion free right R-module is free, is R necessarily left Bezout?

8.3 Skew polynomial rings

We shall now apply the results of **8.1–2** to the skew polynomial ring

$$R = k[t; S, D], \tag{1}$$

where k is a (skew) field, S an automorphism and D an S-derivation. We recall that R is a principal ideal domain, in fact a Euclidean domain; we therefore have the diagonal reduction of matrices over R, and since R is Euclidean, this reduction can be carried out by elementary transformations alone.

When R is given, t is determined up to a linear transformation; a change of variable changes the automorphism S by an inner automorphism and, when S is fixed, changes D by an inner S-derivation, i.e. a derivation of the form

$$\Delta_c : a \mapsto ac - ca^S.$$

We utilize this choice to simplify the expression (1) for R as follows. Let C be the centre of k and assume first that S does not leave C fixed, say $\gamma^S \neq \gamma$ for some $\gamma \in C^*$. On changing the variable to $t_1 = \gamma t - t\gamma$, we find for any $a \in k$,

$$at_1 = a(\gamma t - t\gamma) = \gamma ta^S + \gamma a^D - ta^S \gamma - a^D \gamma = (\gamma t - t\gamma)a^S,$$

i.e.

$$at_1 = t_1 a^S,$$

so that D is inner, and by a suitable choice of variable may be taken to be 0. Next we assume that S leaves C fixed, but there exists $\gamma \in C$ such that $\gamma^D \neq 0$. Then $\gamma^S = \gamma$ and for any $a \in k$, $a\gamma = \gamma a$, hence $a^D \gamma + a\gamma^D = \gamma^D a^S + \gamma a^D$, i.e. $a\gamma^D = \gamma^D a^S$, and on putting $t_1 = t(\gamma^D)^{-1}$, we have

$$at_1 = ta^S(\gamma^D)^{-1} + a^D(\gamma^D)^{-1} = t(\gamma^D)^{-1}a + a^D(\gamma^D)^{-1} = t_1 a + a^{D'},$$

where D' is an ordinary derivation.

The only remaining possibility is that S leaves C fixed and D maps it to 0. Thus we have

THEOREM 3.1. *Let* $R = k[t; S, D]$ *be a skew polynomial ring over a field* k *with automorphism* S *and* S-*derivation* D. *Then either*

(i) D *is inner and by a suitable choice of* t *may be taken to be* 0, *or*

(ii) S *is inner, and may by a suitable choice of* t *be taken to be* 1, *or*

(iii) S *leaves the centre* C *of* k *fixed and* D *maps it to* 0; *in that case* C *is the centre of* R. ■

Next we determine the invariant elements of R. Let $x \in R$ be right invariant, i.e. $Rx \subseteq xR$. We may take x to be monic; if its degree is n, then for any $c \in k$, $cx = ct^n + \ldots = t^n c^{S^n} + \ldots$, where the dots indicate terms of degree less than n, hence

$$cx = xc^{S^n} \qquad \text{for all} \qquad c \in k. \tag{2}$$

Further, $tx = xu$ and a comparison of highest terms shows that u is monic of degree 1, hence for some $\tau \in k$,

$$tx = x(t + \tau). \tag{3}$$

Thus $\Sigma t^i a_i \mapsto \Sigma (t + \tau)^i a_i^{S^n}$ is an endomorphism of R; in fact it is an automorphism for it is injective, and every element of R can be written as a polynomial in $\tau + t$, therefore it is bijective, i.e. an automorphism. This shows that every right invariant element of R is actually invariant. By symmetry the same holds for left invariant elements. Let us apply the automorphism induced by x to the equation

$$at = ta^S + a^D;$$

we obtain

$$a^{S^n}(t + \tau) = (t + \tau)a^{S^{n+1}} + a^{DS^n}, \quad \text{i.e.}$$

$$ta^{S^{n+1}} + a^{S^nD} + a^{S^n}\tau = ta^{S^{n+1}} + \tau a^{S^{n+1}} + a^{DS^n},$$

hence

$$a\tau - \tau a^S = a^{S-1DS^n} - a^D. \tag{4}$$

If R has non-unit invariant elements, let x be the monic invariant element of least degree in t. Any invariant element f, again taken monic for convenience, may be written

$$f = xq + r \qquad \text{where} \qquad q, r \in R \qquad \text{and} \qquad \deg r < \deg x.$$

Let $cx = xc^\alpha$, $cf = fc^\beta$, then $fc^\beta = cf = c(xq + r) = xc^\alpha q + cr$, hence $x(qc^\beta - c^\alpha q) = cr - rc^\beta$, $\deg(cr - rc^\beta) < \deg f$, hence $cr = rc^\beta$, $qc^\beta = c^\alpha q$ and so q, r are both invariant. Moreover, by the minimality of $\deg x$, r has degree 0, i.e. $r \in k$. Now an induction on the degree of f shows that

$$f = x^m + x^{m-1}c_1 + \ldots + c_m \qquad c_i \in k. \tag{5}$$

Here the highest coefficient is 1 because both x and f are monic. Hence for any $a \in k$,

$$fa^\beta = x^m a^\beta + x^{m-1}c_1 a^\beta + \ldots + c_m a^\beta,$$

$$af = x^m a^{\alpha^m} + x^{m-1}a^{\alpha^{m-1}}c_1 + \ldots + ac_m.$$

Equating coefficients of x^m, we see that $\beta = \alpha^m$; hence on equating powers of x, we find that $c_i a^{\alpha^m} = a^{\alpha^{m-i}} c_i$ $(i = 1, 2, ..., m)$, i.e.

$$ac_i = c_i a^{\alpha^i} \qquad (i = 1, ..., m). \tag{6}$$

Now if x has degree d, then $\alpha = S^d$, $\beta = S^{md}$. Let S^r be the least power that is inner, then S^n is inner if and only if $r | n$, and by (6), $c_i = 0$ unless $r | id$. Put $s = r/(r, d)$, then α^s is the least power that is inner, say $c^{\alpha^s} = ece^{-1}$, hence by (5) and (6), f is a polynomial in $y = x^s e$ with coefficients in k. Here y is in the centralizer of k, by the definition of e. Let $f = \Sigma y^i c_i$ and renormalize f so that $c_0 = 1$, then for any $a \in k$, $af = \Sigma a y^i c_i = \Sigma y^i a c_i$, $fa^\beta = \Sigma y^i c_i a^\beta$, hence $ac_i = c_i a^\beta$. But $c_0 = 1$, hence $a^\beta = a$ and c_i lies in the centre of k.

In particular, if D is inner, we may take it to be 0, and then $x = t$ is invariant. If D is outer but S is inner, we may take $S = 1$. Let $f = t^n + t^{n-1} a_1 + ...$ be invariant, then for any $c \in k$, $cf = t^n c + nt^{n-1} c^D + t^{n-1} ca_1 + ...$; this must equal fc' (for some $c' \in k$), hence $c' = c$, $nc^D = a_1 c - ca_1$. If $n \neq 0$ in k, this shows D to be inner:

$$c^D = \frac{1}{n}(a_1 c - ca_1). \tag{7}$$

Hence in characteristic 0 there can be no non-unit invariant elements. In characteristic p, we find instead of (7) that $p | n$, and as before $cf = fc$ $(c \in k)$. Further, $tf = f(t + \lambda)$ for some $\lambda \in k$, and on expansion we find $t^{n+1} + t^n a_1 + ... = t^{n+1} + t^n \lambda + t^n a_1 + ...$, where dots indicate terms of degree less than n. Hence $\lambda = 0$, and so f is central. Thus we have proved

THEOREM 3.2. *Let $R = k[t; S, D]$ be a skew polynomial ring over a field k with automorphism S and S-derivation D, and denote the centre of k by C. Then any left or right invariant element of R is invariant; if there are any non-unit invariant elements, let x be a monic invariant element of least degree n, say, then*

$$ax = xa^{S^n}, \qquad tx = x(t + \tau), \qquad where \qquad a\tau - \tau a^S = a^{S^{-n} DS^n} - a^D. \tag{8}$$

Moreover, if no power of S is inner, then every invariant element has the form $x^r c$ $(c \in k^*)$ and every element of this form is invariant.

If some power of S is inner on k, let x^r be the least power of x inducing an inner automorphism of k, say $ax^r e = x^r ea$, then every invariant element of R has the form $x^i fc$, where f is a polynomial in $y = x^r e$ with coefficients in C and $c \in k^*$. Further,

   (i) *If D is inner, we may take it to be 0, and then $x = t$ is invariant,*

(ii) *if D is outer but S is inner, we may take S to be* 1. *If there are non-unit invariant elements, then k must be of finite characteristic p, and in that case the invariant elements are precisely the central elements of R.* ∎

Exercises 8.3

1. Let $R = k[x; S, D]$ be a skew polynomial ring and α an automorphism commuting with S. Find conditions under which α can be extended to an automorphism of R. Examine the case $\alpha = S$.

2. Let $R = k[x; S, D]$ be a skew polynomial ring. Find conditions for an anti-automorphism of R to exist which maps x to $-x$ and fixes k.

3. Let k be a commutative field with automorphism α and fixed field C. Determine the centre of the field $k(x; \alpha)$, (i) when α has infinite order, (ii) when α has finite order.

4*. Let A be a square matrix over a skew polynomial ring R and let $\mathrm{diag}(a_1, ..., a_n)$ be its canonical diagonal form. To what extent is the product $a_1 ... a_n$ unique? Use this product to define a notion of determinant; examine the particular case when R is simple.

5°. Obtain an analogue of Theorem 3.1 in case S is not surjective.

8.4 Theory of a single pseudo-linear transformation

Let k be a (skew) field, with an automorphism S and an S-derivation D. Given a right k-space V, a mapping $\theta: V \to V$ is said to be *pseudo-linear* (relative to S, D) if

$$(x + y)\,\theta = x\theta + y\theta, \quad (xa)\,\theta = x\theta \cdot a^S + x \cdot a^D \qquad (x, y \in V, \ a \in k). \quad (1)$$

This includes the linear case when $S = 1$, $D = 0$. Other important examples are the *semilinear* transformations (where $k \in C$, $D = 0$ and S is conjugation) and differential transformations $(k = F(t), S = 1$ and $D = d/dt)$.

To obtain a matrix representation of θ we take a basis $v_1, ..., v_n$ in V. Let

$$v_j\,\theta = \Sigma v_i a_{ij} \qquad (a_{ij} \in k), \quad (2)$$

and write $A = (a_{ij})$, then

$$(\Sigma v_j \xi_j)\,\theta = (\Sigma v_j \theta)\,\xi_j{}^S + \Sigma v_j \xi_j{}^D = \Sigma v_i(\Sigma a_{ij}\xi_j{}^S + \xi_i{}^D),$$

hence on writing $\xi = (\xi_1, ..., \xi_n)^T$, we have

$$\theta: \xi \mapsto A\xi^S + \xi^D. \quad (3)$$

Conversely, every $n \times n$ matrix A over k defines a pseudo-linear transformation θ by (3). Let us write the basis in V as a row vector $v = (v_1, ..., v_n)$, so that the general vector $x \in V$ takes the form $x = v\xi$. If we change the basis in V, say $x = v\xi = v'\xi'$, where $v' = vP$, $\xi' = P^{-1}\xi$ and $P \in \mathbf{GL}_n(k)$, then

$v\theta = vA$, $v'\theta = v'B$, hence $vPB = vP\theta = v\theta P^S + vP^D = v(AP^S + P^D)$, and so we obtain

$$PB = AP^S + P^D. \tag{4}$$

Two matrices A, B that are related as in (4), where $P \in \mathbf{GL}_n(k)$, are said to be (S, D)-*similar*. Thus a given pseudo-linear transformation is represented in different coordinate systems by matrices that are mutually (S, D)-similar, and one of our main tasks will be to obtain a canonical form for a given matrix under (S, D)-similarity.

Let us write $R = k[t; S, D]$, then our space V can be defined as a right R-module by letting $\Sigma t^i a_i$ correspond to $\Sigma \theta^i a_i$; the equations (1) ensure that this is a homomorphism.

Taking again a right k-basis $v = (v_1, ..., v_n)$ of V, we find that V is generated as right R-module by $v_1, ..., v_n$ and satisfies the relations $\Sigma v_i(a_{ij} - t\delta_{ij}) = 0$, or in matrix form

$$v(A - tI) = 0. \tag{5}$$

In fact this is a complete set of defining relations for V as right R-module, as is easily checked. Now by the diagonal reduction in R there exist P, $Q \in \mathbf{GL}_n(R)$ such that

$$P(A - tI)Q = \mathrm{diag}(\lambda_1, ..., \lambda_n) \qquad \lambda_{i-1} \| \lambda_i. \tag{6}$$

The polynomials λ_i in t are called the *invariant factors* of A. It follows that as right R-module, V is isomorphic to

$$R/\lambda_1 R \oplus ... \oplus R/\lambda_n R. \tag{7}$$

We observe that $R/\lambda R = 0$ if and only if $\lambda = 1$ and $R/\lambda R \cong R$ if and only if $\lambda = 0$. Moreover, $\dim_k (R/\lambda R) = \deg \lambda$, hence $\Sigma \deg \lambda_i = n$. The expression (7) shows that V is a direct sum of cyclic R-modules, and as minimal decomposition into cyclic R-modules, this expression is unique up to isomorphism.

We ask: When is V itself cyclic? By (6) this is the case if there is just one invariant factor $\neq 1$. In this case the pseudo-linear transformation θ or also the matrix A representing it, is called *cyclic* (or sometimes *non-derogatory*). For a cyclic pseudo-linear transformation we have the following normal form: let v be a generator of V as right R-module and put $v_i = v\theta^{i-1}$, then V is spanned over k by $v_1, v_2, ...$ Let v_{r+1} be the first of these vectors linearly dependent on the preceding ones over k, say

$$v_{r+1} = v_1 a_1 + ... + v_r a_r. \tag{8}$$

Applying θ, we have

$$v_{r+2} = v_1 a_1 \theta + \dots + v_r a_r \theta$$
$$= v_2 a_1{}^S + v_1 a_1{}^D + \dots + v_{r+1} a_r{}^S + v_r a_r{}^D,$$

which shows that all the v_i are linearly dependent on $v_1, \dots, v_r$. By hypothesis V is spanned by the v_i, hence $r = n$ and $v_1, \dots, v_n$ is a basis for V over k. Relative to this basis θ has the matrix

$$\begin{pmatrix} 0 & 0 & 0 & \dots & 0 & a_1 \\ 1 & 0 & 0 & \dots & 0 & a_2 \\ 0 & 1 & 0 & \dots & 0 & a_3 \\ & & \dots & & \dots & \\ 0 & 0 & 0 & \dots & 1 & a_n \end{pmatrix}. \tag{9}$$

Given a pseudo-linear transformation θ, we can decompose V into a direct sum of spaces on each of which θ is cyclic, therefore θ will have a matrix which is a diagonal sum of pieces (9). This is of course the familiar canonical form for linear transformations; we see that it still applies to pseudo-linear transformations.

In the commutative case every linear transformation satisfies a polynomial of degree n (Cayley–Hamilton theorem) which is a multiple of the minimal polynomial. This will not hold in general: If θ is a pseudo-linear transformation, and V the associated R-module, then the minimal polynomial of θ is the generator of the two-sided ideal of R annihilating V, and this may well be 0, even when V is finite-dimensional over k. In fact, as we see from (7), θ has a minimal polynomial if and only if its last invariant factor is bounded; in that case we say that θ (or also the matrix representing θ) is *bounded*. The bound for the last invariant factor is then the minimal polynomial. For example, the ring of differential transformations (in characteristic 0) is simple, hence the only matrices that are bounded are the invertible matrices.

Exercises 8.4

1. Find the normal form for $(\lambda - a) I$, where a is a non-central element of K.

2. Let $R = k[x; S, D]$ and let $A, B \in R_n$. Give two proofs that A is (S, D)-similar to B if and only if $A - tI$ and $B - tI$ are associated in R_n, one via pseudo-linear transformations and one by using Proposition 0.4.3.

3. Show that a system of linear differential equations in several unknown functions is equivalent to a single linear differential equation. (Hint: Use the fact that the ring of differential operators is simple.)

4. Let θ be a pseudo-linear transformation. Show that θ is bounded if and only if the kernel of the homomorphism $R = k[t; S, D] \to k[\theta]$ is a bounded ideal in R, and express the bound of this ideal in terms of the invariant factors of θ.

5. (Jacobson [37]). If θ is a bounded indecomposable pseudo-linear transformation, show that its minimal polynomial is a power of an I-atom.

6. (Jacobson [37]). Show that a bounded pseudo-linear transformation is completely reducible if and only if its minimal polynomial is a product of distinct I-atoms.

7. Show that every matrix over a ring of the form $K*k\langle X \rangle$ is equivalent (after taking a diagonal sum with a unit-matrix) to a matrix of the form $A + \Sigma A_i x_i$ (where the A's have entries in K). (Hint: Observe that the construction of the matrices in the proof of Theorem 7.1.2 was linear.)

8^0. Let A be a matrix over a field K, and x an indeterminate commuting with the elements of the centre of K. Show that the matrix $A - xI$ assumes non-singular values for some values of x, if K has infinite centre, but not always if K is finite. What happens if K is infinite with finite centre?

8.5 Skew field extensions

As an application of some of the preceding ideas we shall now consider algebraic extensions of (skew) fields.

Let E be a field with subfield k. An element $c \in E$ is said to be *right algebraic* over k in case it satisfies a monic equation with coefficients (on the right) belonging to k. This is the same as the definition of right integral (*cf.* **6**.1), but in the case of fields we use the above terminology which agrees with the terminology in the commutative case. From Proposition **6**.1.3 we know that c satisfies the equation $f = 0$ ($f \in k[x]$) if and only if f has $x - c$ as left factor in the ring $E[x]$. In that case we call c a *left zero* of f. The ring $E[x]$ is a principal ideal domain, as is its subring $k[x]$, and if we take a complete factorization of f in $k[x]$,

$$f = p_1 p_2 \cdots p_r \tag{1}$$

we know that $r \leqslant \deg f$. Moreover, any atomic left factor of f is similar to one of the atomic factors on the right of (1). Now two similar elements in $k[x]$ have the same degree in x; this follows from the Euclidean algorithm, or simply from the fact that the degree of any $f \in R = k[x]$ is characterized as the dimension of the right k-space R/fR. From Proposition 0.4.3 we see that $x - a$ and $x - b$ are similar in R if and only if a and b are conjugate in k. Applying the result to a polynomial of degree n, we obtain

THEOREM 5.1. *The left zeros of a polynomial of degree n, in any field k, lie in at most n different conjugacy classes of k.* ∎

Let f be a polynomial with coefficients in the centre C of k; of course here we do not need to distinguish between left and right zeros in k. Thus f is a central element of $k[x]$. Now the semigroup I of invariant elements of $k[x]$ just consists of associates of central elements; hence if f is an atom in $C[x]$, i.e. irreducible over C, then f is an atom in I. By Proposition 6.5.2, all atoms dividing f are similar, hence if f has a zero in k, then all its zeros in k are conjugate, and we obtain

PROPOSITION 5.2. *Let k be a field with centre C, and f any irreducible polynomial over C. Then all zeros of f in k (if any) are conjugate.* ∎

We now return to the general situation. In order to determine the number of left zeros of f in a given conjugacy class we use Theorem 4.2.5; this tells us that the number of left factors of f similar to $x - a$ (up to associates) is 0, 1 or at least α, where α is the cardinality of the eigenring of $x - a$ in $k[x]$. Now the latter is just the centralizer of a in k (by Theorem 0.4.2, Corollary). Thus we have

THEOREM 5.3. *Let f be a polynomial over a field k and let a be a left zero of f in some extension E of k. Denote by α the cardinality of the centralizer of a in E. Then the number of left zeros of f conjugate to a in E is either 1 or at least α.* ∎

From this result it is easy to deduce that the number of conjugates of a, if greater than 1, is infinite. We need only show that in a field that is non-commutative, every element has an infinite centralizer. Clearly this follows from the more general

LEMMA 5.4. *In a non-commutative field k, every element is contained in an infinite commutative subfield.*

Proof. Let C be the centre of k, then $C \neq k$ and for any $a \in k$, $C(a)$ is a commutative subfield containing both C and a. Clearly it will be enough to find an infinite commutative subfield containing $C(a)$ when $a \notin C$. Suppose that $C(a)$ is finite, then so is C, say $|C| = q$, $|C(a)| = q^r$, where $r > 1$. The mapping $x \mapsto x^q$ is an automorphism of $C(a)$ over C, hence a and a^q satisfy the same irreducible equation over C. By Proposition 5.2, they are conjugate in k, i.e. there exists $b \in k^*$ such that

$$b^{-1} ab = a^q. \tag{2}$$

Consider the field $A = C(a, b^r)$. By (2) b^r induces the mapping $x \mapsto x^{q^r}$ on $C(a)$; this is the identity, so A is commutative. We assert that A is infinite; for if not, then $A(b)$ would be a finite extension of A, and therefore a finite field. But by (2) $A(b)$ is non-commutative, and this contradicts Wedderburn's theorem (which asserts that all finite fields are commutative). ■

Combining this result with Theorem 5.3, we obtain

COROLLARY 1. *Let f be a polynomial over a field k, and let a be a left zero of f in some extension E of k. Then the number of left zeros of f in E conjugate to a is 1 or infinite.* ■

And by combining this result with Theorem 5.1 we find

COROLLARY 2. *The number of left zeros of a polynomial of degree n over a field is either at most n or infinite.* ■

Let us now consider briefly the problem of constructing field extensions E/k. At present it seems a hopeless undertaking to classify all such extensions, partly because there is no satisfactory way of constructing arbitrary field extensions, and this is so even if we confine ourselves to all right algebraic or right finite extensions of k (defined as the right integral and right finite extensions in 6.1). We therefore limit ourselves to a special case, the pseudo-linear extensions.

A *pseudo-linear* extension of a field k is a field E generated (as ring) by k and an element θ satisfying

$$a\theta = \theta a_1 + a_2 \quad \text{for any} \quad a \in k.$$

Leaving aside the trivial case $\theta \in k$, we see that a_1, a_2 are uniquely determined by a and as in 0.8 it follows that $S: a \mapsto a_1$ is an endomorphism of k, while $D: a \mapsto a_2$ is an S-derivation. It follows that every such pseudo-linear extension E is a homomorphic image of the skew polynomial ring $R = k[t; S, D]$; this is slightly more general than the case considered in 8.3 in that here S need not be an automorphism (although this is the most important special case). Since E is a field, it must be a proper homomorphic image of R, and this is obtained by factoring out a non-zero ideal of R. Since R is a principal right ideal domain, this ideal has the form fR, where f is a right invariant polynomial, which may be taken to be monic. Thus

$$E = R/fR.$$

If f has degree n in t, and θ is the residue class of t, then E is a right k-space with basis $1, \theta, ..., \theta^{n-1}$. It follows that for an arbitrary non-zero ideal fR, R/fR is either a field extension of k or it has zero-divisors. For each left multiplication is a linear transformation of a finite-dimensional k-space and so is either invertible or has non-zero kernel. So to obtain all pseudo-linear extensions of k associated with S and D we have to find all right invariant monic polynomials f such that R/fR is an integral domain.

Let us consider the case $D = 0$. Since f is right invariant, we have, for any $c \in k$, $cf = fc^\alpha$, where $c^\alpha \in k$, by a comparison of degrees. Clearly α is an endomorphism of k, and if f is monic of degree n, then by comparing highest terms, we find $\alpha = S^n$. A comparison of constant terms gives $ca_n = a_n c^\alpha$, hence α is an inner automorphism; in particular this shows S to be an automorphism, so that we are in the case of Theorem 3.2. Let r be the least positive integer such that S^r is inner on k, induced by e say, and write $x = t^r e$, then x centralizes k, and if $t \mapsto \theta$, $x \mapsto \xi$ in the homomorphism $R \mapsto R/fR$, we have

$$k \subseteq k(\xi) \subseteq k(\theta). \tag{3}$$

Let us call a pseudo-linear extension *central* if the associated endomorphism S is 1 (and the associated derivation is 0), and the generator satisfies a monic equation with coefficients in the centre C of k. Thus a central extension of k is a simple extension of the form $k \otimes_C F$, where F/C is a commutative simple field extension. For example, the extension $k(\xi)/k$ above is central. Secondly we define a *binomial* or *pure* extension as a pseudo-linear extension (with zero derivation) whose generator satisfies a binomial equation

$$x^n - u = 0,$$

and whose associated endomorphism S is such that S^n is the inner automorphism induced by u^{-1}. Then it is clear that the extension $k(\theta)/k(\xi)$ in (3) is of this form, since θ satisfies the equation $t^r = xe^{-1}$, by the definition of x. Thus we have proved

THEOREM 5.5. *Every pseudo-linear extension with zero derivation can be obtained by taking a central extension, followed by a binomial extension. In particular, every pseudo-linear extension of prime degree (with zero derivation) is either central or binomial.* ∎

We can therefore confine our attention to central and binomial extensions. Consider first central extensions. Given a field k with centre C, any monic polynomial f with coefficients in C defines a ring extension of k, and this will be a field if and only if f is an atom in $k[x]$. This proves

PROPOSITION 5.6. *Let k be a field with centre C, and let $F = C[\xi]$ be a finite commutative extension of C. Then the central extension $k \otimes_C F$ is a field if and only if the minimal polynomial for ξ over C is irreducible over k. Moreover, all central extensions of k are obtained in this way.* ∎

Next consider binomial extensions. Let $R = k[t; S]$ and assume that S^n is the inner automorphism induced by u^{-1}, then the element $f = t^n - u$ of R is invariant. To determine when R/fR is a field we need to know when $t^n - u$ is an atom in R. Now every binomial extension can be obtained as a succession of binomial extensions of prime degrees, for if $k(\theta)/k$ is binomial of degree $n = pm$, then $k(\theta)/k(\theta^p)$ and $k(\theta^p)/k$ are binomial extensions, of degrees p, m respectively. Thus we may take $n = p$, a prime.

Clearly $t^p - u$ is an I-atom, for if not, the automorphism S would be inner on k. If $t^p - u$ is not itself an atom, then by Proposition 6.5.2, it must be a product of similar atoms. Now any two similar atoms have the same degree, so $t^p - u$ is a product of linear factors. Further, two linear factors $t - a$, $t - b$ are similar if and only if they are S-conjugate: if $(t - a)v = u(t - b)$, we may take u, v to be of degree 0, and a comparison of terms shows that

$$b = c^{-1} ac^S \quad \text{for some} \quad c \in k^*. \tag{4}$$

Let $t - a$ be a left factor of $t^p - u$, then

$$t^p - u = (t - a)(t^{p-1} + t^{p-2} b_1 + \ldots + b_{p-1}). \tag{5}$$

Equating coefficients, we find $b_1 = a^{S^{p-1}}$, $b_2 = a^{S^{p-2}} a^{S^{p-1}}$, ..., $b_{p-1} = a^S a^{S^2} \ldots a^{S^{p-1}}$. Therefore

$$u = a a^S \ldots a^{S^{p-1}}. \tag{6}$$

Conversely, given any element u of k of the form (6), for some $a \in k$, we have a factorization (5). Thus $t^p - u$ is an atom if and only if u does not satisfy (6), for any $a \in k$. Summarising, we have

PROPOSITION 5.7. *Let k be a field, S an automorphism of k whose pth power (for some prime p) is inner, induced by u say, where u satisfies no equation of the form (6), for any $a \in k$. Then the equations*

$$t^p - u = 0, \quad at = ta^S \quad (a \in k) \tag{7}$$

define a binomial field extension of k, and conversely, all such extensions of k are obtained by this process, repeated if necessary. ∎

In the commutative case we know that a binomial expression $t^p - u$ of prime degree either is irreducible or splits into linear factors, provided only that the ground field contains a primitive pth root of 1. The same result still holds in the non-commutative case:

PROPOSITION 5.8. *Let p be a prime and k a field containing a primitive pth root of 1, ω say. Further, let S be an automorphism of k whose pth power is inner on k, induced by u, say. Then the polynomial*

$$t^p - u \tag{8}$$

is either an atom in $k[t; S]$, or it splits into linear factors, a decomposition being

$$t^p - u = (t - a)(t - \omega a^S) \dots (t - \omega^{p-1} a^{S^{p-1}}), \tag{9}$$

where a is any left zero of (8).

Proof. We know that the polynomial (8) is an atom in $R = k[t; S]$ if and only if there is no representation of the form (6) for u. More precisely, the proof of Proposition 5.7 shows that an expression (6) is possible whenever $t - a$ is a left factor of $t^p - u$, and conversely. Let $t - a$ be a left factor of (8), then since $at = ta^S$, we have $t^{-1} a = a^S t^{-1}$ in $k(t; S)$. Hence

$$(at^{-1})^p = at^{-1} at^{-1} \dots at^{-1}$$

$$= aa^S \dots a^{S^{p-1}} t^{-p}$$

$$= ut^{-p}.$$

By hypothesis ut^{-1} is central, hence we have the decomposition

$$1 - ut^{-p} = 1 - (at^{-1})^p = (1 - at^{-1})(1 - \omega at^{-1}) \dots (1 - \omega^{p-1} at^{-1});$$

now (9) follows on multiplying up by t^p on the right and taking the commutation formula into account. ∎

Exercises   8.5

1. Let k be a field and p^* an I-atom of length r in a skew polynomial ring over k. Show that $r^2 \mid \deg p^*$.

2. Let k be a field, $R = k[t]$ and f a polynomial with no central factors, i.e. no factors (apart from 1) with coefficients all in the centre of R. Show that the ideal generated by f is the whole of R.

3. Let k be a field and $R = k[t]$. If f is an I-atom in R, show that f has a central zero in an extension of k if and only if f is an atom.

4. Let K/k be a pseudo-linear extension of right dimension n, with an endo-morphism S. Denoting the left dimension by $[K:k]_L$, obtain the formula

$$[K:k]_L = 1 + [k:k^S]_L + [k:k^S]_L{}^2 + \dots + [k:k^S]_L{}^{n-1}.$$

Deduce that $[K:k]_L \geqslant [K:k]_R$ (the right dimension), with equality if and only if S is an automorphism of K.

5. Given a skew polynomial ring $R = k[x; S, D]$ with non-surjective endo-morphism S, obtain conditions for R to have an ideal of finite right codimension, giving rise to a pseudo-linear extension of k. Why does this not in general provide an example where the inequality of Exercise 4 is strict?

6. (Jacobson [43]). Let k be a field and S an automorphism of order n. If A is a matrix satisfying $AA^S \dots A^{S^{n-1}} = I$, show that there exists a matrix B such that $A = B^{-1}B^S$.

7^0. Let K be a field with a subfield k, and $p \in K*k\langle X \rangle$, but $p \notin K$. Does there always exist an extension field Ω of K such that p vanishes for some values of X in Ω? Consider the same question for rational functions in X.

8^0. Given K and k as in Exercise 7, and $A \in K_n$, investigate whether $A - Ix$ always specializes to a singular matrix over a suitable extension field of K. (An affirmative answer to this question would also provide an answer to Ex. 7. Note that by Theorem 7.6.3, to answer it we must show that the unit-matrix I cannot be written as a determinantal sum of matrices $(A - Ix) \dotplus B$ and non-full matrices.)

8.6 Iterated skew polynomial rings

Let R be a ring with an automorphism S; then we can form the skew polynomial ring $R[x; S]$, but this will not be a principal ideal domain unless R was a field. However, when S is only required to be an endomorphism, the skew polynomial ring is a principal right ideal domain under wider conditions; the exact condition was found by Jategaonkar who used it to give an ingenious construction of 'iterated skew polynomial rings' which form a useful source of counter-examples. This section is devoted to a brief derivation of his results.

Given a ring R, a subring and an ordinal number τ, we shall say that R is an *iterated skew polynomial ring of type* τ *over* K, if R contains an ascending chain of K-rings R_α ($\alpha < \tau$) and has endomorphisms S_γ ($\gamma < \tau$) mapping each R_α into itself, such that, on writing $R_\alpha^* = \bigcup_{\beta < \alpha} R_\beta$, we have

$$R_0 = K, \quad R_\alpha = R_\alpha^*[x_\alpha; S_\alpha], \quad R_\tau^* = R.$$

It is clear from the definition that every element of R can be written uniquely as

$$\Sigma x_{\alpha_1} \dots x_{\alpha_r} a_{\alpha_1 \dots \alpha_r} \quad \text{where} \quad a_{\alpha_1 \dots \alpha_r} \in K, \; \alpha_1 \geqslant \dots \geqslant \alpha_r. \tag{1}$$

We know from Proposition 0.8.4 that any skew polynomial ring over a right Ore ring is again right Ore; hence if K is a right Ore ring, then (by a transfinite induction), so is R. The next result is Jategaonkar's criterion for R to be a principal right ideal domain. If K is any ring, with an endomorphism S, we can form the skew polynomial ring $K[x; S]$; clearly this is an integral domain if and only if K is an integral domain and S is injective.

THEOREM 6.1. *Let K be a ring with an endomorphism S, and put $R = K[x; S]$. Then R is a principal right ideal domain if and only if K is a principal right ideal domain and S maps K^* into $U(K)$.*

Proof. Assume that R is a principal right ideal domain; then so is K, as homomorphic image of R (on putting $x = 0$).

Further, for any $a \in K^*$, let $xR + aR = cR$, then as factor of a, c has degree 0, so $x = c \cdot xd = xc^S d$, i.e. $c^S d = 1$, so c^S is a unit, and hence $xu + a^S v = 1$ for some u, $v \in R$. Put $x = 0$, then we find $a^S v_0 = 1$, hence $a^S \in U(K)$.

Conversely, under the given hypothesis, let $\mathfrak{a}$ be a right ideal of R and let n be the least degree of polynomials occurring in $\mathfrak{a}$. The leading coefficients of polynomials of degree n in $\mathfrak{a}$ form a right ideal in K, generated by a say. Let $f = x^n a + \ldots \in \mathfrak{a}$, then $a^S \in U(K)$ by hypothesis, and $fx(a^S)^{-1} = x^{n+1} + \ldots \in \mathfrak{a}$. Thus $\mathfrak{a}$ contains a monic polynomial of degree $n + 1$ and hence of any degree $> n$. It follows that $\mathfrak{a} = fR$. ∎

COROLLARY. *An iterated skew polynomial ring is a principal ideal domain if and only if each S_γ maps each R_α^* into $U(K)$ and K is a principal right ideal domain.*

This follows once we observe that $U(R_\alpha) = U(K)$ by (1), and check that (in this case), the property of being principal is preserved on taking the direct union. ∎

The main problem is to construct skew polynomial rings of a given type τ. The ordinary skew polynomial rings over a field are of type 2; rings of type 3 may be obtained by an ad hoc construction (*cf.* Exercise 6), but beyond that it is no easier to construct a ring of finite type than one of arbitrary type. In particular it is not possible to give an inductive construction, because the set $U(K)$ must contain a copy of R^*, so that K depends very much on τ. To construct K we shall need an embedding of the free algebra $k\langle X \rangle$ in a field, in which the elements have a simple normal form. Here it is more convenient to use instead of the method of 7.6, an earlier method due to Malcev [48] and Neumann [49], which more generally provides an embedding of the group algebra of any ordered group in a field. We digress to describe this construction, referring to Cohn [65] for proofs.

Let G be a totally ordered group, i.e. a group with a total ordering such that $a_i < b_i$ $(i = 1, 2)$ implies $a_1 a_2 < b_1 b_2$, and let k be any field (not necessarily commutative). Consider the direct power k^G, regarded as k-space. With each $a \in k^G$ we associate a subset $D(a)$ of G, its *support*, defined as

$$D(a) = \{g \in G | a(g) \neq 0\}.$$

The elements with finite support may be written as finite sums $\Sigma a(g) g$, and just constitute the group algebra kG of G over k, with the multiplication rule

$$ab = c, \qquad \text{where} \qquad c(g) = \Sigma a(u) b(u^{-1} g). \qquad (2)$$

Let E be the set of elements of k^G with well-ordered support; here the definition (2) for the product still makes sense, for the sum occurring here is finite, and the resulting element again has well-ordered support. It is not hard to show that E is a ring containing the group algebra as subring. Moreover, E is actually a field; to show this one writes any $a \neq 0$ in the form $a = \lambda u(1 - f)$, where $u \in G$, $\lambda \in k^*$ and $f(g) = 0$ for $g < 1$. Then $a^{-1} = (\Sigma f^n) u^{-1} \lambda^{-1}$ and it remains to verify that the series Σf^n can be rearranged to form an element of E. We refer to Cohn [65], p. 276 for a proof of this fact. In this way the group algebra kG has been embedded in a field, the 'power series field' E.

Now we note that any free group F can be totally ordered: we write any $g \in F$ as a (possibly infinite) power product of the basic commutators: $g = c_1^{\alpha_1} c_2^{\alpha_2} \ldots$, and use the lexicographic ordering of the exponents $(\alpha_1, \alpha_2, \ldots)$ (*cf.* Neumann [49']). Clearly the free algebra $k\langle X \rangle$ can be embedded in kF, where F is the free group on X, and hence, by the above process, $k\langle X \rangle$ has been embedded in a field.

Let $X = \{x_\alpha\}$ $(\alpha < \tau)$ be a sequence of indeterminates and denote by E the field of formal power series constructed in the above way from the free group on X. Let K be the subfield generated over k by the elements

$$u_{\beta \alpha_1 \ldots \alpha_r} = (x_{\alpha_1} \ldots x_{\alpha_r})^{-1} x_\beta x_{\alpha_1} \ldots x_{\alpha_r}, \quad \alpha_1 \geqslant \ldots \geqslant \alpha_r, \quad \alpha_1 > \beta, \quad r \geqslant 1. \quad (3)$$

LEMMA 6.2. *For any* x_γ *the centralizer of* x_γ *in* K *is* k.

Proof. Let G be the subgroup of F generated by the right-hand sides of (3). Each of these generators has odd length in the x's, and in any expression of an element of G as a power product of the u's, cancellation cannot affect the central factor of any u. It follows that each such expression must begin with a factor x_α^{-1} and end in a factor x_β, even after all cancellations have been made, for this is true of the u's and their inverses. In particular it follows that G does not contain x_γ^n, for any $n \neq 0$.

Consider $a \in K$; this is a power series: $a = \Sigma a_u u$, where u runs over G. Conjugation by x_γ maps G into itself; explicitly we have

$$x_\gamma^{-1} u_{\beta \alpha_1 \dots \alpha_r} x_\gamma = u_{\alpha_r \gamma}^{-1} \dots u_{\alpha_i \gamma}^{-1} u_{\beta \alpha_1 \dots \alpha_{i-1} \gamma} u_{\alpha_i \gamma} \dots u_{\alpha_r \gamma}$$

where i is such that $\alpha_{i-1} \geqslant \gamma > \alpha_i$.

Since x_γ commutes only with the powers $x_\gamma{}^n$, it follows that conjugation by x_γ fixes only $1 \in G$ and moves all other elements in infinite orbits; to be precise, the elements in each orbit arise from the positive powers of x_γ, since G admits conjugation by x_γ but not by $x_\gamma{}^{-1}$. Hence $x_\gamma{}^{-1} a x_\gamma = a$ is possible only if $a_u = 0$ for $u \neq 1$, i.e. $a = a_1 \in k$. ∎

Let S_γ be the endomorphism of G induced by conjugation with x_γ. Clearly this can be extended to an endomorphism of K, again denoted by S_γ. Then

$$a x_\gamma = x_\gamma a^{S_\gamma} \qquad (a \in K). \tag{4}$$

Denote by R the subring of E generated by K and the x_α ($\alpha < \tau$). By (4), every element of R can be written as a finite sum

$$\Sigma x_{\alpha_1} \dots x_{\alpha_r} a_{\alpha_1 \dots \alpha_r} \qquad (a_{\alpha_1 \dots \alpha_r} \in K). \tag{5}$$

Fix any term in (5) and let α_i be the last suffix such that $\alpha_i < \alpha_{i+1} \geqslant \dots \geqslant \alpha_r$. Then we can use (3) to pull x_{α_i} through to the right; this will only change the coefficient. Repeating this process if necessary we can ensure that $\alpha_1 \geqslant \dots \geqslant \alpha_r$ in each term (5). We assert that under this condition the expression (5) is unique. Thus let

$$\Sigma x_{\alpha_1} \dots x_{\alpha_r} a_{\alpha_1 \dots \alpha_r} = 0 \qquad (\alpha_1 \geqslant \dots \geqslant \alpha_r) \tag{6}$$

and assume that not all coefficients are 0. Let α be the highest suffix such that x_α occurs in (6), then (6) may be written as

$$\Sigma x_\alpha{}^i c_i = 0, \tag{7}$$

where each c_i is a polynomial in the x_β ($\beta < \alpha$) with right coefficients in K, and not all the c_i vanish. On conjugating by x_α we obtain

$$\Sigma x_\alpha{}^i c_i^{S_\alpha} = 0,$$

where the coefficients now lie in K, hence x_α is right algebraic over K. Take a monic equation of least degree satisfied by x_α:

$$\Sigma x_\alpha{}^i b_i = 0.$$

This equation is unique; but conjugating with x_α we obtain another equation of the same degree, with coefficients $b_i{}^{S_\alpha}$. Hence conjugation must fix each b_i and so (by the lemma), $b_i \in k$. Thus x_α is right algebraic over k, which is a contradiction. Thus we have established the uniqueness of (5), and we see that R is a skew polynomial ring of type τ.

By Theorem 6.1, R is a principal right ideal domain; however it is not atomic if $\tau > 2$, for then we have

$$x_2 = x_1 x_2 u_{12}{}^{-1} = x_1{}^n x_2 u_{12}{}^{-n},$$

so x_2 has factorizations of arbitrary length.

In any iterated skew polynomial ring, the monomials occurring in the normal form (1) are totally ordered by right divisibility. If we assign ordinals to them according to this ordering, we get a function on these monomials, which we can extend to the polynomials (1) by taking the value of the highest term occurring with a non-zero coefficient. In this way we obtain a function on R with ordinal numbers as values, and it is not hard to verify that R satisfies the transfinite division algorithm relative to this function (*cf.* **2.9**).

A closely related class of rings has been studied by Brungs [69']; he considers rings in which all right ideals are well-ordered by inclusion. Such a ring contains a unique maximal proper right ideal $J = J(R)$ which is also the Jacobson-radical, hence two-sided, and R/J is a field because it has no non-trivial right ideals. Thus R is a local ring with maximal ideal J. More precisely, we have the following structure theorem due to Brungs:

THEOREM 6.3. *Let R be any ring in which the right ideals are well-ordered by inclusion. Then all right ideals of R are principal and are in fact two-sided ideals.*

Proof. Let $\mathfrak{a}$ be any non-zero right ideal, $\mathfrak{a}'$ the maximal right ideal properly contained in $\mathfrak{a}$, and $c \in \mathfrak{a}$, $c \notin \mathfrak{a}'$. Then $cR \subseteq \mathfrak{a}$, $cR \nsubseteq \mathfrak{a}'$ hence $\mathfrak{a}' \subset cR \subseteq \mathfrak{a}$, and so $cR = \mathfrak{a}$. Thus all right ideals are principal. We assert that each right ideal is two-sided (i.e. R is right invariant). If not, let aR be a maximal right but non-left ideal, then there exists $b \in R$ such that $baR \supset aR$, hence $a = bac$, where c is a non-unit, and so lies in J. But baR is two-sided, by the maximality of aR, and so $b . ba = ba . b'$, hence

$$ba = b . bac = bab'c.$$

This shows that $ba(1 - b'c) = 0$, where $c \in J$, and so $1 - b'c$ is a unit. Hence $ba = 0$, a contradiction. ∎

In order to examine these rings more closely we note that for any ring R with the property of Theorem 6.3,

$$\text{if}\quad a, b \in R,\ ab \in cR,\ b \notin cR\quad \text{implies}\quad a^n \in cR\quad \text{for some}\quad n.$$

For let $\bigcap a^n R = dR$, then $adR = dR$, hence $d = adu$ for some $u \in R$. Now either $bR \subseteq dR$, say $b = dv$, then $b = dv = aduv = advu' = abu' \in cR$, or $dR \subset bR$, then $a^n R \subseteq bR$ for some n, and $a^{n+1} \in abR \subseteq cR$.

We now define elements p_α $(0 \leqslant \alpha)$ of R as follows: $p_0 R$ is the maximal right ideal of R, and for any $\alpha > 0$, p_α is defined (up to a right unit factor) by

$$p_\alpha R = \begin{cases} \bigcap_{\beta < \alpha} p_\beta R & \text{if } \alpha \text{ is a limit ordinal,} \\ \bigcap p_{\alpha-1}^n R & \text{otherwise.} \end{cases} \qquad (8)$$

Clearly if $\alpha > \beta$, then $p_\alpha R \subset p_\beta R$; moreover, $p_\beta p_\alpha R \subseteq p_\alpha R$, and by definition of p_α, $p_\alpha R \subset p_\beta R$, say $p_\alpha = p_\beta c$. Since $p_\beta{}^n \notin p_\alpha R$ for all n, we have $c \in p_\alpha R$, so $p_\alpha \in p_\beta p_\alpha R$, and therefore $p_\alpha R = p_\beta p_\alpha R$, i.e.

$$p_\beta p_\alpha = p_\alpha u_{\alpha\beta} \quad \text{for} \quad \beta < \alpha, \quad \text{where} \quad u_{\alpha\beta} \in U(R). \qquad (9)$$

We assert that each $a \in R^*$ is expressible uniquely as

$$a = p_{\alpha_1} \dots p_{\alpha_r} u \quad \text{where} \quad u \in U(R),\ \alpha_1 \geqslant \dots \geqslant \alpha_r. \qquad (10)$$

For let α be the least ordinal such that $a \notin p_\alpha R$, then α cannot be a limit ordinal, say $\alpha = \alpha_1 + 1$ and $a \in p_{\alpha_1}{}^n R$ for some n. Taking n as large as possible, we have

$$a = p_{\alpha_1}^{n_1} a_1, \quad \text{where} \quad a_1 \notin p_{\alpha_1} R.$$

Repeating this process on a_1, we find by induction, that a has the form (10) as claimed, and this expression is unique, from the way it is formed.

Let the order type of the sequence of right ideals of R be σ, and express σ in the form

$$\sigma = \omega^{\tau_1} n_1 + \omega^{\tau_2} n_2 + \dots + \omega n_{k-1} + n_k + 1, \qquad (11)$$

where $\tau_1 > \tau_2 > \dots > \tau_{k-2} > 1$, then the p_α are indexed by all $\alpha < \tau_1$ and (11) corresponds to the relation

$$p_{\tau_1}^{n_1} p_{\tau_2}^{n_2} \dots p_1^{n_{k-1}} p_0^{n_k} = 0. \qquad (12)$$

It only remains to construct a ring with these properties. To do this we take an iterated skew polynomial ring of type τ_1 over any field, and localize at the set of all polynomials with non-zero constant term (using Proposition 3.3.5). Now add the relation (12) (where the x_α are to be replaced by the p_α).

Exercises 8.6

1. Show that any iterated skew polynomial ring of type τ has a transfinite division algorithm and determine its range of values.

2. (Jategaonkar [69]). Let P be the localization of an iterated skew polynomial ring of type τ. Show that the Jacobson radical of P is $J = x_1 P$; if transfinite powers of J are defined by the equations $J^{\alpha+1} = J^\alpha J$, $J^\alpha = \bigcap_{\beta < \alpha} J^\beta$ (at a limit ordinal α), show that $J^\alpha \supseteq x_\alpha P$. Deduce that $J^\alpha \neq 0$ if and only if $\alpha < \tau$.

3. (Jategaonkar [69]). Let R be an iterated skew polynomial ring of type τ. Show that the elements $1 + x_\alpha$ ($\alpha < \tau$) are left linearly independent over R. If τ is a limit ordinal, show that every non-zero right ideal contains an ideal of the form $x_\alpha R$. Show also that $\mathfrak{a} = \Sigma R(1 + x_\alpha)$ is a proper left ideal, and that no maximal left ideal containing $\mathfrak{a}$ can contain a non-zero ideal. Deduce that in this case R is left but not right primitive.

4. Determine the strong prime ideals in an iterated skew polynomial ring of given type.

5. (Brungs [69']). Let R be a ring whose right ideals are well-ordered of type σ. Show that R is an integral domain if and only if $\sigma = \omega^\tau + 1$. Show that R is left Noetherian if and only if $\sigma \leqslant \omega + 1$.

6. Construct a commutative field k with an endomorphism α such that k contains an element t transcendental over k^α. Let K be the subring of $k(y)$ consisting of all fractions $f(1 + yg)^{-1}$, where $f, g \in k[y]$. Show that α can be extended to K by letting $y \mapsto t$, and verify that the resulting endomorphism maps K^* into $U(K)$. Show that $R = K[[x; \alpha]]$ is a principal right ideal domain in which the right ideals are well ordered, and determine the order type of its chain of right ideals.

Notes and comments on Chapter 8

There are many papers on the diagonal reduction of matrices. A weak form of Theorem 1.1 was proved for (non-commutative) Euclidean domains by Wedderburn [32], and the full form by Jacobson [37]; this was generalized to principal ideal domains by Teichmüller [37], and a uniqueness statement added by Nakayama [38]. Our account follows Jacobson [43], with some simplifications (*cf.* Amitsur [63]). For a general study of elementary divisor rings see Kaplansky [49].

Section 8.4 follows Jacobson [37]; the applications to skew field extensions in the first half of 8.5 are for the most part due to Gordon–Motzkin [65], generalizing earlier results by Richardson [27] and Herstein [56]. Our account follows Cohn [70]. For Theorem 5.5 and an application to free products see Cohn [61'']; Proposition 5.8 is well known. Theorem 6.1 and the iterated skew polynomial ring are due to Jategaonkar [69, 69']; the presentation given here is new. For Theorem 6.3 see Brungs [69'].

Appendix

This appendix gives a brief summary of the main facts needed from attice theory and homological algebra, with references to proofs (or the proofs themselves, where no convenient reference was available). In each section some reference books are listed, with an abbreviation which is used in quoting them in the appendix.

1. Lattice theory

References

LT. G. Birkhoff, "Lattice Theory," 3rd ed. (Providence R.I. 1967),
UA. P. M. Cohn, "Universal Algebra," (New York, London, Tokyo 1965).

We recall that a *lattice* is a partially ordered set in which any pair of elements a, b has a supremum (least upper bound), also called *join* and written $a \vee b$, and an infimum (greatest lower bound), also called *meet* and written $a \wedge b$. If we reverse the ordering, we again obtain a lattice, in which meet has been replaced by join and vice versa. This is the basis of the principle of duality in lattice theory, by which we obtain from each theorem about lattices, another one which is its dual.

Given any elements $x \leqslant y$ in a lattice L, the set $\{z \in L | x \leqslant z \leqslant y\}$ is a sublattice (i.e. closed under the operations of join and meet in L). It is denoted by $[x, y]$ and called the *interval* from x to y.

Nearly all the lattices we deal with will be *modular*, i.e. they satisfy the *modular law*:

$$(x \vee y) \wedge z = x \vee (y \wedge z) \quad \text{for all} \quad x, y, z \quad \text{such that} \quad x \leqslant z. \quad (1)$$

An important example of a modular lattice, which will much occupy us here (and which incidentally is responsible for the name 'modular') is the set $\text{Lat}_R(M)$ of submodules of an R-module M, partially ordered by inclusion. A criterion for modularity is provided by

PROPOSITION A.1. *A lattice is modular if and only if it contains no sublattice isomorphic to the 5-element lattice of Fig.* 1. (*cf.* LT, *p.* 13, UA, *p.* 66). ∎

314

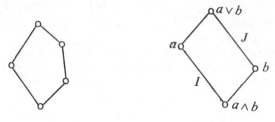

FIGURE 1.   FIGURE 2.

If a, b are any two elements of a modular lattice L, there is an isomorphism between the intervals $I = [a \wedge b, a]$ and $J = [b, a \vee b]$, given
by the mapping $x \mapsto x \vee b$, with inverse $y \mapsto y \wedge a$. This fact is often referred
to as the *parallelogram law* (the proof is an easy exercise, see LT, p. 13,
UA, p. 69). Two intervals related in this way are said to be *perspective*;
more generally, two intervals are said to be *projective* if we can pass from the
one to the other by a series of perspectivities.

The parallelogram law shows that projective intervals are isomorphic as
lattices. This may not tell us much (e.g. when the intervals refer to simple
modules), but in the module case a stronger assertion can be made: by the
Noether isomorphism theorem, perspective intervals give isomorphic module
quotients: $(a + b)/b \cong a/(a \wedge b)$, hence by induction, so do projective intervals.

Many basic theorems on modules are lattice-theoretic in nature, in the
sense that the results can be stated and proved in terms of lattices. This is
true of the next group of theorems, which are all used in the text
(mainly Chapter 3).

SCHREIER REFINEMENT THEOREM. *In a modular lattice, two finite chains
between the same end-points have refinements that are isomorphic, in the
sense that their intervals can be paired off in such a way that corresponding
intervals are projective.* (UA, p. 70). ■

The result shows in particular that any two maximal chains between
given end-points in a modular lattice have the same length. By the *length*
of a lattice L one understands the supremum of the lengths of chains in L,
the length of a finite chain being the number of its links. A lattice of finite
length necessarily has a greatest element, written 1, and a least element,
written 0. The length of the interval $[0, a]$ is also called the *height* of a. A
modular lattice is of finite length whenever it satisfies the ascending and
descending chain conditions, for then all its maximal chains are finite and

of equal length. As in group theory, the Schreier refinement theorem has the
following consequence:

JORDAN–HÖLDER THEOREM. *In a modular lattice of finite length, any chain
can be refined to a maximal chain, and any two maximal chains are
isomorphic* (UA, *p.* 70). ∎

A lattice L is said to be *distributive*, if

$$(x \vee y) \wedge z = (x \wedge z) \vee (y \wedge z) \qquad \text{for all} \qquad x, y, z \in L. \tag{2}$$

This is easily seen to be equivalent to its dual (obtained by interchanging
$\vee$ and $\wedge$) and to imply (1), so that every distributive lattice is modular.
Like modularity, distributivity can be characterized by the non-existence of
a certain sublattice:

PROPOSITION A.2. *A lattice L is distributive if and only if it is modular and
contains no 5-element sublattice of length* 2. (*cf.* LT, *p.* 14, UA, *p.* 69). ∎

FIGURE 3.

There is precisely one 5-element modular lattice of length 2, up to
isomorphism; it is shown in Fig. 3.

Distributive lattices are much more special than modular ones, as is clear
from Chapters **3** and **4**. This is also apparent from the special form taken
by the next two theorems in the distributive case. An element a in a lattice
is said to be expressed as a *join of independent elements*: $a = a_1 \vee \ldots \vee a_n$,
if $a_i \neq 0$ and

$$a_i \wedge \left(\bigvee_{j \neq i} a_j \right) = 0 \qquad \text{for} \qquad i = 1, \ldots, n.$$

KRULL–SCHMIDT THEOREM. *In a modular lattice of finite length, let*

$$1 = a_1 \vee \ldots \vee a_m$$

and

$$1 = b_1 \vee \ldots \vee b_n$$

*be two representations of 1 as a join of independent elements. Then $m = n$
and for some permutation $i \mapsto i'$ of $(1, 2, ..., n)$, $[0, a_i]$ is projective with
$[0, b_{i'}]$. In the case of a distributive lattice projectivity may be replaced by
equality*: $a_i = b_{i'}$ (LT, p. 168, UA, p. 73). ■

A finite decomposition $a = a_1 \vee ... \vee a_n$ is called *irredundant*, if no a_i can
be omitted. If no irredundant decomposition of a with more than one
element exists, a is called *join-irreducible*.

KUROŠ–ORE THEOREM. *In a modular lattice L, let*

$$c = p_1 \vee ... \vee p_r$$

and

$$c = q_1 \vee ... \vee q_s$$

*be two irredundant decompositions of c into join-irreducible elements. Then
$s = r$ and the p_i may be exchanged against the q_j, i.e. after suitable
renumbering of the q's we have*

$$c = q_1 \vee ... \vee q_t \vee p_{i+1} \vee ... \vee p_r \qquad (i = 1, ..., r).$$

If moreover, L is distributive, the p's and q's are equal except for their order.
(LT, p. 75, UA, p. 76 f.). ■

It is clear how an algebraic notion like 'homomorphism' is to be interpreted for lattices, namely as join-and-meet-preserving mapping. A homomorphism of a modular lattice L which collapses an interval I (i.e.
identifies its end-points) will clearly collapse all intervals perspective with I,
and hence all intervals projective with I. Conversely, if we collapse all
intervals projective to a given one, I, we obtain a homomorphic image of L.
Thus each congruence on L (i.e. each collection of inverse image sets of a
homomorphic image) is a union of projectivity classes of intervals of L.

The *direct product* of lattices L_i is the Cartesian product ΠL_i on which
the operations $\vee$, $\wedge$ are defined component-wise. A sublattice of ΠL_i
which projects onto each factor L_i is called a *subdirect product* of the L_i.

If L can be written as a subdirect product of the family (L_i) where none
of the projections $L \to L_i$ is an isomorphism, L is said to be *subdirectly
reducible*, otherwise *subdirectly irreducible*.

BIRKHOFF'S REPRESENTATION THEOREM. *Every modular lattice L is a subdirect
product of a family (possibly infinite) of subdirectly irreducible modular
lattices, which are homomorphic images of L. A corresponding result holds
for distributive lattices.* (LT, p. 193, UA, p. 100). ■

2. Categories and homological algebra

References:

TF: R. Godement, Théorie des Faisceaux (Paris 1958).

FR: I. Kaplansky, Fields and Rings (Chicago 1969).

H: S. MacLane, Homology (Berlin–Göttingen–Heidelberg 1963)

We shall not give a formal definition of a category here; let us just recall that if one merely looks at maps (by identifying objects with their identity maps) a category is a class (i.e. a big set) with a multiplication not everywhere defined, but where defined it is associative, and left and right neutrals for multiplication always exist. In diagrams each map is represented by an arrow going from the *source* to the *target*. The anti-isomorph of a category $\mathscr{C}$ (obtained by reversing all arrows) is again a category, called the *opposite* of $\mathscr{C}$ and denoted by $\mathscr{C}^o$.

The most important category for us in this book is $\mathscr{M}_R$, the category of right R-modules and homomorphisms, for a given ring R. Similarly $_R\mathscr{M}$ denotes the category of left R-modules.

Let $\mathscr{C}$ be any category and write $\mathscr{C}(X, Y)$ for the set of all maps $X \to Y$ with source X and target Y. A subcategory $\mathscr{B}$ is defined (as in algebra) as a subclass closed under multiplication when defined, and containing the left and right neutral of each map. If for any objects X, Y in $\mathscr{B}$, we have $\mathscr{B}(X, Y) = \mathscr{C}(X, Y)$, $\mathscr{B}$ is said to be *full*. Clearly a full subcategory is determined by its objects alone.

In any category a map is called an *isomorphism* if it has a 2-sided inverse, and two objects are *isomorphic* if there is an isomorphism between them. An object X_0 in a category $\mathscr{C}$ is said to be *initial* if there is precisely one map $X_0 \to Y$ for each Y; clearly any two initial objects are isomorphic. Dually an object Y_0 is *final* if there is just one map $X \to Y_0$ for each X. An object which is both initial and final is called a *zero object* and is written 0.

In any category consider a diagram consisting of two maps with the same target: $\underset{\rightarrow}{\alpha} \downarrow \beta$. The different ways of completing this figure to a commutative square $\downarrow \underset{\rightarrow}{\ } \downarrow$ form themselves the objects of a category, in which the 'maps' are maps between the objects added to get a 'commutative wedge'. A final object in this category is called the *pullback* of α and β. Thus the pullback consists of a pair of maps α', β' with the same source, such that $\alpha' \beta = \beta' \alpha$, and universal with this property; in other words, the pullback of α and β is just their least common left multiple (when it exists). *Pushouts* are defined dually, as the least common right multiple of two maps with the same source.

A category $\mathscr{A}$ is said to be *additive* if (i) $\mathscr{A}(X, Y)$ is an abelian group such that composition is distributive when defined: $\alpha(\beta + \beta') = \alpha\beta + \alpha\beta'$, $(\alpha + \alpha')\beta = \alpha\beta + \alpha'\beta$, (ii) there is a zero-object, and (iii) to each pair of objects X_1, X_2 there corresponds an object S and maps

$$X_1 \overset{i_1}{\underset{p_1}{\rightleftarrows}} S \overset{i_2}{\underset{p_2}{\rightleftarrows}} X_2 \tag{3}$$

with $i_1 p_1 = 1$, $i_2 p_2 = 1$, $p_1 i_1 + p_2 i_2 = 1$. The object S is called the *direct sum* of X_1 and X_2 with injections i_ν and projections p_ν. E.g. $\mathscr{M}_R$ is an additive category, in which the direct sum has its usual meaning.

Let $\mathscr{A}$ be an additive category; with each map $f: X \to Y$ we can associate a new category whose objects are maps α of $\mathscr{A}$ with target X and satisfying $\alpha f = 0$, and whose maps are commutative triangles in $\mathscr{A}$, as shown in the figure; thus $\lambda: \alpha \to \alpha'$ is such that $\alpha = \lambda\alpha'$. A final object μ in this category (when one exists) is called a *kernel* of f, written $\ker f$; it is always a *monomorphism* or *monic*, i.e. $\xi\mu = \xi'\mu$ implies $\xi = \xi'$, hence its source is unique up to isomorphism, and we shall sometimes use $\ker f$ to indicate this source. The dual notion is the *cokernel* of f, written $\operatorname{coker} f$; it is an *epimorphism* or *epic*, i.e. it can be cancelled whenever it appears as left-hand factor.

Given $f: X \to Y$, we define $\operatorname{im} f = \ker \operatorname{coker} f$, $\operatorname{coim} f = \operatorname{coker} \ker f$; of course they need not exist, but when they do we have the following picture, where we have identified $\ker f$ and $\operatorname{im} f$ with their source and $\operatorname{coker} f$, $\operatorname{coim} f$ with their target:

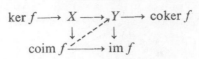

Here the map $\text{coim} f \to \text{im} f$ follows from the definition of im and coim. If the additive category $\mathscr{A}$ is such that every map has a kernel and cokernel and the natural map $\text{coim} f \to \text{im} f$ is an isomorphism, the category $\mathscr{A}$ is said to be *abelian*. We note that the notion of an abelian category is again self-dual; observe also that in an abelian category a map which is both epic and monic is necessarily an isomorphism. The category of all modules over a ring is an abelian category, as is easily seen; of course not every abelian category is of the form $\mathscr{M}_R$, for example if R is any non-zero ring, then the opposite of $\mathscr{M}_R$ is not of the form $\mathscr{M}_S$ for any ring S.

We note the following easy consequence of the definitions:

PROPOSITION A.3. *Let $\mathscr{A}$ be an abelian category and $\mathscr{B}$ a full subcategory; then $\mathscr{B}$ is abelian if and only if it has finite direct sums, kernels and cokernels.* ∎

A sequence of maps in an abelian category

$$\ldots \to X_{i-1} \xrightarrow{f_{i-1}} X_i \xrightarrow{f_i} X_{i+1} \to \ldots$$

is *exact at* X_i if $\ker f_i = \text{im} f_{i-1}$; if it is exact at each object, it is said to be *exact*. An exact sequence beginning and ending in 0 cannot have just one non-zero term, and it it has two, they must be isomorphic. Thus the first non-trivial case is that of a 3-term exact sequence.

$$0 \to A' \to A \to A'' \to 0, \tag{4}$$

also called a *short exact sequence*. E.g. in the case of modules, (4) represents an extension: A is an extension of A' by A''.

The following two results are easily proved for modules by a diagram chase (and this is all we need), but in fact they hold in any abelian categories:

FIVE-LEMMA. *Given any commutative diagram with exact rows*

$$\begin{array}{ccccccccc}
\longrightarrow & & \longrightarrow & & \longrightarrow & & \longrightarrow & & \\
\alpha_1\downarrow & & \alpha_2\downarrow & & \alpha_3\downarrow & & \alpha_4\downarrow & & \alpha_5\downarrow \\
\longrightarrow & & \longrightarrow & & \longrightarrow & & \longrightarrow & &
\end{array}$$

if α_1 is epic and α_2, α_4 are monic, then α_3 is monic; dually, if α_5 is monic and α_2, α_4 are epic, then α_3 is epic. In particular, if α_1, α_2, α_4, α_5 are isomorphism, then so is α_3. (H, p. 14). ∎

Three–By–Three Lemma. *Given a commutative diagram with exact rows and columns*

$$
\begin{array}{ccccccccc}
 & & 0 & & 0 & & 0 & & \\
 & & \downarrow & & \downarrow & & \downarrow & & \\
0 & \to & A' & \to & A & \to & A'' & \to & 0 \\
 & & \downarrow & & \downarrow & & \downarrow & & \\
0 & \to & B' & \to & B & \to & B'' & \to & 0 \\
 & & \downarrow & & \downarrow & & \downarrow & & \\
0 & \to & C' & \to & C & \to & C'' & \to & 0 \\
 & & \downarrow & & \downarrow & & \downarrow & & \\
 & & 0 & & 0 & & 0 & & \\
\end{array}
$$

there is a unique way of filling in the column $0 \to A' \to B' \to C' \to 0$ *so as to keep the diagram commutative and then the new column is exact too.* (H, *p.* 49). ∎

With every algebraic system there is associated the notion of homomorphism, namely a structure-preserving mapping. In the special case of categories one speaks of a functor. Thus a *functor* $F : \mathcal{A} \to \mathcal{B}$ is a mapping from one category $\mathcal{A}$ to another, $\mathcal{B}$ which preserves neutrals and satisfies

$$F(\alpha\beta) = F\alpha \cdot F\beta. \tag{5}$$

If instead of (5) we have

$$F(\alpha\beta) = F\beta \cdot F\alpha, \tag{6}$$

F is said to be *contravariant*, in contrast to the sort defined before, which are also called *covariant*. Thus a contravariant functor from $\mathcal{A}$ to $\mathcal{B}$ may also be regarded as a covariant functor from $\mathcal{A}^\circ$ to $\mathcal{B}$, or from $\mathcal{A}$ to $\mathcal{B}^\circ$. If F, G are two covariant functors from $\mathcal{A}$ to $\mathcal{B}$, a *natural* or *functorial transformation* $t : F \to G$ is a function which assigns to each $\mathcal{A}$-object X a $\mathcal{B}$-map $t_X : FX \to GX$ such that for any map $\alpha : X \to Y$ the square

$$
\begin{array}{ccc}
FX & \xrightarrow{\; F\alpha \;} & FY \\
t_X \downarrow & & \downarrow t_Y \\
GX & \xrightarrow[\; G\alpha \;]{} & GY
\end{array}
$$

commutes. If each t_X is an isomorphism, t is called a *natural equivalence*, or also a *functorial isomorphism*.

We shall mainly be concerned with additive (in particular abelian) categories; in that case all functors are assumed to be *additive*, i.e. $F(\alpha + \beta) = F\alpha + F\beta$. E.g. $\mathrm{Hom}_R(-, -)$ and $- \otimes -$, the tensor product, are additive functors. In what follows, all functors are tacitly assumed to be additive.

Any functor F transforms an exact sequence

$$\overset{\alpha}{\to} A \overset{\beta}{\to} \tag{7}$$

into a sequence

$$\overset{F\alpha}{\to} FA \overset{F\beta}{\to} \tag{8}$$

with composition zero: $F\alpha . F\beta = 0$, but in general (8) will not be exact. If (7) is *split exact*, i.e. if im $\alpha = \ker \beta$ is a direct summand of A, then so is (8). The functor F is said to be *exact*, if it transforms any exact sequence into an exact sequence.

Exact functors are rare; thus $\mathrm{Hom}_R (P, -)$ is exact if and only if P is projective; this may be taken as the definition of a projective module, and it can then be proved that P is projective if and only if it is a direct summand of a free module. Similarly $\mathrm{Hom}_R (-, I)$ is exact if and only if I is injective. The minimal injective containing a given module M is unique up to isomorphism and is called the *injective hull* of M. There is an alternative definition of injective module and injective hull, as a sort of algebraic closure, but this will not be needed (*cf.* UA, p. 261).

However, most functors have a certain partial exactness property, which we now describe:

A covariant functor F is *left exact*, if the sequence

$$0 \to FA' \to FA \to FA'' \to 0 \tag{9}$$

obtained by applying F to a short exact sequence (4) is exact except possibly at FA'', and F is *right exact* if (9) is exact except possibly at FA'. For a contravariant functor $D : \mathscr{A} \to \mathscr{B}$ these notions are defined by applying the definitions just given to the associated covariant functor from $\mathscr{A}^o$ to $\mathscr{B}$. With these definitions one verifies without difficulty that $\mathrm{Hom}_R (-, -)$ is left exact in each of its arguments.

With each functor F we can associate a series of derived functors F^i ($i \in \mathbf{Z}$) which measure the departure of F from exactness. When F is exact, $F^i = 0$ for $i \neq 0$ and F^0 is naturally equivalent to F, in symbols $F^0 \cong F$. For a left exact functor F, $F^i = 0$ for $i < 0$ and (i) $F^0 \cong F$, (ii) $F^n I = 0$ for $n > 0$ and I injective, and (iii) to each short exact sequence (4) there corresponds a long exact sequence

$$0 \to F^0 A' \to F^0 A \to F^0 A'' \overset{\Delta}{\to} F^1 A' \to F^1 A \to \dots \tag{10}$$

with a 'connecting' homomorphism Δ which is a natural transformation. Moreover, F^i is uniquely determined up to natural equivalence by these three properties.

As an example let us take $FA = \operatorname{Hom}_R (M, A)$. The derived functor in this case is denoted by Ext. Thus $\operatorname{Ext}_R{}^0 (M, N) \cong \operatorname{Hom}_R (M, N)$, $\operatorname{Ext}_R{}^i (M, I) = 0$ for $i > 0$ and I injective, and we have an exact sequence

$$\begin{cases} 0 \to \operatorname{Hom}_R (M, A') \to \operatorname{Hom}_R (M, A) \to \operatorname{Hom}_R (M, A'') \overset{\Delta}{\to} \operatorname{Ext}_R{}^1 (M, A') \to \\ \operatorname{Ext}_R{}^1 (M, A) \to \operatorname{Ext}_R{}^1 (M, A'') \overset{\Delta}{\to} \operatorname{Ext}_R{}^2 (M, A') \to \cdots \end{cases} \quad (11)$$

Next let us take $\operatorname{Hom}_R (M, N)$, regarded as a functor in M. This is contravariant, and we have to replace injectives in (ii) above by projectives. The derived functor is the same as before, namely Ext. Thus we have $\operatorname{Ext}_R{}^i (P, N) = 0$ for $i > 0$ and P projective, and from any short exact sequence (4) we obtain

$$\begin{cases} 0 \to \operatorname{Hom}_R (A'', N) \to \operatorname{Hom}_R (A, N) \to \operatorname{Hom}_R (A', N) \overset{\Delta}{\to} \operatorname{Ext}_R{}^1 (A'', N) \to \\ \operatorname{Ext}_R{}^1 (A, N) \to \operatorname{Ext}_R{}^1 (A', N) \overset{\Delta}{\to} \operatorname{Ext}_R{}^2 (A'', N) \to \cdots \end{cases} \quad (12)$$

Whenever R is a field, or more generally, when R is any semisimple Artinian ring, $\operatorname{Ext}^i = 0$ for all $i \neq 0$, so in this case Hom is exact in both arguments. Most of the rings considered in this book are hereditary, which amounts to saying that Ext^i vanishes identically for $i > 1$. Therefore (11) and (12) reduce to 6-term sequences in this case. For any module M, the *projective* (*injective*) *dimension* is defined as the least n such that $\operatorname{Ext}_R{}^{n+1} (M, -) = 0 \, (\operatorname{Ext}_R{}^{n+1} (-, M) = 0)$, or ∞ if no such n exists.

Let us briefly note the connexion with module extensions. In (11) put $M = A''$ and consider the image of $1 \in \operatorname{Hom}_R (A'', A'')$ under Δ; this is an element θ of $\operatorname{Ext}_R{}^1 (A'', A')$. It can be shown (H, Chapter 3) that two short exact sequences (4) and

$$0 \to A' \to B \to A'' \to 0 \quad (13)$$

give rise to the same θ if and only if the extensions A, B are isomorphic, in the sense that there is an isomorphism $f: A \to B$ making the diagram

$$0 \to A' \to A \to A'' \to 0$$
$$\downarrow 1 \quad \downarrow f \quad \downarrow 1$$
$$0 \to A' \to B \to A'' \to 0$$

commute (note that even a homomorphism $f: A \to B$ with this property must be an isomorphism, by the 5-lemma). Moreover, every element of $\operatorname{Ext}_R{}^1 (A'', A')$ gives rise to an extension in this way; so the isomorphism classes of extensions of A' by A'' may be classified by $\operatorname{Ext}_R{}^1 (A'', A')$. We remark that the element θ can also be obtained as the image of 1 under Δ by putting $N = A'$ in (12). This element θ is zero if and only if the extension (4) splits. In particular, (4) splits if A'' is projective (or if A' is injective); of course this can easily be checked directly.

Since every free module is projective, every module can be written as a homomorphic image of a projective module. Thus for any R-module M, we have a short exact sequence

$$0 \to Q \to P \to M \to 0, \qquad \text{where } P \text{ is projective.} \qquad (14)$$

This is called a *presentation* of M; different presentations of M are compared in

SCHANUEL'S LEMMA. *Let M be any module; given two presentations of M, say* (14) *and* $0 \to Q' \to P' \to M \to 0$, *where P' is projective, then*

$$P \oplus Q' \cong P' \oplus Q.$$

(FR, *p.* 167). ∎

We now turn to tensor products. Given modules U_R, $_R V$ over a ring R, $T = U \otimes_R V$ is defined as an abelian group with a mapping $\phi : U \times V \to T$ which is *biadditive*, i.e. additive in each argument, and *balanced* in R: $\phi(ur, v) = \phi(u, rv)$ for all $r \in R$, and is universal with this property, i.e. every biadditive balanced mapping of $U \times V$ into an abelian group A can be factored uniquely by ϕ. More generally, given rings Q, R, S and bimodules $_Q U_R$, $_R V_S$, the tensor product T defined as above, has a (Q, S)-bimodule structure. The definition shows that for any modules U_R, $_R V_S$, W_S we have a natural isomorphism

$$\text{Hom}_R \left(U, \text{Hom}_S \left(V, W \right) \right) \cong \text{Hom}_S \left(U \otimes V, W \right). \qquad (15)$$

This is sometimes expressed by saying that $- \otimes V$ and $\text{Hom} \, (V, -)$ are *adjoint* to each other.

$$U \times V \xrightarrow{\ \phi\ } T$$

Suppose that U is a free right R-module with basis (u_λ); in particular, such a basis always exists if R is a field. Then every element of $U \otimes V$ can be written in the form

$$f = \Sigma u_\lambda \otimes a_\lambda \qquad (a_\lambda \in V),$$

and moreover, the coefficients a_λ are uniquely determined by f. In general, no such convenient normal form exists, but one has the following description of the relations in $U \otimes V$. The proof depends on the fact (easily deduced from (15)) that the tensor product is right exact in each argument. It is included below (*cf.* Cohn [59'] for a special case; see also Bourbaki [61], p. 41).

LEMMA. *Let R be any ring, and U_R, $_R V$ modules over R; assume that V is generated by a family $(e_\beta, \beta \in B)$ with defining relations $\Sigma a_{\alpha\beta} e_\beta = 0$ $(\alpha \in A)$. Given any family (x_β), of elements in U indexed by B and almost all 0, if*

$$\Sigma x_\beta \otimes e_\beta = 0 \quad \text{in} \quad U \otimes V,$$

then there exist elements $y_\alpha \in U$, almost all 0, such that

$$x_\beta = \Sigma y_\alpha a_{\alpha\beta}. \tag{16}$$

Proof. By hypothesis V has a presentation

$$0 \to G \xrightarrow{\lambda} F \xrightarrow{\mu} V \to 0,$$

where F is free on a family (f_β) and G is the submodule of F spanned by the elements $\Sigma a_{\alpha\beta} f_\beta$. Tensoring with U and observing that this operation is right exact, we obtain the exact sequence

$$U \otimes G \xrightarrow{\lambda'} U \otimes F \xrightarrow{\mu'} U \otimes V \to 0$$

By hypothesis, $(\Sigma x_\beta \otimes f_\beta)\mu' = 0$, hence exactness shows that

$$\Sigma x_\beta \otimes f_\beta = (\Sigma y_\alpha \otimes a_{\alpha\beta} f_\beta) \lambda',$$

for some elements $y_\alpha \in U$, almost all 0. Now λ' is the homomorphism induced by the inclusion $G \to F$, and F is free on the f's, so on equating coefficients we obtain (16). ∎

A module U is said to be *flat* if $U \otimes -$ is an exact functor. E.g. any free module, and more generally, any projective module is flat. The following property of flat modules is often useful:

Given modules U_R, $_R V$, where U is flat, if

$$\Sigma u_i \otimes v_i = 0 \quad \text{in} \quad U \otimes V, \tag{17}$$

then there exist elements $x_h \in U$ and $a_{hi} \in R$ such that

$$u_i = \Sigma x_h a_{hi}, \qquad \Sigma a_{hi} v_i = 0. \tag{18}$$

For if V' is the submodule generated by the v_i, let $\Sigma a_{\alpha i} v_i = 0$ $(\alpha \in A)$ be a set of defining relations for V'. Since U is flat, the relation (17), valid in $U \otimes V$, also holds in $U \otimes V'$ and by the lemma there exists a family (x_α) of elements of U, almost all 0, such that $u_i = \Sigma x_\alpha a_{\alpha i}$. If we renumber the non-zero x's as $x_1, \ldots, x_m$ and the $a_{\alpha i}$ accordingly, we obtain (18). ∎

At first sight the reader might be tempted to think that (18) holds without any hypothesis on U, but this is not so; in fact if (18) holds for $V = R$, then U must be flat, so the above property is actually characteristic of flat modules.

Since $- \otimes -$ is right exact in each argument, we can again form a derived functor. This is written Tor_i^R and has the following properties (which also characterize it):

(i) $\mathrm{Tor}_0^R (U, V) = U \otimes V$, $\mathrm{Tor}_i^R (U, V) = 0$ for $i > 0$ if U or V is projective,

(ii) each short exact sequence (4) gives rise to a derived sequence

$$
\begin{cases}
\ldots \to \mathrm{Tor}_2^R (M, A'') \to \mathrm{Tor}_1^R (M, A') \to \mathrm{Tor}_1^R (M, A) \to \\[2mm]
\mathrm{Tor}_1^R (M, A'') \to M \otimes A' \to M \otimes A \to M \otimes A'' \to 0.
\end{cases}
\tag{19}
$$

A similar sequence exists when the second argument in Tor is held fixed and the first ranges over the terms of (4). For a semisimple Artinian ring we again have $\mathrm{Tor}_i^R = 0$ for $i \neq 0$, while $\mathrm{Tor}_i^R = 0$ for $i > 1$ when R is hereditary.

Suppose that $\mathrm{Tor}_1^R (M, C) = 0$ for some fixed M and all finitely presented cyclic modules C. By induction (on the number of generators) it follows from (19) that $\mathrm{Tor}_1^R (M, A) = 0$ for all finitely presented modules A. Now Tor commutes with direct limits (essentially because $\otimes$ does), and since every module can be written as a direct limit of finitely presented modules, it follows that $\mathrm{Tor}_1^R (M, -) = 0$.

Bibliography and author index

Apart from listing works referred to in the text, the bibliography includes papers related to the main topics, although there has been no attempt at complete coverage of such general topics as principal ideal domains. The page references at the end of the entries indicate the places in the text where the entry is quoted; other references to an author are listed after his name.

ABHYANKAR, S. S. [247

ALBRECHT, F.
61. On projective modules over semihereditary rings. *Proc. Amer. Math. Soc.* **12** (1961), 638–639. [41

AMITSUR, S. A.
48. On unique factorization in rings (Hebrew). *Riveon Lematematika* **2** (1948), 28–29. [117
48′. A generalization of a theorem on differential equations. *Bull. Amer. Math. Soc.* **54** (1948), 937–941.
50. Finite differential polynomials (Hebrew) *Riveon Lematematika* **4** (1950), 1–8.
54. Differential polynomials and division algebras. *Ann. of Math.* **59** (1954), 245–278.
58. Commutative linear differential operators. *Pacif. J. Math.* **8** (1958), 1–10. [41
63. Remarks on principal ideal rings. *Osaka Math J.* **15** (1963), 59–69. [313
66. Rational identities and applications to algebra and geometry. *J. Algebra* **3** (1966), 304–359. [287

ASANO, K.
38. Nichtkommutative Hauptidealringe. *Act. Sci. Ind.* No. 696 (Paris 1938). [145

BASS, H. [174
64. Projective modules over free groups are free. *J. Algebra* **1** (1964), 367–373. [41
68. "Algebraic K-theory" New York 1968. [41

BAUMSLAG, B. and BAUMSLAG, G.
71. On ascending chain conditions. *Proc. London Math. Soc.* (3) **22** (1971), 681–704. [63

BEAUREGARD, R. A.
69. Infinite primes and unique factorization in a principal right ideal domain. *Trans. Amer. Math. Soc.* **141** (1969), 245–254.
71. Right LCM domains. *Proc. Amer. Math. Soc.* **30** (1971), 1–7.

BEAUREGARD, R. A. AND JOHNSON, R. E.
70. Primary factorization in a weak Bezout domain. *Proc. Amer. Math. Soc.* **25** (1970), 662–665.

BEHRENS, E.-A.
65. "Algebren" Mannheim, 1965.   [146, 150

BERGMAN, G. M.   [106, 128, 174, 243, 258
67. "Commuting elements in free algebras and related topics in ring theory". Thesis, Harvard University 1967. [30, 41, 63, 71f., 93, 111, 145, 156, 174, 179, 187, 192, 197, 209f., 246f.
69. Centralizers in free associative algebras, *Trans. Amer. Math. Soc.* **137** (1969), 327–344.   [247
69'. Ranks of tensors and change of base field. *J. Algebra* **11** (1969), 613–621.
70. Skew fields of noncommutative rational functions, after Amitsur, *Séminaire Schützenberger–Lentin–Nivat, Année* 1969/70, No. 16. Paris 1970.   [287
71. Groups acting on hereditary rings. *Proc. London Math. Soc.* (3) **23** (1971) 70–82.   [95
71'. Hereditary commutative rings and centers of hereditary rings. *Proc. London Math. Soc.* (3) **23** (1971), 214–236.
72. Infinite multiplication of ideals in $\aleph_0$-hereditary rings. *J. Algebra*, To appear.
72''. Hereditarily and cohereditarily projective modules, Park City Symposium 1971. To appear.   [41
72''. Boolean rings of projection maps. *J. London Math. Soc.* To appear.   [210
72'''. Notes on epimorphisms of rings. To appear.

BERGMAN, G. M. AND COHN, P. M.
69. Symmetric elements in free powers of rings. *J. London Math. Soc.* (2), **1** (1969), 525–534.
71. The centres of hereditary rings and 2-firs. *Proc. London Math. Soc.* (3), **23** (1971), 83–98.   [225, 247

BIRKHOFF, G.
67. "Lattice Theory", (3rd ed.) Providence, 1967.   [58, 151

BLUM, E. K.
65. Free subsemigroups of a free semigroup. *Mich. Math. J.* **12** (1965), 179–182.

BOKUT, L. A.
64. Some examples of rings without zero-divisors. (Russian). *Algebra i Logika* **3**, No. 5–6 (1964), 5–28.
65. Factorization theorems for certain classes of rings without zero-divisors. (Russian) I. *Algebra i Logika* **4**, No. 4 (1965), 25–52, II. *ibid.* No. 5 (1965), 17–46.
69. On Malcev's problem (Russian). *Sibirsk. Math. Zh.* **10** (1969), 965–1005. [286

BOURBAKI, N.
61. "Algèbre commutative". Ch. 1–2 (Paris 1961).   [325
61'. "Algèbre commutative". Ch. 3–4 (Paris 1961).   [97
65. "Algèbre commutative". Ch. 7 (Paris 1965).   [63, 218

BOWTELL, A. J.
67. "The multiplicative semigroup of a ring and the embedding of rings in skew fields". Thesis, London University, 1967.   [63, 127, 145
67'. On a question of Malcev. *J. Algebra* **6** (1967), 126–139.   [111, 286

BOWTELL, A. J. AND COHN, P. M.
71. Bounded and invariant elements in 2-firs. *Proc. Cambridge Phil. Soc.* **69**, (1971), 1–13. [247

BRENNER, J. L.
55. Quelques groupes libres de matrices. *C. R. Acad. Sci. Paris* **241** (1955), 1689–1691. [92

BRUMER, A.
66. Pseudocompact algebras, profinite groups and class formations. *J. Algebra* **4** (1966), 442–470.

BRUNGS, H. H.
69. Ringe mit eindeutiger Faktorzerlegung. *J. reine angew. Math.* **236** (1969), 43–66. [145
69'. Generalized discrete valuation rings. *Can. J. Math.* **21** (1969), 1404–1408. [311, 313

BURMISTROVIČ, I. E.
63. On the embedding of rings in skew fields (Russian). *Sibirsk. Mat. Zh.* **4** (1963), 1235–1240.

CARCANAGUE, J.
71. Idéaux bilatères d'un anneau de polynomes non commutatifs sur un corps. *J. Algebra* **18** (1971), 1–18. [247

CHASE, S. U.
61. A generalization of the ring of triangular matrices. *Nagoya Math. J.* **18** (1961), 13–25. [55
62. On direct sums and products of modules. *Pacif. J. Math.* **12** (1962), 847–854. [132

CLARK, W. E. [243

CLAUS, H. J.
55. Über die Partialbruchzerlegung in nicht notwendig kommutativen euklidischen Ringen. *J. reine angew. Math.* **194** (1955), 88–100.

COCKCROFT, W. H. AND SWAN, R. G.
61. On the homotopy type of certain two-dimensional complexes. *Proc. London Math. Soc.* (3), **11** (1961), 193–202. [41

COHEN, I. S. [224

COHEN, J. M. AND GLUCK, H.
70. Stacked bases for modules over principal ideal domains. *J. Algebra* **14** (1970), 493–505.

COHN, P. M.
59. Simple rings without zero-divisors and Lie division rings. *Mathematika* **6** (1959), 14–18. [258
59'. On the free product of associative rings. *Math. Zeits.* **71** (1959), 380–398. [325
60. On the free product of associative rings II. The case of skew fields. *Math. Zeits.* **73** (1960), 433–456. [111

61. On a generalization of the Euclidean algorithm. *Proc. Cambridge Phil. Soc.* **57** (1961), 18–30. [111

61'. On the embedding of rings in skew fields. *Proc. London Math. Soc.* (3), **11** (1961), 511–530. [287

61''. Quadratic extensions of skew fields. *Proc. London Math. Soc.* (3), **11** (1961), 531–556. [313

62. Factorization in non-commutative power series rings. *Proc. Cambridge Phil. Soc.* **58** (1962), 452–464. [111, 145, 247

62'. On subsemigroups of free semigroups. *Proc. Amer. Math. Soc.* **13** (1962), 347–351. [247

63. Noncommutative unique factorization domains. *Trans. Amer. Math. Soc.* **109** (1963), 313–332, Correction *ibid.* **119** (1965), 552. [63, 145

63'. Rings with a weak algorithm. *Trans. Amer. Math. Soc.* **109** (1963), 332–356. [111, 247

64. Subalgebras of free associative algebras. *Proc. London Math. Soc.* (3) **14** (1964), 618–632. [247

64'. Free ideal rings. *J. Algebra* **1** (1964), 47–69, Correction *ibid.* **6** (1967), 410. [63

65. "Universal Algebra", New York, 1965. [5, 20, 30, 32, 82, 286, 308f.

66. Some remarks on the invariant basis property. *Topology* **5** (1966), 215–228. [7, 41, 46, 63, 72, 111, 197, 258f., 280

66'. On the structure of the GL_2 of a ring. *Publ. Math. I.H.E.S. No. 30 Paris* (1966), 5–53. [88, 111

66''. A remark on matrix rings over free ideal rings. *Proc. Cambridge Phil. Soc.* **62** (1966), 1–4. [63

66'''. Hereditary local rings. *Nagoya Math. J.* **27–1** (1966), 223–230. [132

67. Torsion modules over free ideal rings. *Proc. London Math. Soc.* (3), **17** (1967), 577–599. [41, 63, 209

68. Bezout rings and their subrings. *Proc. Cambridge Phil. Soc.* **64** (1968), 251–264. [63, 198

68'. On the free product of associative rings III. *J. Algebra* **8** (1968), 376–383, Correction *ibid.* **10** (1968), 123.

69. Free associative algebras. *Bull. London Math. Soc.* **1** (1969), 1–39. Correction *ibid.* p. 218. [63, 111, 145, 247

69'. Rings with a transfinite weak algorithm. *Bull. London Math. Soc.* **1** (1969), 55–59. [111

69''. Dependence II. The dependence number. *Trans. Amer. Math. Soc.* **135** (1969), 267–279. [46, 63, 111, 136

69'''. Progrès recent dans l'étude des algèbres associatives libres, *Séminaire Dubreil–Pisot*, 22e année 1968/69 No. 16.

70. Factorization in general rings and strictly cyclic modules. *J. reine angew. Math.* **239/40** (1970), 185–200. [41, 145, 313

70'. Torsion and protorsion modules over free ideal rings. *J. Austral. Math. Soc.* **11** (1970), 490–498. [186

70''. On a class of rings with inverse weak algorithm. *Math. Zeits.* **117** (1970), 1–6. [210

71. The embedding of firs in skew fields. *Proc. London Math. Soc.* (3), **23** (1971). 193–213. [60, 285, 287

71'. Rings of fractions. *Amer. Math. Monthly* **78** (1971), 596–615. [286

71''. Un critère d'immersibilité d'un anneau dans un corps gauche. *C.R. Acad. Sci. Paris, Ser. A,* **272** (1971), 1442–1444.

71'''. The embedding of radical rings in simple radical rings. *Bull. London Math. Soc.* **3** (1971), 185–188. [111, 287

71*. Free ideal rings and free products of rings. Actes, Congrès internat. math., 1970, t.1 (1971), 273–278.

72. Universal skew fields of fractions. *Symposia Math.* VIII (1972). [287

72'. Bound modules over hereditary rings. To appear. [210

72''. Generalized rational identities. "Proc. Park City Conference 1971". [287

72'''. Skew fields of fractions, and the prime spectrum of a general ring. "Tulane Symposium on Rings and Operator algebras". [287

CORNER, A. L. S.

69. Additive categories and a theorem of Leavitt. *Bull. Amer. Math. Soc.* **75** (1969), 78–82.

CZERNIAKIEWICZ, A. [247

72. Automorphisms of a free associative algebra of rank 2. *Trans. Amer. Math. Soc.* To appear.

COZZENS, J. H.

70. Homological properties of the ring of differential polynomials. *Bull. Amer. Math. Soc.* **76** (1970), 75–79.

CURTIS, C. W.

52. A note on non-commutative polynomial rings. *Proc. Amer. Math. Soc.* **3** (1952), 965–969. [41

DICKSON, S. E.

66. A torsion theory for abelian categories. *Trans. Amer. Math. Soc.* **121** (1966), 223–235. [210

DOSS, R.

48. Sur l'immersion d'un semi-groupe dans un groupe. *Bull. Sci. Math.* **72** (1948), 139–150. [22

DUBREIL–JACOTIN, M.-L.

47. Sur l'immersion d'un semigroupe dans un groupe. *C.R. Acad. Sci. Paris* **225** (1947), 787–788.

ELIZAROV, V. P.

69. Rings of fractions (Russian). *Algebra i Logika,* **8**, No. 4 (1969), 381–424. [41

ENGEL, W.

55. Ein Satz über ganze Cremona-Transformationen der Ebene. *Math. Ann.* **130** (1955), 11–19.

62. Über die Derivationen eines Polynomringes. *Math. Nachr.* **24** (1962), 275–280.

ENGSTROM, H. T.

41. Polynomial substitutions. *Amer. J. Math.* **63** (1941), 249–255.

EUCLID, Elements (-300) [110

FELLER, E.

60. Intersection irreducible ideals of a non-commutative principal ideal domain. *Canad. J. Math.* **12** (1960), 592–596. [145

FIELDS, K. L.

69. On the global dimension of skew polynomial rings. *J. Algebra* **13** (1969), 1–4, Addendum, *loc. cit.* **14** (1970), 528–530.

FISHER, J. L.

72. Embedding free algebras in skew fields. *Proc. Amer. Math. Soc.*   [258

FITTING, H.

35. Primärkomponentenzerlegung in nichtkommutativen Ringen. *Math. Ann.* **111** (1935), 19–41.   [41

36. Über den Zusammenhang zwischen dem Begriff der Gleichartigkeit zweier Ideale und dem Äquivalenzbegriff der Elementarteilertheorie. *Math. Ann.* **112** (1936), 572–582.   [128, 145

FLANDERS, H.

55. Commutative linear differential operators, Technical Report No. 1 (ONR) University of California, Berkeley 1955.

FLIESS, M.

70. Transductions algébriques, R.I.R.O. R–1 (1970), 109–125.   [111

70′. Inertie et rigidité des séries rationnelles et algébriques. *C.R. Acad. Sci. Paris, Ser. A* **270** (1970), 221–223.

70″. Sur le plongement de l'algèbre des séries rationnelles non commutatives dans un corps gauche. *C.R. Acad. Sci. Paris* **271** (1970), 926–927.

FOX, R. H.

53. Free differential calculus I. Derivation on the free group ring. *Ann. Math.* **57** (1953), 547–560.

GELFAND, I. M. AND KIRILLOV, A. A.

66. Sur les corps liés aux algèbres enveloppantes des algèbres de Lie. *Publ. Math. I.H.E.S.* No. 31 (1966), 5–19.

GENTILE, E. R.

60. On rings with a one-sided field of quotients. *Proc. Amer. Math. Soc.* **11** (1960), 380–384.   [29, 41

62. Singular submodule and injective hull. *Indag. Math.* **24** (1962), 426–433.

GERSTEN, S. M.

65. Whitehead groups of free associative algebras. *Bull. Amer. Math. Soc.* **71** (1965), 157–159.

GOLDIE, A. W.

58. The structure of rings under ascending chain conditions. *Proc. London Math. Soc.* (3), **8** (1958), 589–608.   [41

GORDON, B. AND MOTZKIN, T. S.

65. On the zeros of polynomials over division rings. *Trans. Amer. Math. Soc.* **116** (1965), 218–226, Correction *ibid.* **122** (1966), 547.   [313

GREGER, K.

53. Multilinear forms and elementary divisors in a principal ideal ring. *Fysiogr. Sällsk. Lund Förhdl.* **23** (1953), 131–137.

GUAZZONE, S.
62. Sui Λ-moduli liberi i alcuni teoremi di C. J. Everett, *Rend. Sem. Mat. Univ. Padova* **32** (1962), 304–312.

HASSE, H.
28. Über eindeutige Zerlegung in Primelemente oder in Primhauptideale in Integritätsbereichen, *J. reine angew. Math.* **159** (1928), 3–12.   [69

HELMER, O.
40. Divisibility properties of integral functions. *Duke Math. J.* **6** (1940), 345–356.
43. The elementary divisor theorem for certain rings without chain condition. *Bull. Amer. Math. Soc.* **49** (1943), 225–236.

HENSEL, K.
27. Über die eindeutige Zerlegung in Primelemente. *J. reine angew. Math.* **158** (1927), 195–198.
29. Über Systeme in einfachen Körpern. *J. reine angew. Math.* **160** (1929), 131–142.
33. Elementare Rechenoperationen in Ringen. *J. reine angew. Math.* **169** (1933), 67–70.

HERSTEIN, I. N.
56. Conjugates in division rings. *Proc. Amer. Math. Soc.* **7** (1956), 1021–1022. [313

IKEDA, M.
69. Über die maximalen Ideale einer freien assoziativen Algebra. *Hamb. Abh.* **33** (1969), 59–66.   [86

JACOBSON, N.
34. A note on non-commutative polynomials. *Ann. Math.* **35** (1934), 209–210. [111
37. Pseudo-linear transformations. *Ann. Math.* **38** (1937), 484–507.   [301, 313
43. "Theory of Rings". Providence, 1943.   [121, 145, 234, 247, 313
64. "Structure of Rings", rev. ed. Providence 1964.   [13, 16, 230

JAFFARD, P.
60. "Les systèmes d'idéaux". Paris, 1960.   [63

JATEGAONKAR, A. V.
69. A counter-example in homological algebra and ring theory. *J. Algebra* **12** (1969), 418–440.   [107, 313
69'. Rings with a transfinite left division algorithm. *Bull. Amer. Math. Soc.* **75** (1969), 559–561.   [313
69''. Ore domains and free algebras. *Bull. London Math. Soc.* **1** (1969), 45–46. [33
70. Left principal ideal rings. "Springer Lecture Notes in Mathematics", No. 123. Berlin, 1970.

JENSEN, C. U.
63. On characterizations of Prüfer rings. *Math. Scand.* **13** (1963), 90–98. [146, 150
66. A remark on semihereditary local rings. *J. London Math. Soc.* **41** (1966), 479–482.
69. Some cardinality questions for flat modules and coherence. *J. Algebra* **12** (1969), 231–241.   [63

JOHNSON, R. E.

65. Unique factorization in a principal right ideal domain. *Proc. Amer. Math. Soc.*
16 (1965), 526–528. [145

67. The quotient domain of a weak Bezout domain. *J. Math. Sci.* **2** (1967), 21–22.

JOOSTE, T. DE W.

71. "Derivations in free power series rings and free associative algebras". Thesis,
London University, 1971. [105

JUNG, H. W. E.

42. Über ganze birationale Transformationen der Ebene. *J. reine angew. Math.*
184 (1942), 161–174. [247

KAPLANSKY, I.

49. Elementary divisors and modules. *Trans. Amer. Math. Soc.* **66** (1949), 464–491.
[290, 313

58. Projective modules. *Ann. Math.* **68** (1958), 372–377. [13, 17, 41

70. "Commutative Rings". Boston, 1970. [247

KAZIMIRSKII, P. S.

62. A theorem on elementary divisors for the ring of differential operators,
(Ukrainian, Russian summary), *Dopovidi Akad. Nauk Ukrain. SSR* (1962),
1275–1278.

KLEIN, A. A.

67. Rings nonembeddable in fields with multiplicative semigroups embeddable
in groups. *J. Algebra* **7** (1967), 100–125. [286

69. Necessary conditions for embedding rings into fields. *Trans. Amer. Math. Soc.*
137 (1969), 141–151. [59f., 63

70. A note about two properties of matrix rings. *Israel J. Math.* **8** (1970), 90–92.

70′. Three sets of conditions on rings. *Proc. Amer. Math. Soc.* **25** (1970), 393–398.

72. On conditions for the embeddability of rings into fields. *J. Algebra.* [197

KNIGHT, J. T.

70. On epimorphisms of non-commutative rings. *Proc. Cambridge Phil. Soc.* **68**
(1970), 589–600.

KNUS, M. A.

68. Homology and homomorphisms of rings. *J. Algebra* **9** (1968), 274–284.

KOŠEVOI, E. G.

66. On the multiplicative semigroup of a class of rings without zero-divisors
(Russian). *Algebra i Logika* **5**, No. 5 (1966), 49–54. [132, 145

70. On certain associative algebras with transcendental relations (Russian).
Algebra i Logika **9**, No. 5 (1970), 520–529. [33

71. On pure subalgebras of free associative algebras (Russian). *Algebra i Logika*
10, No. 2 (1971), 183–187. [247

KRULL, W.

24. Die verschiedenen Arten der Hauptidealringe. *S.-B. Heidelberg. Akad. Wiss.*
(1924), 6. Abhandlung.

31. Über die Zerlegung der Hauptideale in allgemeinen Ringen. *Math. Ann.* **105**
(1931), 1–14.

54. Zur Theorie der kommutativen Integritätsbereiche. *J. reine angew. Math.* **192** (1954), 230–252.

LANDAU, E.

02. Ein Satz über die Zerlegung homogener linearer Differentialausdrücke in irreduzible Faktoren. *J. reine angew. Math.* **124** (1902), 115–120.   [145

LANG, S.

58. "Introduction to Algebraic Geometry". New York, 1958.   [217, 239

LEAVITT, W. G.

57. Modules without invariant basis number. *Proc. Amer. Math. Soc.* **8** (1957), 322–328.   [41

LEAVITT, W. G. AND WHAPLES, G.

49. On matrices with elements in a principal ideal ring. *Bull. Amer. Math. Soc.* **55** (1949), 117–118.

LEWIN, J.

68. On Schreier varieties of linear algebras. *Trans. Amer. Math. Soc.* **132** (1968), 553–562.
69. Free modules over free algebras and free group algebras: the Schreier technique. *Trans. Amer. Math. Soc.* **145** (1969), 455–465.   [86, 111

LEWIN, J. AND LEWIN, T.

68. On ideals of free associative algebras generated by a single element. *J. Algebra* **8** (1968), 248–255.

LOEWY, A.

03. Über reduzible homogene Differentialausdrücke. *Math. Ann.* **56** (1903), 549–584.   [145
11. Über lineare homogene Differentialausdrücke derselben Art. *Math. Ann.* **70** (1911), 550–560.
12. Zur Theorie der linearen homogenen Differentialausdrücke. *Math. Ann.* **72** (1912), 203–210.
20. Begleitmatrizen und lineare homogene Differentialausdrücke. *Math. Zeits.* **7** (1920), 58–128.

LOPATINSKII, Y. B.

45. Linear differential operators (Russian). *Mat. Sbornik* **17** (59) (1945), 267–288.
46. A theorem on bases, (Russian). *Trudy Sek. Mat. Akad. Nauk Azerb. SSR* **2** (1946), 32–34.
47. On some properties of rings of linear differential operators (Russian), *Nauch. Zap. L'vov Univ. I Ser. fiz.-mat.* **2** (1947), 101–107.

McCONNELL, J. C. AND ROBSON, J. C.

72. Homomorphisms and extensions of modules over A_1 and related rings. *J. Algebra.* To appear.

MACDONALD, I. G.

68. "Algebraic Geometry: Introduction to Schemes". New York, 1968.   [287

MacLane, S.

63. "Homology". Berlin, 1963.   [57

Magnus, W., Karrass, A. and Solitar, D.

66. "Combinatorial Group Theory". New York, 1966.

Malcev, A. I.

37. On the immersion of an algebraic ring into a field. *Math. Ann.* **113** (1937), 686–691.

39. Über die Einbettung von assoziativen Systemen in Gruppen (Russian, German summary) I. *Mat. Sbornik* **6** (48) (1939), 331–336, II. *ibid.* **8** (50) (1940), 251–264.   [286

48. On the embedding of group algebras in division algebras (Russian). *Doklady Akad. Nauk* **60** (1948), 1499–1501.   [308

Moh, T.-T.   [247

Motzkin, T. S.

49. The euclidean algorithm. *Bull. Amer. Math. Soc.* **55** (1949), 1142–1146. [110f.

Nagata, M.

57. A remark on the unique factorization theorem. *J. Math. Soc. Japan* **9** (1957), 143–145.   [145

Nakayama, T.

38. A note on the elementary divisor theory in non-commutative domains. *Bull. Amer. Math. Soc.* **44** (1938), 719–723.   [313

Neumann, B. H.

49. On ordered division rings. *Trans. Amer. Math. Soc.* **66** (1949), 202–252.   [308
49'. On ordered groups. *Amer. J. Math.* **71** (1949), 1–18.   [309

Nivat, M.

68. Transduction des langages de Chomsky. *Ann. Inst. Fourier* **18**, 1 (1968), 339–456.   [111

69. Séries rationnelles et algébriques en variables non commutatives. *Cours du DEA* 1969/70.   [287

Niven, I.

41. Equations in quaternions. *Amer. Math. Monthly* **48** (1941), 654–661.

Nöbeling, G.

68. Verallgemeinerung eines Satzes von Herrn Specker. *Inventiones Math.* **6** (1968), 41–55.   [210

Noether, E. and Schmeidler, W.

20. Moduln in nichtkommutativen Bereichen, insbesondere aus Differential- und Differenzenausdrücken. *Math. Zeits.* **8** (1920), 1–35.   [174

Nuñez, P. Libro de Algebra (Antwerp 1567).   [110

Ore, O.

31. Linear equations in non-commutative fields. *Ann. Math.* **32** (1931), 463–477.
                                                                      [41

32. Formale Theorie der linearen Differentialgleichungen. *J. reine angew. Math.* **167** (1932), 221–234, II. *ibid.* **168** (1932), 233–252.   [40f., 145

33. Theory of non-commutative polynomials. *Ann. Math.* **34** (1933), 480–508.
                                                                      [41, 145

OSTMANN, H.-H.
50. Euklidische Ringe mit eindeutiger Partialbruchzerlegung. *J. reine angew. Math.* **188** (1950), 150–161.

POINCARÉ, H. [145

PONTRJAGIN, L.
39. "Topological Groups". Princeton, 1939. [210

RABINOW, D. G.
37. Independent sets of postulates for abelian groups and fields in terms of the inverse operation. *Amer. J. Math.* **59** (1937), 211–224. [279

RAINVILLE, E. D.
41. A discrete group arising in the study of differential operators. *Amer. J. Math.* **63** (1941), 136–140.

RICHARDSON, A. R.
27. Equations over a division algebra. *Messenger of Math.* **57** (1927), 1–6. [313

RINEHART, G. S.
62. Note on the global dimension of a certain ring. *Proc. Amer. Math. Soc.* **13** (1962), 341–346.

RITT, J. F.
22. Prime and composite polynomials. *Trans. Amer. Math. Soc.* **23** (1922), 51–66.

ROBSON, J. C. [50
72. Idealizers and hereditary Noetherian prime rings. *J. Algebra.* [20

ROOS, J.-E.
67. Locally distributive spectral categories and strongly regular rings. "Reports of the Midwest Category Seminar, Springer Lecture Notes in Mathematics". No. 47. Berlin, 1967. [174

SAMUEL, P. [69
68. Unique factorization. *Amer. Math. Monthly* **75** (1968), 945–952. [110

SANOV, I. N.
67. Euclidean algorithm and one-sided prime factorization for matrix rings (Russian). *Sibirsk. Mat. Zh.* **8** (1967), 846–852. [68

SCHLESINGER, L.
97. "Handbuch der Theorie der Differentialgleichungen". Leipzig, 1897. [145

SCHUR, I.
04. Über vertauschbare lineare Differentialausdrücke, *Berl. Math. Ges. Sitzber.* **3** (Archiv d. Math. Beilage (3) 8) (1904), 2–8. [246

SCHÜTZENBERGER, M. P.
59. Sur certains sous-demi-groupes qui interviennent dans un problème de mathématiques appliquées. *Publ. Sci. Univ. d'Alger Ser. A* **6** (1959) 85–90. [247
62. On a theorem of R. Jungen. *Proc. Amer. Math. Soc.* **13** (1962), 885–890. [287

338   BIBLIOGRAPHY AND AUTHOR INDEX

SCHWARZ, L.
47. Zur Theorie der nichtkommutativen rationalen Funktionen. *Berichte d. Math. Tagung Tübingen* 1946 (1947), 134–136.
49. Zur Theorie des nichtkommutativen Polynombereichs und Quotientenrings. *Math. Ann.* **120** (1947/49), 275–296.   [145

SERBIN, H.
38. Factorization in principal ideal rings. *Duke Math. J.* **4** (1938), 656–663.

SHODA, K.
29. Über die mit einer Matrix vertauschbaren Matrizen, *Math. Zeits.* **29** (1929), 696–712.

SILVESTER, J. R.   [55

SKOLEM, T.
39. Eine Bemerkung über gewisse Ringe mit Anwendung auf die Produktzerlegung von Polynomen. *Norske Mat. Tidsskrift* **21** (1939), 99–107.

SKORNYAKOV, L. A.
65. On Cohn rings (Russian). *Algebra i Logika* **4**, No. 3 (1965). 5–30.   [111
67. The homological classification of rings (Russian). *Mat. Vesnik* **4** (19) (1967), 415–537.

SMITS, T. H. M.
68. Nilpotent *S*-derivations. *Indag. Math.* **30** (1968), 72–86.
68′. Skew polynomial rings. *Indag. Math.* **30** (1968), 209–224.   [41
69. The free product of a quadratic number field and semifield. *Indag. Math.* **31** (1969), 145–159.

SPECHT, W.
50. Gesetze in Ringen, I. *Math. Zeits.* **52** (1950), 557–589.

SPECKER, E.
50. Additive Gruppen von Folgen ganzer Zahlen. *Portugal. Math.* **9** (1950), 131–140.   [210

STENSTRÖM, B.
71. Rings and modules of quotients. Springer Lecture Notes in Mathematics, No. 237. Berlin 1971.   [210

STEPHENSON, W.   [174

STEVIN, S.   Arithmétique 1585 (vol. II of collected works, 1958).   [110

STEWART, B. M.
49. A note on *LCMs*. *Bull. Amer. Math. Soc.* **55** (1949), 587–591.

TAMARI, D.
53. On the embedding of Birkhoff–Witt rings in quotient fields. *Proc. Amer. Math. Soc.* **4** (1953), 197–202.

TARASOV, B. V.
67. On free associative algebras (Russian). *Algebra i Logika* **6**, No. 4 (1967), 93–105.

TAZAWA, M.

33. Eine Bemerkung über den Elementarteilersatz. *Proc. Imper. Acad. Japan* **9** (1933), 468–471.

TEICHMÜLLER, O.

37. Der Elementarteilersatz für nichtkommutative Ringe. *S.-B. Preuss. Akad. Wiss.* (1937), 169–177. [313

UTUMI, Y.

63. On rings of which any one-sided quotient rings are two-sided. *Proc. Amer. Math. Soc.* **14** (1963), 141–147.

WAGNER, E.

44. Über Shodasche Matrizen und Polynome in einer Matrix. *Math. Zeits.* **49** (1944), 517–537.

WEBBER, D. B.

70. Ideals and modules in simple Noetherian rings. *J. Algebra* **16** (1970), 239–242.
[4

WEDDERBURN, J. H. M.

14. On continued fractions in non-commutative quantities. *Ann. Math.* (2) **15** (1913–14), 101–105.
32. Non-commutative domains of integrity. *J. reine angew. Math.* **167** (1932), 129–141. [313

WILLIAMS, R. E.

68. A note on weak Bezout rings. *Proc. Amer. Math. Soc.* **19** (1968), 951–952. [63
68'. Sur une question de P. M. Cohn. *C.R. Acad. Sci. Paris, Ser. A* **267** (1968), 79–80.
69. A note on rings with weak algorithm. *Bull. Amer. Math. Soc.* **75** (1969), 959–961 (*cf.* correction in *Math. Rev.* **40** (1970) No. 7308).
69'. On the free product of rings with weak algorithm. *Proc. Amer. Math. Soc.* **23** (1969), 596–597.

WOLF, L. A.

36. Similarity of matrices in which the elements are real quaternions. *Bull. Amer. Math. Soc.* **42** (1936), 737–743.

WOLF, M. C.

36. Symmetric functions of noncommuting elements. *Duke Math. J.* **2** (1936), 626–637.

ZAKS, A.

70. Hereditary local rings. *Mich. Math. J.* **17** (1970), 267–272.

Subject Index

A

α-fir, 47
Abelian category, 320
Abelianizing a ring (or group), 33, 286
Abstract atomic factor, 166
ACC = ascending chain condition, 48
ACC_n = ACC on n-generator
    submodules, 48
ACC_{ds} = ACC on finitely generated
    submodules with dense inclusions,
    205
Additive category, 319
Additive functor, 321
Adjoint functor, 324
Affine scheme, 286
Algebraic element (left, right), 301
Artinian ring = ring with (left, right)
    DCC, 15, 25
Atom, atomic, xv
Augmentation ideal, 32

B

Balanced, 324
Bergman's centralizer theorem, 211, 246f
Bezout domain (left, right), 45
Bezout identity, 63
Bezout ring, weak, 57
Biadditive, 324
Bicentral action, 294
Binomial extension, 304
Birkhoff's representation theorem, 317
Bound module, 175
Bound, strongly 181
Bound of an element, 227
Bounded element, 226f
Bounded linear transformation, 300
Bounded module, 226

C

$\mathscr{C}$-indecomposable, 121, 141
$\mathscr{C}$-module, 117
$\mathscr{C}$-simple, 121
Cancellation semigroup, 235
Capacity, 231
Card (I) = cardinality of I, 5
Category, 318
Cayley–Hamilton theorem, 300
Central extension, 304
Characteristic of a module, 182
Cleavage, cleft, 143
Closed (K-closed) submodule, 192, 197
Closure of a submodule, 192
Cohen's theorem, 224
Coherent family of matrix groups, 51
Cokernel, 319
Column rank, 194
Comaximal, xv
Comaximal relation, 89
Comaximal transposition, 134
Comaximally transposable, 130
Commensurable, xv
Complete direct decomposition, 141
Completely primary ring, 121
Completely reducible module, 142
Completion (of a filtered ring), 97
Complex-skew polynomial ring, 36
Composition series, 186
Conductor, 214
Conical semigroup, 25, 112
Conjugate elements, xv
Conjugate mappings, 212
Connected negatively filtered k-ring, 102
Connecting homomorphism, 322
Conservative 2-fir, 157, 215
Continuant polynomial, 89
Contravariant functor, 321
Coprime relation, 126